T0342273

Spacecraft Lithium-Ion Battery Power Systems

Spacecraft Lithium-Ion Battery Power Systems

Edited by
Thomas P. Barrera

Registered Offices
John Wiley & Sons, Inc., 111 River Street, Hoboken, NJ 07030, USA
John Wiley & Sons Ltd, The Atrium, Southern Gate, Chichester, West Sussex, PO19 8SQ, UK

Editorial Office
The Atrium, Southern Gate, Chichester, West Sussex, PO19 8SQ, UK

For details of our global editorial offices, customer services, and more information about Wiley products visit us at www.wiley.com.

Wiley also publishes its books in a variety of electronic formats and by print-on-demand. Some content that appears in standard print versions of this book may not be available in other formats.

Library of Congress Cataloging-in-Publication Data applied for
Hardback ISBN: 9781119772149

Cover Design: Wiley
Cover Image: © NASA

Set in 9.5/12.5pt STIXTwoText by Straive, Pondicherry, India
Printed and bound by CPI Group (UK) Ltd, Croydon, CR0 4YY

C9781119772149_161122

Dedicated to the Memory of Mrs. Sara Thwaite
January 13, 1972 – May 31, 2022

Contents

About the Editor

Thomas P. Barrera, PhD, is President, LIB-X Consulting (Long Beach, CA, USA), where he provides engineering and educational services in the broad area of lithium-ion battery power systems. Previously, Tom was a Technical Fellow for The Boeing Co., Satellite Development Center (El Segundo, CA), where he led multidisciplinary teams in systems engineering of advanced space electrical power subsystem technologies. During his 19-year Boeing career, Tom provided mission operations support for the NASA Space Shuttle and International Space Station programs as well as battery expertise for the CST-100 Starliner, Space Launch System, numerous commercial and government LEO/GEO satellite systems, and various high-value proprietary satellite programs. Before joining The Boeing Co., Tom served as a space battery R&D test engineer at The Aerospace Corporation and electrical power systems engineer at the NASA Lyndon B. Johnson Space Center. Tom currently serves as an industry member of the NASA Engineering and Safety Center and is on the advisory board for South 8 Technologies. Frequently invited to speak and lecture at domestic and international conferences, Tom has over 50 combined conference presentations and publications, including 3 US patents in the area of aviation battery safety. Dr. Barrera earned his PhD in chemical engineering from the University of California, Los Angeles (UCLA) with a minor in atmospheric chemistry and physics. He also served as a Postdoctoral Research Fellow in the department of materials science and engineering at UCLA. Tom earned his MS in industrial engineering and management sciences from Northwestern University, BS in chemical engineering cum laude and BA in mathematics–economics both from University of California, Santa Barbara. He is also an AIAA Associate Fellow, member-at-large for the Battery Division of the ECS, and member of the MRS and Tau Beta Pi. Tom and his wife Joan reside in Long Beach, California where they actively enjoy a Southern California lifestyle and traveling abroad.

About the Contributors

Thomas P. Barrera, PhD, is President, LIB-X Consulting, Long Beach, CA, USA. He has over 35 years of relevant aerospace experience in fuel cell, battery, solar array, power electronics, and harness technologies. During his 19-year career at The Boeing Co., Dr. Barrera served as a Technical Fellow where he led systems engineering teams in advanced satellite electrical power system development, commercial and government spacecraft launch campaigns, and on-orbit spacecraft mission operations activities. Previously, he was a space battery R&D test engineer at The Aerospace Corporation and electrical power systems engineer at the NASA Lyndon B. Johnson Space Center. Tom has over 50 combined conference presentations and publications, including 3 US patents in the area of aviation battery safety. He earned his PhD in chemical engineering from UCLA and is a member of the ECS, AIAA, and MRS.

Yannick Borthomieu, PhD, is Product Manager and Saft Fellow, at Saft Defense and Space Division, Poitiers, France. He has worked 33 years in the space battery field acquiring a worldwide recognized expertise in more than 700 Ni-Cd, Ni-H_2 and Li-ion combined satellite battery programs in collaboration with all the worldwide satellite prime contractors and operators. He has authored or co-authored 200 publications, 3 book chapters, and 5 patents. Yannick earned his PhD from CNRS in Bordeaux, France.

Ratnakumar V. Bugga, PhD, is Technical Fellow, Lyten, Inc., San Jose, CA, USA. Previously, Dr. Bugga spent 34 years at the NASA Jet Propulsion Laboratory as a Principal Technologist conducting research on various advanced battery technologies for NASA planetary missions. He provided support to several NASA/JPL planetary missions, including the first major NASA missions to implement Li-ion batteries, 2003 Mars Rovers (Spirit and Opportunity) and 2011 Mars Curiosity Rover as rover battery cognizant engineer and overall battery lead. He was awarded the NASA exceptional service award for his contributions to Mars missions, the JPL Magellan award, and NASA Innovative and Advanced Concepts grants for Venus mission power solutions. He has authored over 100 peer-reviewed journal articles, granted over 25 US patents, authored four book chapters, and presented in numerous national and international conferences. Kumar earned his PhD in electrochemistry from the Indian Institute of Science. He is member of ECS, MRS, ACS and AIAA.

Keith Chin, PhD, is Sr. Research Technologist, NASA Jet Propulsion Laboratory, California Institute of Technology, Pasadena, CA, USA. During his 21 years at NASA/JPL, he has led numerous flight and research technology programs in the broad areas of energy storage, energy generation, and in-situ electrochemical instrumentation. He also has more than 15 years of system engineering experience on spacecraft power system design and operation for CubeSat to Flagship missions and served on Grace (2002), MER (2003), Deep Impact (2005), Phoenix (2007), MSL (2011), and Psyche (2022) mission teams. Keith is the recipient of numerous NASA Awards including the MSL Mission System Development Team and Li-Ion Battery Tech Dev. Team Awards. Keith earned his PhD and MS in chemical engineering from the University of Southern California and UCLA, respectively.

Steven E. Core served as Associate Technical Fellow, The Boeing Company, El Segundo, CA, USA. Prior to his retirement, Steve was regional manager for the Boeing Satellite Development Center's Customer Operations Support Center. With over 36 years of aerospace engineering experience, Steve led teams for 7 international customers to maintain their fleet of 16 satellites, resolve on-orbit anomalies, and perform decommissioning operations. He is an expert in propulsion systems engineering, executing programs for 26 satellites up through AI&T operations, including 20 launch campaigns and mission operations. He holds 2 patents and has MS degrees in both Environmental and Aerospace Engineering from University of Southern California, and a BS in Aero/Astro Engineering from Ohio State University.

David Curzon is Customer Relationship Manager, EnerSys/ABSL Space Products, Abingdon, England. David joined the ABSL Space Products team in 1998. Previously, David served as Program Manager with the NASA GSFC customer on numerous key space battery programs such as ST-5, LRO, SDO and GPM. David has worked with major space agencies including ESA, Indian Space Research Organization, Mohammed bin Rashid Space Centre and Korea Aerospace Research Institute as well as the majority of European Spacecraft Primes. Major projects and achievements have included Proba-1, Galileo and KSLV-2. David earned his Class #1 honors degree in Mechanical Engineering from Birmingham University.

Penni J. Dalton served as Electrical Power System Battery Subsystem Manager, International Space Station, NASA Glenn Research Center, Cleveland, OH, USA. Prior to her retirement, she had over 40 years of experience designing, building, testing, and operating spacecraft batteries, including the first launch of the NASA ISS Ni-H_2 batteries in 2000 through the latest on-orbit replacement of those original batteries with Li-ion batteries in 2021. She is a two-time recipient of the NASA Exceptional Service Medal, the Spaceflight Awareness Award, and the Silver Snoopy Award and has authored numerous publications on the ISS battery operations. She has a MS in Chemistry from Indiana University and a BS in Chemistry from the University of California, Santa Barbara.

Eric C. Darcy, PhD, is Battery Technical Discipline Lead, NASA, Johnson Space Center, Houston, TX, USA. During his 35-year NASA career, Eric has led battery design, verification, and safety assessments for various manned spacecraft applications. He has pioneered the development of safe, high performing lithium-ion battery systems with a focus on understanding, preventing, and mitigating cell latent defects that can lead to catastrophic hazards. He is co-inventor of a cell implantable internal short circuit device and fractional thermal runaway calorimeter providing significant design insights into cell responses during thermal runaway and enabled valid battery propagation assessments. He earned his PhD in chemical engineering at the University of Houston.

Aakesh Datta is Lead EPS engineer and battery expert at OHB System AG, Bremen, Germany. He is responsible for the development of satellite EPS architectures and Li-ion battery technologies for on-orbit applications varying from telecommunications to scientific exploration. In this role, he established the Li-ion battery in-flight management for the Galileo FOC constellation ground operational team. Previously, Aakesh served four and a half years as a battery development engineer for Li-ion technologies at Saft where he was responsible for technical management of the Saft VES16 Li-ion cell industrialization process and quality control. Aakesh earned a MS in Aeronautics and Aerospace Engineering from the Institut Polytechnique des Sciences Advancées.

Scott Hull is Lead Orbital Debris Engineer, NASA Goddard Space Flight Center, Greenbelt, MD, USA. He has over 20 years of experience performing orbital debris assessments, consulting on decommissioning planning, and providing related studies for over 60 spacecraft missions. Mr. Hull has also supported multiple orbital debris related studies for the NASA Engineering and Safety Center, and authored or co-authored over a dozen conference papers and contributed to two textbooks. Previous experience includes failure analysis, electronic parts engineering, and space mission operations support. He graduated from Drexel University with a BS in Materials Engineering.

Eloi Klein is a battery engineering specialist at Thales Alenia Space, Cannes, France. Eloi joined Thales Alenia Space in 2006, where he has been involved in the development of Li-ion cells and batteries for various spacecraft, such as the Exomars Trace Gas Orbiter, the Iridium Next constellation, and the Space Inspire commercial satellite product line. Eloi served first at Saft, working on the mechanical and electrical design of the first Li-ion batteries for GEO and LEO satellite missions. He earned his M.Sc. in Mechanical and Power Engineering from Institut National des Sciences Appliquées de Hauts-de-France.

David J. Reuter is the space battery subject matter expert, Northrop Grumman Strategic Space Systems Division, Redondo Beach, CA, USA. David has over 25 years of end-to-end spacecraft battery design, test, integration, launch and on-orbit operating experience. He has led multi-disciplined teams developing high performance, long-life energy storage systems for many commercial and government spacecraft programs with expertise in Ni-H_2, Li-ion, and thermal battery power systems. Previously, Dave was a responsible engineering authority for space batteries at Hughes Space and Communications and a lead systems engineer at Raytheon Space and Airborne Systems. Dave earned a BS in mechanical engineering from California State Polytechnic University, Pomona.

Samuel P. Russell is Project Manager and Systems Engineer, NASA Lyndon B. Johnson Space Center, Houston, TX, USA. With more than 20 years of experience in developing and testing human space flight hardware, Sam has managed the development of the first Li-ion battery for critical human space application, and after nearly a decade of successful performance, the first generation of propagation resistant Li-ion batteries for human spaceflight. Sam is a PhD candidate in Systems Engineering at Stevens Institute of Technology and holds a BS in Environmental Engineering and a MS in Materials Engineering from the New Mexico Institute of Mining and Technology.

Shriram Santhanagopalan, PhD, is Team Leader for the Battery Materials, Analysis and Diagnostics Group, National Renewable Energy Laboratory, Golden Co, USA. Before joining NREL, Shriram was a Senior Scientist at Celgard, LLC, where he was responsible for developing new characterization tools and Li-2 components for automotive batteries. Over the last 20 years Shriram has developed models, materials and methodology to address performance and safety limitations of automotive batteries. He actively participates in committees on battery standards. He has authored over 70 peer-reviewed journal articles, several book chapters and three books. He earned his PhD in Chemical Engineering from the University of South Carolina.

Marshall C. Smart, PhD, is Principal Member of the Technical Staff, NASA Jet Propulsion Laboratory, California Institute of Technology, Pasadena, CA, USA. During his 28-year NASA career, Dr. Smart's major research focus has been the development of low temperature electrolytes for aerospace and automotive lithium-ion battery applications. Dr. Smart's low temperature electrolyte technology was successfully used on the MER, Phoenix Lander, MSL, M2020, and InSight missions to Mars. He is currently the Cognizant Engineer of the Li-ion Batteries for the Europa Clipper project and the Mars 2020 Ingenuity Helicopter. Dr. Smart has authored or co-authored 59 publications in peer-reviewed journal articles, over 280 conference proceedings papers and presentations, and has 13 US patents. In 2015, Dr. Smart was awarded a NASA Exceptional Achievement Award for the "Infusion of Li-ion Battery Electrolytes into Flight Missions". Dr. Smart earned a PhD in Organic Chemistry from the University of Southern California, a MS in Chemistry from California State University, Los Angeles, and BA degrees in Philosophy and Chemistry from University of California, Santa Barbara. He is also a member of the ACS, ECS and AAAS.

Samuel Stuart is Chief Scientist, Power & Energy Division, Naval Surface Warfare Center, Crane, IN, USA. Sam has over 37 years of experience with commercial and military batteries, working extensively with test, evaluation, failure analysis, and manufacturing technologies across the battery engineering lifecycle. He earned a BS and MS in Chemical Engineering from New Mexico State University. He is a member of the AIAA, AIChE, ASNE, and ECS.

Sara Thwaite was a Battery Technical Specialist, EnerSys/ABSL Space Products, Abingdon, England. For much of her early childhood she and her brother lived in Northern Argentina where their parents worked with the Wichi people of the Chaco region. Returning to England in December 1980 she lived and was educated near and in Exeter, Devon. She read Theology at Exeter College Oxford, graduating in 1993. After graduation she studied a range of engineering disciplines through the Open University. She originally started work in the space industry in 2000 and worked in a variety of roles from technical to product assurance before moving into defense development work on hybrid fuel cell and battery systems. Returning to her real passion of Space in 2010, she continued in a senior Product Assurance team management role until 2018 when she took on her final role of Technical Specialist, providing system engineering and knowledge development support to the business based on her long-standing experience in the industry. In addition to her strong interest in Space and developing technologies, Sara had a love of nature, regularly hiking and cycling in the UK. She was also a keen photographer of flora and fauna – both in the UK and in various African countries she visited on safaris.

Lloyd Zilch, PhD, is Chief Engineer, Power & Energy Division, Naval Surface Warfare Center, Crane, IN, USA. Lloyd has over 13 years of experience in military manned and unmanned energy storage systems for aviation, missile, and space applications. In this role, Lloyd serves as a technical advisor to the aerospace lithium-ion battery life cycle test facility. Lloyd earned his PhD in Chemistry from Indiana University and a MS and BS in Chemistry and Biochemistry from Brigham Young University.

List of Reviewers

Valerie Ang
The Aerospace Corporation
El Segundo, CA, USA

Christopher R. Ashley
Ashley Technical Services, LLC
Redondo Beach, CA USA

François Bausier, PhD
European Space Agency, European Space Research and Technology Centre
Noordwijk, The Netherlands

Gary Bayles, PhD
Science Applications International Corporation, Crane, IN, USA

Erik J. Brandon, PhD
Jet Propulsion Laboratory, California Institute of Technology, Pasadena, CA, USA

Robert M. Button
NASA, Glenn Research Center
Cleveland, OH, USA

Gregory A. Carr
Jet Propulsion Laboratory, California Institute of Technology, Pasadena, CA, USA

Boyd Carter, PhD
The Aerospace Corporation
El Segundo, CA, USA

Jeff Case
Space Power Systems Engineering
Chantilly, VA, USA

David Delafuente, PhD
NASA, Johnson Space Center
Houston, TX, USA

Jan Geder, PhD
VDE Renewables Asia, Singapore

Rob Gitzendanner, PhD
EaglePicher Technologies, East Greenwich, RI, USA

Joshua Lamb, PhD
Sandia National Laboratories
Albuquerque, NM, USA

Michel Lannes
EaglePicher Technologies, East Greenwich, RI, USA

Jamal Mardini, PhD
The Boeing Co., El Segundo, CA, USA

Allen Muroi
The Boeing Co. (Retired), Huntington Beach, CA, USA

David W. Olsen
NASA, John F. Kennedy Space Center, FL, USA

Eugene R. Schwanbeck IV
NASA, Johnson Space Center
Houston, TX, USA

Evelyne Simon, PhD
European Space Agency, Harwell
Campus, UK

Pinakin M. Shah, PhD
PS Battery Tech Consulting, Spring, TX, USA

Dick (Richard) Shaw
Lockheed Martin Space (Retired)
Littleton, CO, USA

Ramanathan Thillaiyan, PhD
Teledyne Energy Systems, Inc. Hunt
Valley, MD, USA

Carl Thwaite
EnerSys/ABSL, Abingdon, Oxfordshire, UK

Joe Troutman
EnerSys/ABSL, Longmont, CO, USA

Margot L. Wasz, PhD
The Aerospace Corporation
El Segundo, CA, USA

Mark J. Welch
NASA, Johnson Space Center
Houston, TX, USA

Eric J.M. Young
NASA, Goddard Space Flight Center
Greenbelt, MD, USA

Foreword by Albert H. Zimmerman and Ralph E. White

A man emerges from the shadows. Silhouetted in the moonlight, he looks furtively over his shoulder and cradles a glowing coal in his arms before hurrying ahead towards a darkened town. You may recognize this as the mythological story of Prometheus, who stole the technology of fire from the gods and gave it to humanity. This story has come to represent striving for and developing new knowledge, as well as the risks that often come with new technologies. Control over fire is indeed a technology that has energized and altered the course of civilization. The development of lithium-ion batteries (LIB) has also created a technology that when combined with modern microelectronics, has transformed the course of civilization in the twenty-first century. New LIB-powered machines emerge every day and have given people instantaneous connection to both each other and to limitless information.

This book extends the LIB narrative to the power systems used in today's machines of space, the satellites, the space exploration craft, the planetary explorers, probes, and rovers. The space environment includes challenges not commonly encountered on Earth, temperature extremes of hot and cold, radiation, meteorites, as well as the more familiar day/night cycle that governs charging and discharging of batteries (like the Promethean cycle of torment and rebirth for stealing fire). One additional challenge: the battery systems that power these space vehicles must perform without failure over the vehicle lifetime; battery replacement is usually not an option. This book provides a series of nine chapters in which technology practitioners describe LIB technologies and practices that are utilized to make cells, batteries, and power systems for space vehicles, to integrate these power systems into satellites or planetary spacecraft, to verify safety, risks, reliability, and performance, and finally to discuss experiences operating and managing some representative space systems over their life cycle. The information in these chapters make it abundantly clear that successfully fielding a space LIB power system requires much work and attention to a lot of detail along the way.

This book outlines an overall process that, if followed, should result in the successful design and operation of a space LIB power system. The process is intended to be flexible enough to accommodate cell technology advances and to be applicable to the full range of today's spacecraft types, ranging from inexpensive CubeSats to the largest and most expensive satellite or deep space mission. From where does this process come from? The answer to this question gets to the heart of why we design space systems as we do today. Simply put, the process is based on experience. This experience is documented in standards and other publications that are referenced in each chapter, and by this means captures both the good and not so good lessons learned from the past. This outstanding book provides a gold standard, which can be followed to optimize the likelihood of power system success, but a standard without any flexibility has limited usefulness. Flexibility is achieved because the critical portions of the standards that are adopted by each space vehicle can be tailored to meet its specific cost, schedule, and reliability needs. This book provides an excellent starting point for the battery scientist and power system engineer who wants to leverage the experience of the past into a space power system for the future.

Albert H. Zimmerman, PhD
Technical Fellow, The Aerospace Corporation

Ralph E. White, PhD
Professor, Chemical Engineering, University of South Carolina

Preface

Whether serving as an educator in a university classroom environment or as a mentor to early career professionals, I learned that we all benefit from sharing relevant experiences, lessons-learned and proven best practices. To that end, the vision for this book was initially formed while teaching graduate-level spacecraft power system and LIB short-courses at USC, UCLA, and NASA. During this time, it was evident that a single-source comprehensive treatment of the cradle-to-grave "life cycle" of spacecraft LIB power systems did not yet exist. Instead, the majority of space Li-ion cell and LIB engineering requirements, test data, and program experience were widely distributed across hard-to-find or difficult-to-access public domain resources. Within the spacecraft designer community, this reality had seemingly created challenging knowledge gaps in a common understanding of space LIB power system fundamentals.

After establishing this growing space industry need, a systematic plan for authoring the first-ever spacecraft LIB technical reference book was socialized within the aerospace community. This included lengthy conversations with industry colleagues at various professional conferences, careful reviews of student course evaluations, and an extensive evaluation of the existing literature. Based on a significant amount of constructive industry feedback, and enthusiastic encouragement from my family, the book project had the needed traction and motivation to move forward to completion. To best serve the space community, this book was intentionally authored for a wide spectrum of practicing industry professionals with varying degrees of experience. As such, early career professionals new to spacecraft LIB power systems, practitioners who require additional technical details to increase their understanding of space LIB applications, and experienced subject matter experts who need a centralized technical LIB reference source will greatly benefit from this book. Since LIBs are an enabling technology, an important objective was to meet the intent of applicable data protection laws, US export controls, and international traffic and arms regulations. As a result, in lieu of LIB design "how-to" or detailed process explanations, this book provides the basic fundamentals of LIB power systems complemented by detailed public-domain citations readily available for further learning opportunities.

This ambitious book project would not have been completed without the sage guidance, keen focus, and understanding of the world-class John Wiley & Sons UK-based publishing team. A very special thanks to Sandra Grayson (Commissioning Editor) for her encouragement and creative ideas from the beginning of this journey; Becky Cowan (Editorial Assistant) for her patience and attention-to-detail in organizing the book contributor team and copyright processing; and to Patricia Bateson for her detailed copy editing of the book manuscript text, figures, and tables. Most of all I gratefully thank Juliet Booker (Managing Editor) for her continuous candid feedback, flexibility with evolving book content dynamics, and welcome focus on execution.

My special gratitude goes to Thomas Evans (NASA-GSFC) and Dr. Margot Wasz (The Aerospace Corporation) for providing early feedback to the book proposal; Lynn Hitschler for her constant encouragement and genuine interest in this project; Dennis Devine (EaglePicher Technologies) for technical book writing lessons-learned and invaluable advice; Dr. Dan Doughty (Sandia National Laboratories, retired) and Dr. Robert Spotnitz (Battery Design, LLC) for their willingness to co-teach LIB courses with me at UCLA; Dr. Chris Iannello (NASA-KSC) for sponsoring a series of NESC LIB short-courses which helped develop a detailed book vision; Michael Butler (Johns Hopkins Applied Physics Lab) for his technical expertise on spacecraft EPS architecture; and Stephen Ottaway and Danny Montgomery (Thermal Hazard Technology) for their expertise with accelerated rate calorimetry test and analysis of LIBs. I also thank all my past mentors, friends, and colleagues at the NASA-Johnson Space Center, The Aerospace Corporation, and The Boeing Co., who gave me the opportunity to learn-by-doing in dynamic technical work environments.

A book of this detail would not have been possible without the core expertise and early candid feedback from the major space Li-ion cell and battery manufacturers. I am therefore forever grateful to Dr. Robert Gitzendanner (EaglePicher Technologies), Joe Troutman (EnerSys/ABSL), Curtis Aldrich and Tom Pusateri (GS Yuasa Lithium Power, Inc.), and Dr. Yannick Borthomieu (Saft) for their interest in the project, willingness to collaborate by sharing relevant data, and creative ideas for increasing book quality. The biggest challenge in completing a technical book of this scope is finding the quality time needed to write while meeting full-time work commitments. I will always be in debt to each book contributor who selflessly gave their time to complete the project by enduring through my countless virtual meetings, demands for more data, and unrelenting reminders to meet deadlines. This book is a tribute to your dedication and commitment toward achieving our mutual objectives. In addition, each industry peer reviewer who enthusiastically volunteered their valuable time to provide constructive feedback to the final manuscript deserves special credit and recognition. The book team is also grateful to Julianna Calin for her graphics design expertise and guidance with the book chapter artwork.

Our entire book contributor team is forever blessed for the opportunity to work closely with Sara Thwaite throughout the duration of this book project. Sara's technical expertise in space LIB design and test was only exceeded by her ability to foster team good will and camaraderie. More importantly, Sara's ability to listen to different points of view and offer alternate explanations to complex ideas had a welcome calming effect on the team dynamics. Brought up with an Amerindian tribe in Northern Argentina where her parents were missionary aid workers, Sara started life without the modern technology most of us take for granted. Nevertheless, on her return to the UK, she developed a love and innate ability for science and engineering, which led her to the space industry she worked in for the majority of her adult life. Upon reflection from her time in Argentina to her career at EnerSys/ABSL, appropriately one of Sara's favorite sayings was "*From Stone Age to Space Age*." Despite the high technology work focus, Sara retained a passion for nature and travel in her private life and never lost sight of her roots. She married her husband Carl in 1994, and they remained inseparable until the end, working together for over 20 years as well as sharing their hobbies and interests. Sara sadly passed away in May 2022, but had remained working until her last month and was honored to be a part of creating this publication to help share knowledge of battery systems for new engineers in the industry.

Finally, I thank my loving wife Joan who always listened when I needed it most, kept Bunke and Wolfy busy so I could write (or delete) a few more words, and encouraged me to invest the quality time needed to cross the finish line. May we all be blessed with that one person in our life who unconditionally makes us better each and every day.

Thomas P. Barrera

Acronyms and Abbreviations

AC	Assembly Complete
ADCS	Attitude Determination and Control System
AFSPC	Air Force Space Command
Ag-Cd	Silver Cadmium
Ag-Zn	Silver-Zinc
Ah	Ampere-hour
AIAA	American Institute of Aeronautics and Astronautics
AI&T	Assembly, Integration, and Test
ANSI	American National Standards Institute
AOS	Acquisition of Signal
AP	Adaptor Plate
APL	Johns Hopkins University Applied Physics Laboratory
ARC	Accelerating Rate Calorimetry
BCB	Battery Control Boards
BCDU	Battery Charge–Discharge Unit
BIU	Battery Interface Unit
BMS	Battery Management System
BOL	Beginning of Life
BOM	Bill of Materials
BRC	Battery Risk Classification
BSCCM	Battery Signal Conditioning and Control Module
BTA	Bends Treatment Adaptor
C	Nameplate capacity
CAN	Controller Area Network
CID	Current Interrupt Device
CM	Command Module
COMSAT	Commercial Satellite
COTS	Commercial Off-The-Shelf
COVE	CubeSat On-Board Processing Validation Experiment
CPV	Common Pressure Vessel
CSLI	CubeSat Launch Initiative
CSUN	California State University at Northridge
CT	Computed Tomography
CVL	Charge Voltage Limit

DC	Direct Current
DEC	Diethyl Carbonate
DET	Direct Energy Transfer
DHMR	Dry Heat Microbial Reduction
DLR	German Aerospace Center
DMC	Dimethyl Carbonate
DMSP	Defense Meteorological Satellite Program
DoD	Department of Defense
DOD	Depth of Discharge
DOE	Department of Energy
DoT	Design-of-Test
DOT	Department of Transportation
DPA	Destructive Physical Analysis
DRC	Design Reference Case
EBA	Energy Balance Analysis
EBOT	EVA Battery Operations Terminal
EC	Ethylene Carbonate
EDL	Entry, Descent, and Landing
EEE	Electrical, Electronic, and Electromechanical
EGSE	Electrical Ground Support Equipment
EIA	Electrical Interface Assembly
ELV	Expendable Launch Vehicle
EOM	End of Mission
EPS	Electrical Power System
ESA	European Space Agency
ESC	External Short Circuit
EVA	Extravehicular Activity
EM	Engineering Model
EMC	Electromagnetic Compatibility
EMC	Ethyl Methyl Carbonate
EMF	Electromotive Force
EMI	Electromagnetic Interference
EMU	Extravehicular Mobility Unit
EOC	End-of-Charge
EOCV	End-of-Charge Voltage
EODV	End-of-Discharge Voltage
EOL	End of Life
EOM	End of Mission
ER	Eastern Range
ESD	Electrostatic Discharge
ESEO	European Student Earth Orbiter
ESS	Energy Storage Systems
FAA	Federal Aviation Administration
FAR	Federal Acquisition Regulations
FIT	Failure in Time
FM	Flight Model
FMECA	Failure Mode Effects and Criticality Analysis

FOD	Foreign Object Debris
FTS	Flight Termination System
GEO	Geosynchronous Orbit
GLONASS	Global Navigation Satellite System
GNC	Guidance Navigation, and Control
GNSS	Global Navigation Satellite System
GOES	Geostationary Operational Environmental Satellite
GPM	Global Precipitation Measurement
GPS	Global Positioning System
GRACE	Gravity Recovery and Climate Experiment
GRAIL	Gravity Recovery and Interior Laboratory
GSE	Ground Support Equipment
GSFC	Goddard Space Flight Center
GTO	GEO Transfer Orbit
HAR	Hazard Analysis Report
HCM	Hand Controller Module
HEO	Highly Elliptical Orbit
HST	Hubble Space Telescope
HTV	H-II Transfer Vehicle
HVAC	Heating, Ventilation, Air Conditioning
HVI	Hypervelocity Impact
IADC	Inter-Agency Space Debris Coordination
ICD	Interface Control Document
IEA	Integrated Equipment Assembly
IEC	International Electrochemical Commission
IEEE	Institute of Electrical and Electronics Engineers
IESD	Internal Electrostatic Discharge
IOT	In Orbit Test
IOV	In-Orbit Validation
IPV	Individual Pressure Vessel
ISARA	Integrated Solar Array and Reflectarray Antenna
ISC	Internal Short Circuit
ISO	International Organization for Standardization
ISS	International Space Station
ITAR	International Traffic in Arms Regulations
IVA	Intra-Vehicular Activity
JAXA	Japan Aerospace Exploration Agency
JPL	Jet Propulsion Laboratory
JSC	Johnson Space Center
JWST	James Webb Space Telescope
KOH	Potassium Hydroxide
LAT	Lot Acceptance Test
LBB	Leak Before Burst
LCC	Launch Commit Criteria
LCO	Lithium Cobalt Oxide
LCROSS	Lunar Crater Observation and Sensing Satellite
LCT	Life-Cycle Test

LDI	Local Data Interface
LEO	Low Earth Orbit
LIB	Lithium-Ion Battery
LFP	Lithium Iron Phosphate
LiPo	Lithium Polymer
Li-$SOCl_2$	Lithium Thionyl Chloride
LLB	Long Life Battery
LM	Lunar Module
LMO	Lithium Manganese Oxide
LPGT	Li-ion Pistol-Grip Tool Battery
LREBA	Li-ion Rechargeable EVA Battery Assembly
LRO	Lunar Reconnaissance Orbiter
LV	Launch Vehicle
M&P	Materials and Processing
MAPTIS	Materials and Processes Technical Information System
MAVEN	Mars Atmosphere and Volatile EvolutioN
MBSE	Model-Based Systems Engineering
M-Cubed	Michigan Multipurpose Minisat
MEO	Medium Earth Orbit
MGS	Mars Global Surveyor
MLI	Multi-Layer Insulation
MMOD	Micrometeoroid and Orbital Debris
MMRTG	Multi-Mission Radioisotope Thermoelectric Generator
MMS	Magnetospheric Multiscale
MMU	Manned Maneuvering Unit
MOC	Mission Operations Center
MP	Medium Power
MRB	Material Review Board
MSL	Mars Science Laboratory
MTBF	Mean Time Between Failures
NACA	National Advisory Committee for Aeronautics
Na-S	Sodium Sulfur
NASA	National Aeronautics and Space Administration
NCA	Nickel Cobalt Aluminum Oxide
NCO	Nickel Cobalt Oxide
NESC	NASA Engineering Safety Center
Ni-Cd	Nickel Cadmium
Ni-H_2	Nickel Hydrogen
Ni-MH	Nickel Metal Hydride
NMC	Nickel Manganese Cobalt Oxide
NOAA	National Oceanic and Atmospheric Administration
NOD	Native Objects Debris
NSI	NASA Standard Initiator
NTGK	Newman, Tiedemann, Gu, and Kim
OCV	Open-Circuit Voltage
OEM	Original Equipment Manufacturer
ORU	Orbital Replacement Unit

$Pb\text{-}SO_4$	Lead Acid
PC	Propylene Carbonate
PCB	Printed Circuit Board
PCU	Power Control Unit
PGT	Pistol-Grip Tool
PLSS	Portable Life Support System
PMAD	Power Management And Distribution
PPR	Passive Propagation Resistance
PPT	Peak Power Tracker
PSE	Power System Electronics
PTC	Positive Temperature Coefficient
PV	Photovoltaic
PXI	PCI (Peripheral Component Interconnect) Extensions for Instrumentation
R&D	Research and Development
RCC	Range Commanders Council
RCCA	Root Cause and Corrective Action
RF	Radio Frequency
ROM	Rough Order of Magnitude
ROSA	Roll-Out Solar Array
RTCA	Radio Technical Commission for Aeronautics
RTG	Radioisotope Thermoelectric Generator
SAE	Society of Automotive Engineers
SAFER	Simplified Aid for EVA Rescue
SAR	Solar Array Regulator
SAW	Solar Array Wing
SCXI	Signal Conditioning Extensions for Instrumentation
SDO	Solar Dynamics Observatory
SEI	Solid Electrolyte Interphase
SMAP	Soil Moisture Active-Passive
SME	Subject Matter Expert
SOA	State of the Art
SOC	State of Charge
SOE	Sequence of Events
SOH	State of Health
SoP	State of Practice
SOW	Statement of Work
SPCE	Servicing, Performance, and Checkout Equipment
SSC	Space Systems Command
SSE	Solid-State Electrolyte
SysML	Systems Modeling Language
T&C	Telemetry and Command
TAU	Telemetry Acquisition Unit
TCS	Thermal Control System
TLI	Trans Lunar Injection
TLYF	Test-Like-You-Fly
TR	Thermal Runaway
TRL	Technology Readiness Level

TRP	Temperature Reference Point
TRR	Test Readiness Review
TVAC	Thermal Vacuum
UL	Underwriters Laboratories
UN	United Nations
UPS	Uninterruptible Power Supply
USAF	United States Air Force
USSF	United States Space Force
V	Potential in Volts
VAB	Vertical Assembly Building
VC	Vinylene Carbonate
VDA	Voltage Drop Analysis
WCCA	Worst-Case Circuit Analysis
Wh	Watt-hour
WR	Western Range

1

Introduction

Thomas P. Barrera

1.1 Introduction

The role of advanced lithium-ion battery (LIB) power systems has been significant in enabling the widespread decarbonization of fossil-fuel consuming modes of transportation. The development of advanced hybrid and all-electric applications of land, aviation, marine, and rail transport systems benefit greatly from the diverse performance characteristics of LIB technologies. Continued research and development (R&D) into advanced LIB technologies remains a top priority to advancing next-generation electric vehicles, all-electric passenger aircraft, and future spacecraft applications. In contrast to the revolutionary changes that characterize advances in commercial LIB applications, the unique performance requirements, stressful environments, high reliability, and cost constraints of spacecraft battery power systems contribute to a more conservative approach in adopting new commercial LIB developments into traditional spacecraft system applications. As such, the space LIB industry has traditionally been viewed as a niche battery marketplace driven by unique spacecraft customer-driven requirements. These unique space LIB requirements are characterized by electrical, mechanical, and thermal performance operating conditions specific to a diverse set of harsh space mission environments and extended service life needs. In addition, rigorous ground performance and safety test requirements drive additional costs into qualifying space LIB-based electrical power systems (EPSs). As a critical component to next-generation spacecraft EPSs, LIB energy storage technologies must continue to realize performance, safety, and reliability design improvements to keep pace with increasing spacecraft power and service life demands.

1.2 Purpose

The purpose of this book is to capture and transfer existing expert technical knowledge associated with the requirements, design, manufacturing, test, safety, deployment, and operation of spacecraft LIB power systems. The level of treatment is based on a practical, not theoretical, approach to characterizing the technical aspects of the entire life cycle of spacecraft LIBs. The book chapters provide comprehensive technical details which enable industry practitioners who are directly involved with new or existing programs to develop advanced spacecraft LIB-based power systems. In addition, academics engaged in R&D, classroom teaching, hands-on learning, or other educational environments will greatly benefit from the depth and breadth of the book content. This book

Spacecraft Lithium-Ion Battery Power Systems, First Edition. Edited by Thomas P. Barrera.

is also suitable for both undergraduate and graduate students requiring an industry-oriented treatment of lithium-ion (Li-ion) cells and battery applications. A significant portion of the book scope is directly applicable to adjacent LIB markets such as portable commercial electronics, electric vehicles, electrified passenger aircraft, stationary grid energy storage, marine, and other markets requiring reliable LIBs. To accomplish these objectives, a broad system engineering approach toward implementing an LIB power system into a spacecraft application is emphasized. Finally, the book provides the latest state of practice (SoP) information and knowledge, based on relevant space industry experience, which can be used to solve today's challenges facing the safe and reliable deployment of LIBs in spacecraft power system applications.

1.2.1 Background

Across multiple global marketplace applications, Li-ion cell technologies have enabled increased performance advantages over more traditional rechargeable battery cell design options. In Earth-orbiting and planetary mission spacecraft applications, incorporation of LIB designs onto EPS platforms have resulted in numerous improvements to on-orbit capability. The most significant impact of on-orbit spacecraft LIB applications are mass and volume savings to the EPS design architecture. When compared to the same power level of heritage space battery power systems, LIB mass and volume savings translate into increased spacecraft system performance capability. More recently, new commercial space partnerships are driving paradigm shifts away from traditional Department of Defense (DoD) and National Aeronautics and Space Administration (NASA) program business models. These partnerships include academic institutions, US Government agencies, and international entities focused on more affordable spacecraft systems and missions. As such, the recent upsurge in commercial space contracts resulting from the DoD-industry partnerships and the NASA Space Act Agreement has enabled commercial space engineering organizations to have a significant impact on the future of space exploration. However, LIB technology gaps have already emerged where new spacecraft types and mission applications (such as missions to Mars, commercial cargo, and lunar exploration) have driven increasing demands for new energy storage solutions.

To meet these needs, today's spacecraft applications have greatly benefited from leveraging advances in Li-ion cell technology led by the growing commercial electronics and electrified transportation industries. However, spacecraft missions experience stressing electrical, thermal, and mechanical environments, as well as needs for long mission lifetimes, not typically required by terrestrial commercial market LIB applications. In addition, the reliability and resilience requirements for long on-orbit mission durations make implementing a space LIB power system design solution a significant technical challenge. Thus, the more stringent quality, performance, and safety characteristics of spacecraft LIBs require a different set of design solutions not currently found in commercial electronics or electrified transportation LIB applications. This book addresses the demanding and challenging needs of spacecraft LIB power systems while offering LIB technical solutions to meet those challenges.

1.2.2 Knowledge Management

Emerging spacecraft markets and changing commercial procurement paradigms have created new opportunities for Li-ion cell and battery manufacturers to provide more innovative products for their space industry customers. However, the rapid space market growth rate has outpaced LIB knowledge capture and transfer across the broader space LIB community of designers and users. The cause of these widening knowledge gaps can be traced to a number of factors, including space industry barriers toward sharing relevant data, shortfalls in customer dissemination of spacecraft lessons-learned from on-orbit experience, and uncontrolled growth of undocumented organizational tribal knowledge. Furthermore, as the number of on-orbit LIB-based satellites approaching the end of their design service life increases, additional end of mission (EOM) LIB performance data will be collected and stored. Analyses of these new on-orbit trend data will enable future spacecraft EPS architectures, create innovative LIB design improvement opportunities, and aid with developing future spacecraft EOM passivation strategies.

The space-based LIB market has achieved a wide and growing geographically diverse community. North America remains a dominant region for the space LIB market due to the presence of NASA, US DoD, and various well-funded commercial spacecraft manufacturers. This presence has increased investments into domestic space technologies with high priorities on commercialization and national security initiatives. Home to the 22-member state European Space Agency (ESA), Europe also provides a world-leading space LIB market due to the presence of numerous satellite manufacturers, high-value partnerships with NASA, and significant investments into electrified transportation LIB technologies. However, the Pacific rim commercial LIB market also has a rapidly growing space LIB market sector due to the region's growing electrified transportation marketplace. More specifically, China, India, Japan, and South Korea continue to make significant investments in manned and unmanned space exploration programs. Since all spacecraft require rechargeable LIB technologies, the global increase in space exploration is accelerating needs for Li-ion cell and battery knowledge management strategies.

1.3 History of Spacecraft Batteries

Over the past eight decades of space exploration, mission requirements for reduced EPS mass, higher power satellite payload capability, increased mission reliability, and reduced cost have dictated the evolutionary direction of space battery cell technologies. More specifically, the history of spacecraft rechargeable battery cells is a pursuit of cell chemistry and design improvements for increased cell-level specific energy, extended mission life cycle performance, and reduced cost. In addition, the unique chemistries and materials of construction needed for space-qualified battery cell manufacturing require a reliable and cost-effective supply chain marketplace environment.

The need to lower space systems cost has also impacted spacecraft cell and battery development in terms of minimizing the use of expensive raw materials, adopting leaner cell manufacturing processes, and reducing acceptance and qualification test burdens on program stakeholders. Historically, government agency and commercial industry investments into R&D of improved rechargeable space battery cell technologies have had the most significant impacts on transformational battery cell technology improvements. Prior to the introduction of rechargeable Li-ion cell technology into spacecraft applications in late 1990s, aqueous alkaline battery systems, such as silver-cadmium (Ag-Cd), nickel-cadmium (Ni-Cd), and nickel-hydrogen (Ni-H_2), were the only space-qualified battery cell technologies used for main power supporting on-orbit spacecraft EPS during eclipse or peak power periods. Other well-known rechargeable cell technologies, such as nickel-metal hydride (Ni-MH), lead-acid (Pb-SO_4), and sodium-sulfur (Na-S) chemistries, were only used in limited specialty space applications or as flight technology experiments.

1.3.1 The Early Years – 1957 to 1975

The evolution of space battery technologies begins with the earliest spacecraft applications requiring reliable energy storage for short duration missions. By today's standards, early spacecraft EPS energy storage demands were low due to short mission durations, which ranged from a few days to several weeks. Early spacecraft were also faced with stringent mass and volume constraints that required energy storage solutions with high specific energy and volumetric energy densities. Additional challenges such as requirements to survive and operate in severe launch vehicle (LV) and on-orbit environments dictated first-generation space battery power system design solutions [1].

1.3.1.1 Silver-Zinc

The first application of an Earth-orbiting spacecraft battery power system was achieved in 1957 when Sputnik I completed a three-week technology demonstration mission in an elliptical low Earth orbit (LEO) environment. The Sputnik I EPS utilized three silver-zinc (Ag-Zn) batteries to power a radio transmitter and internal thermal management control system [2]. Although the Sputnik I Ag-Zn batteries were not recharged on-orbit, the application was successful in demonstrating an on-orbit energy storage capability under stressing launch and space environments. Throughout the 1960s, the NASA Mercury, Gemini, and Apollo manned space programs primarily utilized non-rechargeable (single discharge) and rechargeable (limited cycle life) Ag-Zn battery technologies in main, standby, and pyrotechnic (squib) power applications. Manufactured by EaglePicher Technologies, the requirements established for the Apollo command module (CM), service module, and lunar module (LM) Ag-Zn batteries were well within the existing battery technology state-of-the-art (SOA) design capabilities. As such, no in-flight Ag-Zn battery safety, reliability, or performance problems were experienced during the short life cycle of these Apollo space battery power applications [3]. In addition to supporting nominal CM and LM power needs during the Apollo program, the use of the LM high-capacity Ag-Zn batteries to provide emergency contingency lifeboat power and to recharge the CM re-entry batteries is credited with enabling the safe return of the Apollo 13 crew to Earth [4]. Today, the most common use of space Ag-Zn batteries is for LV avionics, thrust vector control, pyrotechnics, propulsion subsystem, and flight termination system (FTS) power (Figure 1.1).

1.3.1.2 Silver-Cadmium

During this same time, first-generation rechargeable aqueous alkaline electrolyte silver-cadmium (Ag-Cd) and nickel-cadmium (Ni-Cd) battery power systems were employed for unmanned Earth-orbiting satellite energy storage. In 1960, the NASA-Goddard Space Flight Center (GSFC) initiated satellite programs to study radiation and magnetic fields in near-Earth orbits. Thus, these missions required that all satellite components be constructed from non-magnetic materials to minimize interference with

Figure 1.1 Ag-Zn batteries for Atlas and Delta launch vehicle applications manufactured by EaglePicher Technologies. *Source:* Courtesy of EaglePicher Technologies.

onboard spacecraft magnetometers. As a result, rechargeable Ag-Cd cell technology was chosen over Ni-Cd to meet the non-magnetic and cycle life requirements of these unique spacecraft missions. Launched in August 1961, the NASA Explorer XII spacecraft was the first on-orbit space application of Ag-Cd battery cell technology. Enclosed by an outer all-aluminum battery case, the Explorer XII non-magnetic Ag-Cd battery architecture was composed of 13 prismatic 5Ah Ag-Cd cells (manufactured by Yardney Electric Co.) electrically connected in series. From 1961 to 1973, over 15 NASA spacecraft missions demonstrated that sealed Ag-Cd battery designs were a reliable energy storage device option for short-term space missions operating in ultra-clean magnetic environments [5].

1.3.1.3 Nickel-Cadmium

The first on-orbit application of rechargeable Ni-Cd battery-based EPS technology was aboard the NASA Explorer VI spacecraft. Launched on August 6, 1959, the Explorer VI EPS battery was composed of 14 cylindrical 5 Ah Ni-Cd F-cells (manufactured by Sonotone Corp.). The Ni-Cd battery operated at a shallow depth-of-discharge (DOD) of 2% in a highly elliptical orbit (HEO) [6]. Collaboration between government agency stakeholders and space industry Ni-Cd cell manufacturers (such as General Electric, Saft, and EaglePicher Technologies) enabled design solutions to overcome early technical challenges, which included understanding the root causes of cell anomalous behavior, quantifying the relationships between mission-specific environments and operating conditions, and devising new strategies for reliable charge–discharge monitoring and control (Figure 1.2). In the 1970s, these developments resulted in the widely used NASA standard prismatic 20 and 50 Ah Ni-Cd designs for Earth geosynchronous missions [7, 8]. Over the next 20 years, the evolution of space Ni-Cd cell technology included improvements to the cell terminal seals, separator material, and positive electrode manufacturing processes [9]. These cell-level design improvements extended mission life for numerous early LEO and geosynchronous Earth orbit (GEO) mission spacecraft applications.

1.3.2 The Next Generation – 1975 to 2000

Further demands for higher power and lower mass spacecraft EPS necessitated improvements to Ni-Cd cell specific energy, cycle life, and reliability, which created new development opportunities

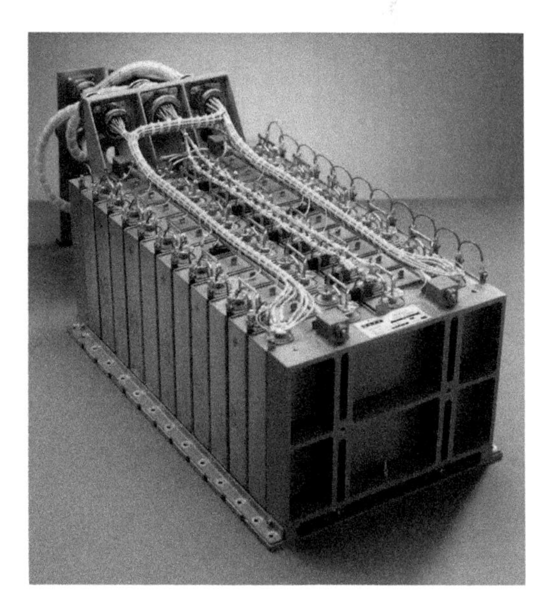

Figure 1.2 Launched in March 1998, the SPOT 4 commercial high-resolution optical Earth imaging satellite was powered by a 28 V 40 Ah Ni-Cd battery power system. *Source:* Courtesy of Saft.

for next-generation space battery cell technologies. The earliest efforts to develop rechargeable Ni-H_2 cell technology for space applications started in 1970 when the Communications Satellite Corporation laboratories developed the first hermetically-sealed Ni-H_2 cell design for GEO satellite applications [10]. Also, an aqueous alkaline-based cell technology, rechargeable Ni-H_2 cell chemistry was a hybrid technology derived from combining the proven nickel-hydroxide positive electrode from space Ni-Cd cells along with the platinum negative electrode commonly used in space-qualified hydrogen-oxygen alkaline fuel cells [11].

1.3.2.1 Nickel-Hydrogen

Similar to the early development of Ni-Cd cells for space battery applications, R&D funding for Ni-H_2 cell development emanated from industry and government entities with an interest in promoting secondary battery technology for GEO and LEO space applications. These organizations included commercial industry (such as AT&T Bell Laboratories, International Telecommunications Satellite Consortium, Hughes Aircraft Co., and EaglePicher Technologies), NASA, US Air Force (USAF), and US Naval Research Laboratory government agencies (Figure 1.3). In part supported by USAF funding, The Aerospace Corporation pioneered early R&D into Ni-H_2 cell chemistry, fundamental electrode analysis techniques, and life cycle testing (LCT) critical to maturing Ni-H_2 battery cell technology for space applications [12].

Launched on June 23, 1977, the US Navy Navigation Technology Satellite (NTS)-2 was the first on-orbit demonstration of the feasibility of Ni-H_2 battery technology in a space application. The NTS-2 Ni-H_2 battery consisted of 14 hermetically-sealed 35 Ah cells (manufactured by EaglePicher Technologies) electrically connected in series and separated into two 7-cell modules [13]. NTS-2 successfully met its primary mission objectives of operating for five years in a polar 12-hour orbit between 40 and 60% DOD [14]. Shortly after the success of NTS-2, a deliberate transition from heritage Ni-Cd battery cell technology was supported by various US government agencies and the rapidly growing commercial satellite manufacturing industry. In 1979, Ford Aerospace (which later became Maxar Technologies) replaced the Ni-Cd batteries baselined for the Intelsat V GEO telecommunications satellite system with 30 Ah Ni-H_2 batteries. Four years later, Intelsat V Flight 6 (F-6) became the world's first commercial satellite to use Ni-H_2 batteries for an on-orbit main EPS application. The higher specific energy, longer LEO cycle life, and inherent safety characteristics of Ni-H_2 battery cells were responsible for their use in the NASA Hubble Space Telescope (HST) and International Space Station (ISS) spacecraft EPS (Figure 1.4). From the launch of NTS-2 up to the mid-2000s, nearly all

Figure 1.3 28V Ni-H_2 satellite battery manufactured by EaglePicher Technologies. *Source:* Courtesy of EaglePicher Technologies.

Figure 1.4 Launched in 1990, the NASA Hubble Space Telescope utilizes six $Ni\text{-}H_2$ batteries (23 cells per battery) enclosed in two separate modules. Each astronaut-replaceable module contains three 88 Ah $Ni\text{-}H_2$ batteries manufactured by EaglePicher Technologies [15]. Credit: NASA.

commercial and government agency satellite programs employed $Ni\text{-}H_2$ batteries for EPS energy storage [16]. During this same time period, numerous deep-space planetary missions (such as 2001 Mars Odyssey, Messenger and the Mars Reconnaissance Orbiter) also utilized $Ni\text{-}H_2$ batteries.

1.3.2.2 Sodium-Sulfur

High-temperature rechargeable sodium-sulfur (Na-S) cell technology was developed in the 1980s and 1990s for emerging spacecraft, electric vehicle, and electric utility energy storage applications. Advantages such as high cell-level specific energy, long cycle life, and high energy efficiency were attractive characteristics when compared to other competing rechargeable battery cell technologies. In the late 1980s, the USAF began funding the development of Na-S cell technologies for spacecraft applications. Na-S batteries operate at high temperatures between 300 and 350 °C and were expected to be a lower mass and more cost-effective alternative to space-qualified $Ni\text{-}H_2$ cells. Furthermore, ground testing of Na-S cells demonstrated a threefold increase in specific energy versus comparable $Ni\text{-}H_2$ cell chemistries. Although the technology readiness of Na-S cell technology was mature, spacecraft thermal integration risks caused this technology to never be used in a spacecraft EPS application. However, in late 1997, an on-orbit technology demonstration experiment aboard the NASA Space Shuttle (STS-87), named the Sodium Sulfur Battery Experiment, successfully characterized the performance of four 40 Ah Na-S cylindrical cells (manufactured by EaglePicher Technologies) in a microgravity environment [17]. Today, Na-S battery cell technologies are commonly used in megawatt-scale power grid applications, including load leveling and standby energy storage power sources for renewable energy resources [18].

1.3.2.3 Transition to Lithium-Ion

Despite the proliferation of space $Ni\text{-}H_2$ batteries in spacecraft applications, the extremely high cost of space $Ni\text{-}H_2$ cells was a significant market-entry barrier for terrestrial commercial applications. $Ni\text{-}H_2$ space cell cost was dominated by high costs associated with manufacturing touch-labor, critical raw materials availability, low-quantity production runs, limited cell manufacturing base, and supply chain constraints. By the mid-2000s, Ni-H_2 cell manufacturers (such as EaglePicher Technologies, The Boeing Co., and Saft) were faced with critical raw materials obsolescence issues and a degrading subtier supplier base, which created significant cost and manufacturing risk to continued space Ni-H_2 battery cell production. For example, heritage Dupont Teflon T-30 emulsion used for producing Ni-H_2 platinum negative electrodes was being phased out due to a worldwide ban on the industrial use of perfluorooctanoic acid [19]. In addition, other critical raw materials, such as positive electrode nickel powder and Zircar separator materials, were at risk of becoming obsolete. Although engineering alternatives were proposed to mitigate the risk of Ni-H_2 cell obsolescence, the cost of maintaining Ni-H_2 technology was found to be significant when compared to the long-term benefits of investing in a transition toward higher specific energy LIB technologies [20]. In addition, the significant increases in specific energy, energy density, and lower forecasted costs of Li-ion cell technology further accelerated Ni-H_2 space cell manufacturers to cease production. In 2016, US domestic suppliers of Ni-H_2 space batteries discontinued production of Ni-H_2 cells for spacecraft applications. Launched on August 18, 2017, the Tracking and Data Relay (TDRS)-M GEO satellite was the final launch of a NASA spacecraft utilizing a Ni-H_2 battery-based EPS.

1.3.3 The Li-ion Revolution – 2000 to Present

In the mid-1990s as the consumer electronics market began a transition to rechargeable LIB technologies, the emerging market for LIB-powered electric vehicles was also accelerated by large R&D investments into new Li-ion cell designs. Existing battery companies such as Japan Storage Battery (now GS Yuasa), SAFT, and BlueStar Advanced Technology Corp (later acquired by EaglePicher Technologies) led the industry in providing commercial satellite manufacturers with first-generation custom Li-ion cells for test and evaluation [21]. At the same time, numerous commercial off-the-shelf (COTS) 18650 Li-ion cell designs were under consideration for first-generation small satellite technology demonstration programs. These early technology transitions were initially based on real-time and accelerated LCT demonstrations performed in controlled laboratory environments using relevant spacecraft EPS operating conditions. In addition to LCT, space-industry relevant electrical, thermal, and mechanical testing further indicated that Li-ion technology would meet GEO and LEO EPS space mission requirements [22–30].

1.3.3.1 First Space Applications

NASA's first on-orbit application of Li-ion cell technology was battery power for portable commercial camcorder video cameras and laptops safety-certified for astronaut crew member use on the NASA Space Shuttle and ISS. The video camera (2s2p design) and laptop LIBs (4p3s design) consisted of first-generation Sony and Panasonic 18650 COTS LIB cells, respectively [31, 32]. Soon afterwards, the world's first on-orbit space application of rechargeable LIBs was launched on October 22, 2001 aboard the ESA-funded Project for On-Board Autonomy (PROBA)-1 LEO technology demonstration spacecraft. Figure 1.5 shows the PROBA-1 EPS 9Ah LIB (manufactured by EnerSys/ABSL), which consisted of 36 Sony 18650HC COTS Li-ion cells electrically connected in a 6s6p design configuration [33]. Originally designed as a one-year mission, the PROBA-1 spacecraft continues to operate after over 20 years (>100 000 LEO cycles) to date of continuous on-orbit service (Figure 1.6).

In the Summer of 2003, NASA launched the twin Mars rovers, Spirit and Opportunity, which utilized custom prismatic Li-ion 10 Ah cells with a low temperature electrolyte in an 8s2p design

Figure 1.5 PROBA-1 6s6p 9 Ah LIB manufactured by EnerSys/ABSL. Credit: Prime Contractor QinetiQ Space.

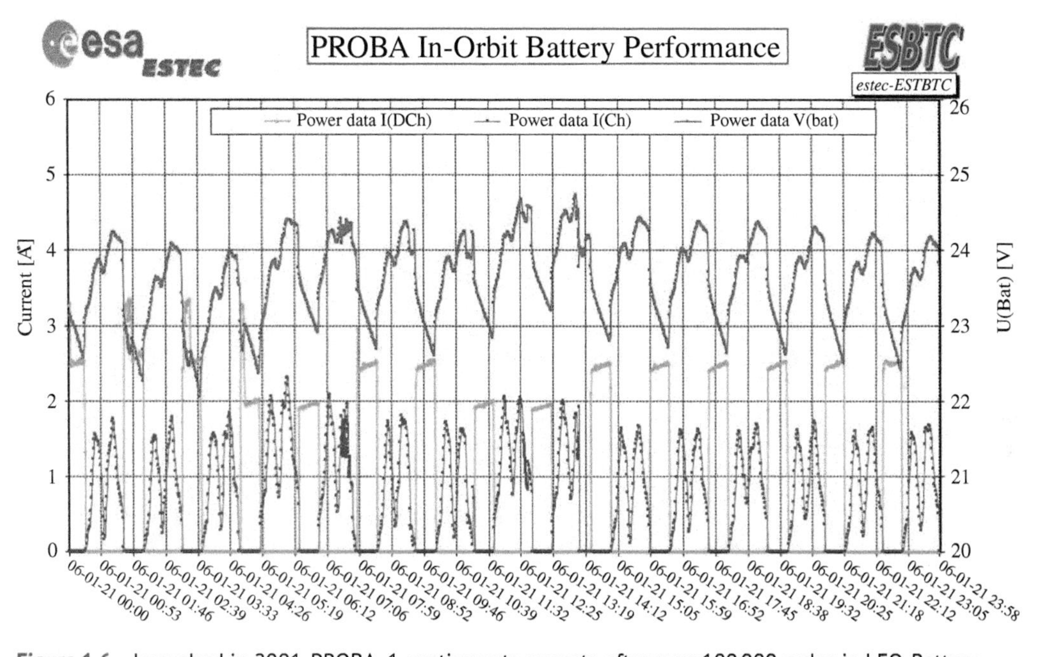

Figure 1.6 Launched in 2001, PROBA-1 continues to operate after over 100 000 cycles in LEO. Battery performance data are shown for a single orbit on January 6, 2021. *Source:* Adapted from Prime Contractor QinetiQ Space.

configuration (manufactured by Yardney Technical Products, later acquired by EaglePicher Technologies). Designed for a 90 Martian solar day mission, both Spirit and Opportunity operated in deep space for over 6 and 14 Earth-years, respectively. Launched in June 2003, Mars Express was the first ESA space exploration science mission to utilize a COTS Li-ion 18650 cell-based LIB (manufactured by EnerSys/ABSL). Still in operation today, Mars Express represents the longest surviving planetary mission with LIBs. Launched later in May 15, 2004, the Astrium (later acquired by Airbus) W3A satellite was the first commercial telecommunications satellite to use an LIB-based EPS in a GEO orbit using custom Li-ion VES140 cells (manufactured by Saft) [34].

Figure 1.7 Japan's resupply ship, the HTV-9 attached to the ISS Harmony module. Stowed inside the Japanese space freighter is the HTV-8 cargo pallet that was brought up to the ISS on a previous resupply mission. The pallet contains old Ni-H_2 batteries for disposal that were replaced by new LIBs during a series of spacewalks in 2019. Credit: NASA.

To date, the most complex spacecraft application of LIB cell technology has been the on-orbit replacement of the NASA ISS Ni-H_2 batteries with new high specific energy LIBs [35]. Completed in February 2021, the NASA ISS required multiple LIB on-orbit deliveries, via the Japanese H-II Transfer Vehicle (HTV) cargo spacecraft, to replace the aging Ni-H_2 batteries (Figure 1.7). Fourteen extravehicular activity (EVA) astronaut spacewalks conducted over four years were needed to replace the Ni-H_2 batteries with new long-life LIBs qualified to power the ISS EPS through to 2030. Due to the higher specific energy of Li-ion cell technology, only 24 LIBs were needed to replace 48 Ni-H_2 batteries across all eight ISS power channels.

1.3.3.2 Advantages and Disadvantages

In practice, it is commonplace for detailed engineering trade studies to analyze Li-ion cell design, performance, and safety data relative to spacecraft LIB performance and safety requirements. Space industry ground and on-orbit experience with Li-ion cell technology has shown that there are characteristic advantages and disadvantages that impact LIB implementation into spacecraft applications. Table 1.1 lists the major advantages and disadvantages of Li-ion cells as compared to heritage space-qualified Ni-Cd and Ni-H_2 battery cell technologies [36]. Stringent battery safety requirements for crewed spacecraft require additional hazard controls for maintaining fault tolerance to protect against LIB hazards, such as those that may lead to catastrophic cell-to-cell propagating thermal runaway (TR). LIB intolerance to overcharge and overdischarge conditions requires additional battery management system (BMS) functionality, which increases mass and cost to most spacecraft EPS designs. Current COTS and custom Li-ion cell technologies are also challenged to meet future planetary science spacecraft energy storage requirements. For example, future missions to the outer planets (such as the Ocean Worlds and Ice Giants), inner planets (Venus and Mercury), and Mars all have unique energy storage needs that require higher specific energy (>500 Wh/kg), longer calendar life (>15 year), and longer cycle life (>50 000 cycles). Development of next-generation rechargeable batteries supporting outer and inner planetary missions will also be required to operate in extreme hot (> +60 °C) and cold (< −180 °C) temperature environments [37].

Table 1.1 Advantages and disadvantages of Li-ion cells and batteries for spacecraft applications.

Advantages	Disadvantages
High specific energy (Wh/kg) and energy density (Wh/L)	Lack of procurement controls on COTS cell suppliers
High coulombic and energy efficiency	Irreversible degradation above +65 °C
Long cycle and calendar life characteristics	Cell degradation at high charge rates and low temperatures
Low self-discharge rate	Intolerant to overcharge and overdischarge
Low heat generation during charge and discharge	Added mass and cost due to need for cell-to-cell charge balancing electronics (if required)
High operating cell-level voltage	Added mass and cost due to battery management systems (if required)
SOC can be estimated from cell voltage	Cell and battery safety hazard controls are required
Wide variety of COTS cell designs enables battery design flexibility	Severity and consequences of thermal runaway hazards
Low cost of COTS cell designs	Propagation resistant battery designs add mass, volume, and cost
No need for reconditioning due to no memory effect	Need to comply with Li-ion battery transportation regulations
Various cell geometries enable battery packaging efficiencies	Supply chain risks due to critical raw materials availability
Ability to electrically connect cells in parallel to create a module of higher capacity	
Operating temperature range (−10 to +40 °C) reduces thermal management requirements	
Competitive and diverse manufacturing base	

1.4 State of Practice

Due to the significant number of on-orbit spacecraft utilizing LIB-based EPS and the wide diversity of successful on-orbit mission applications, the space Li-ion cell and battery market sector is approaching maturity. However, the SoP for spacecraft LIBs remains a continuous balance to achieve an optimal combination of cell cost, mass, and safety for the intended on-orbit application. Since the late 1990s, commercial portable electronics and electric vehicle market needs for advanced Li-ion cell technologies have mostly dictated the trajectory of the space battery market sector. These adjacent commercial market applications share a common Li-ion cell design trade space with spacecraft applications, albeit at different thresholds of requirements compliance. These market dynamics are especially relevant for the accelerating use of COTS Li-ion cell designs in space LIB applications. Custom space-qualified Li-ion cell designs are also dependent on commercial supply chains, manufacturing equipment and tooling, and key critical personnel skills to maintain competitiveness within the niche space LIB marketplace.

1.4.1 Raw Materials Supply Chain

Historically, US space Ni-Cd and Ni-H_2 battery cell manufacturers depended on a secure domestic industrial base for a reliable supply of raw materials for battery cell production. Maintaining a sustainable domestic supplier industrial base increased the ability to control material quality, lowered the risk to materials obsolescence, and enabled compliance to US government restrictions on the use of foreign subtier suppliers. Similarly, stable and reliable sources for critical raw materials used to manufacture space Li-ion cells are necessary to meet spacecraft program cost, schedule, and technical requirements. For example, in order to avoid cell re-qualification costs and increase on-orbit LIB reliability, spacecraft manufacturers commonly require a stable and traceable Li-ion cell raw materials supply over the lifetime of the cell product line.

Today, the US and foreign-based space LIB manufacturers rely mainly on international markets for supplying critical Li-ion cell raw materials. As a result, the complexities of the global Li-ion cell raw materials supply chain have created a significant supply chain risk to the niche space LIB manufacturing industry [38]. Assuring a stable source of high-quality Li-ion cell active raw materials has caused custom space Li-ion cell manufacturers to align with commercial market demands for raw materials and diversify supplies from secondary sources [39]. However, the relatively low-production volume of custom space Li-ion cells, coupled with stringent materials requirements, are insufficient to influence the trajectory of commercial Li-ion cell market demands. This risk has been partially mitigated by the use of high-production volume COTS Li-ion cells commonly used in high-demand commercial portable electronics applications. However, maintaining visibility into cell design changes and advance notice of changes to critical cell raw materials used in COTS Li-ion cells is not available without establishing special supplier contractual agreements that may have prohibitive cost impacts to LIB procurements. To avoid materials obsolescence risks, custom Li-ion space cell manufacturers have successfully employed strategic risk mitigation approaches, such as stockpiling of critical cathode and/or anode materials until replacements are qualified for space applications, as well as funding investments into domestic suppliers to develop next-generation Li-ion cell critical raw materials [40].

1.4.2 COTS and Custom Li-ion Cells

Space industry experience has demonstrated that the use of COTS or custom Li-ion cell designs for space LIB applications is a continuous engineering trade between mission-relevant cost, schedule, and technical requirements. Cell-level specific energy, energy density, safety, reliability, and modularity continue to be key technical requirements for cell selection trade studies [41]. More recently, the hazard severity risks of catastrophic cell-to-cell propagating TR has become a key discriminator for LIB safety requirements. Other cell-level trade study requirements include impacts to EPS mass and cost due to LIB cell-to-cell balancing electronics, if needed. Although custom space Li-ion cell design solutions have always been challenged to compete with COTS Li-ion cell designs on cost and specific energy, satellite size remains a significant space market discriminator for trades between large-format custom and small-format COTS Li-ion cell designs. For example, COTS Li-ion cell designs are generally favored for small satellite applications, new LV designs, and a growing number of planetary mission applications. As such, the share of Earth-orbiting satellites using COTS Li-ion cells is growing with increasing market cost pressures.

1.4.3 Hazard Safety and Controls

Recent industry-wide LIB power system safety incidents have significantly impacted how Li-ion cell and battery suppliers design, manufacture, test, store, transport, and operate LIB-based energy storage systems. These shifting industry safety paradigms are reflected in new industry LIB safety requirement standards applicable to the entire Li-ion cell and battery life cycle. A major focus of the new LIB safety standards is mitigating the risk to catastrophic LIB TR hazards impacting passenger modes of transportation, such as ground transportation, aviation, and marine applications. Similarly, the impact of a catastrophic LIB safety hazard on a manned (crewed) on-orbit spacecraft mission may result in loss of the mission and/or crew members. However, in an unmanned

Earth-orbiting satellite or robotic planetary spacecraft mission, the consequences of a catastrophic LIB safety incident are less severe when compared to a similar safety incident on a crewed spacecraft application. As a result, the implementation of TR propagation-resistant LIB designs for crewed or unmanned spacecraft applications remains focused on safety hazard risk assessments, impacts to EPS mass and cost, and compliance to reliability requirements.

1.4.4 Acquisition Strategies

Present Earth-orbiting satellite acquisition methods are commonly challenged since the duration between concept and launch can take from six to eight years, resulting in constellations where each satellite is unique in structure and capability, due to its relatively slow replacement rate and the continuing need for system modernization [42]. The space industry is developing faster, more agile, acquisition strategies to lower system costs and procurement cycle times so that more current technology can be launched. In addition, there is intense interest in developing the means to quickly field a space system that delivers a basic capability, such as broad-band communication, over a large number of small satellites in lieu of a small constellation of highly reliable, but expensive, large spacecraft. The increased use of the high-reliability associated with high-volume production COTS Li-ion cell-based LIBs may provide a means to satisfy agile acquisition strategy goals without sacrificing traditional reliability objectives for spacecraft LIBs [43]. New spacecraft EPS designs are forecasted to continue benefitting from leveraging technical advances led by the consumer electronics, electric ground transportation, and grid LIB energy storage marketplace. For satellite systems, a significant barrier that challenges the rapid adoption of next-generation LIBs is the five to ten years of ground service life demonstrations required by some high-value programs. This necessitates that real-time ground LCT be performed before a new Li-ion battery cell design can be incorporated on to a spacecraft with long on-orbit mission service life requirements [44].

1.5 About the Book

Space industry experience has demonstrated that the principles of spacecraft LIB power systems are best understood by considering a systematic approach to understanding the entire LIB life cycle. This approach ensures that spacecraft customer top-level requirements related to performance, safety, reliability, and operating conditions are sufficiently decomposed and translated into the applicable spacecraft EPS design solution. The result is a functionally and operationally compliant spacecraft LIB power system that meets spacecraft mission requirements.

Historically, the rechargeable space battery design process has relied on a semi-empirical approach to deliver compliant flight hardware design solutions to meet specification requirements. This balanced design approach is dictated by spacecraft program cost, schedule, and technical requirements associated with the relevant mission-specific objectives. As such, the most common LIB requirement specification compliance methods are test and analysis. Hence, the combination of electrical, thermal, and mechanical analyses, coupled with relevant test verification methodologies, are commonplace in qualifying spacecraft LIB power systems to meet on-orbit service life demands. Although the fundamental engineering principles of heritage space battery life cycle characteristics are well known, the unique performance characteristics of LIB cell technologies have required significant changes to heritage space battery design, analysis, and test best practices. Similarly, LIB ground processing and mission operation approaches have also been modified as needed from heritage ground and on-orbit battery management approaches. Additional technical challenges with increasing LEO and GEO satellite power levels, deep space missions to outer planets with extreme environments, and growing needs for mitigating LIB safety hazards require new design solutions beyond today's knowledge and experience base.

1.5.1 Organization

In this book, a space LIB power system is defined as a multi-cell LIB that provides reliable energy storage capability to a spacecraft EPS. In order to best understand the major process steps associated with deploying a space-qualified LIB, the book is organized with a series of detailed chapters describing the LIB product life cycle. The book chapters are sequenced in chronological order according to the major hardware integration steps in the LIB life cycle. Since all space LIBs are integrated into the spacecraft EPS, an integrated systems engineering approach is employed to emphasize the importance of understanding how spacecraft, environmental, and EPS requirements impact space LIB design and test. The significance of LIB ground processing activities is described, beginning with flight battery shipment from the battery manufacturer to the spacecraft EPS integrator, and concluding with arrival at the spacecraft launch site. The role of an integrated flight battery in pre-launch, launch day, and post-launch spacecraft operations is discussed in terms of preparation for on-orbit mission operations. Finally, on-orbit flight LIB management, via trending of LIB and EPS telemetry data during on-orbit mission operations up through EPS passivation at spacecraft EOM, completes the life cycle of the flight LIB.

1.5.2 Space Li-ion Cells and Batteries

The life cycle of a space-qualified LIB begins with understanding the general principles of Li-ion cell and battery engineering, design, test, and operation as it relates to the unique requirements of spacecraft EPS. Chapter 2 provides a detailed overview of Li-ion cell technologies used in today's space-qualified LIBs. The role of cell-level requirements on design features found in custom and COTS Li-ion cell technologies used in space LIBs is described. After successful completion of cell-level acceptance testing, Li-ion cells are selected for integration into the LIB design. This critical next step in the Li-ion cell life cycle is described in Chapter 3, with a special focus on LIB-level requirements, interfaces, and design features representative of Earth-orbiting and planetary mission spacecraft LIBs. A detailed description of LIB testing for acceptance, proto-qualification, and qualification of space LIBs is provided. The importance of a LIB requirement specification, cell selection and matching processes, electrical/mechanical/thermal interface definition, analyses, and safety testing provide the depth and breadth needed for understanding the unique characteristics of space LIB designs.

1.5.3 Electrical Power System

Composed of all the hardware and software required to generate, store, manage, and distribute electrical energy to all spacecraft user loads, the EPS is a mission-critical spacecraft subsystem. The next major life cycle event for the flight LIB energy storage unit is therefore integration into the spacecraft's EPS. Using a system engineering approach to assembly, integration, and test (AI&T), Chapter 4 provides an overview of the key requirements, architecture options, analysis types, and test events associated with an integrated spacecraft EPS. A special focus on the role of the LIB energy storage capability in meeting EPS requirements is emphasized in terms of the solar array and power management and distribution (PMAD) components. Also described are critical

LIB-based EPS functional and operational capabilities, such as architecture, BMS, energy and power budget analyses, and dead bus event characteristics.

1.5.4 On-Orbit LIB Experience

The purpose of a space-qualified LIB-based EPS is to provide reliable power to enable on-orbit spacecraft to meet their intended mission objectives. Describing this next phase in the LIB life cycle is accomplished in terms of discussing the on-orbit experience of a wide variety of successful Earth-orbiting satellite and planetary mission spacecraft LIB power systems. Chapter 5 is focused on the general design features of LIBs used in Earth-orbiting satellite applications, such as the NASA ISS, James Webb Space Telescope (JWST), and astronaut extravehicular mobility unit (EMU) LIBs. Examples from ESA spacecraft using COTS Li-ion cell-based LIBs are also described. Chapter 6 describes the unique LIB power system requirements and designs for planetary robotic mission orbiters, landers, and rovers to different destinations, such as the Moon, Mars, Jupiter, Venus, and various asteroids. Unique planetary mission requirements, such as radiation tolerance, extreme environmental temperature, and planetary protection are described in the context of LIB design solutions.

1.5.5 Safety and Reliability

The high specific energy, flammable cell electrolyte, and intolerance to overcharge–overdischarge conditions of LIBs present unique safety hazard risks to all spacecraft applications. Identifying safety hazards and implementing hazard controls are critical for mitigating the risk of a catastrophic safety incident during LIB ground processing and/or on-orbit operations. The purpose of Chapter 7 is to describe space LIB safety and reliability for manned (crewed) and unmanned spacecraft applications in terms of requirements, analyses, hazard controls, and safe-by-design opportunities for implementing safe LIB power systems. An emphasis on the risk of TR in space Li-ion cells and batteries is provided in terms of incorporating safe-by-design principles to mitigate catastrophic cell-to-cell TR propagation. In addition, practical approaches to predicting space LIB reliability are discussed in order to estimate the likelihood of the LIB meeting EPS performance requirements for the specified mission lifetime.

1.5.6 Life Cycle Testing

Li-ion cell and battery qualification for space applications includes test and analysis methods to demonstrate compliance with LIB performance requirements over the service life of the spacecraft mission. As such, LCT is considered a critical test event in qualifying a LIB for any on-orbit spacecraft application. Chapter 8 details the general aspects of LCT in terms of test requirements, planning, test facility readiness, and test approaches commonly used in qualifying space LIBs for on-orbit service. A special emphasis on real-time and accelerated LCT techniques as they apply to qualifying space LIBs is discussed. In addition, modeling and simulation methodologies are described for analyzing LCT data to predict on-orbit service life.

1.5.7 Ground Processing and Mission Operations

The final stages of the space LIB life cycle include delivery of the flight LIB to the launch-site, followed by the execution of ground processing, launch, and on-orbit operations. With a special focus on the role of the spacecraft LIB-based EPS, Chapter 9 highlights the key pre-launch ground processing, launch day, and on-orbit mission operation phases of the spacecraft. Described are normal on-station operation, trending of on-orbit LIB telemetry, contingency operations, and EOM decommissioning of the spacecraft. A special focus on the role of the spacecraft EPS on safe and reliable passivation operations is also discussed.

1.6 Summary

Rechargeable battery power systems have successfully supported over 75 years of on-orbit spacecraft operation across a technologically diverse set of unmanned and manned mission applications. The success of heritage battery technologies such as Ag-Zn, Ni-Cd, and Ni-H_2 batteries in powering all types of spacecraft missions provided the foundation for a deliberate transition to LIB technologies. Although some of today's heritage expendable LVs rely on Ag-Zn batteries for main power, these legacy battery power systems are being displaced by compliant LIB technology options. Nonetheless, all of today's Earth-orbiting and planetary mission spacecraft utilize LIB power systems to meet spacecraft EPS requirements. Today's space LIB SoP is continuously challenged by new space program requirements to lower cost and increase mass savings while meeting stringent spacecraft operating requirements. The scope of this book provides a framework for understanding the importance of leveraging on-orbit spacecraft experience, engineering best practices, and space industry standards to meet mission-specific spacecraft LIB requirements. To accomplish these objectives, this book emphasizes the key aspects of the entire space LIB power system life cycle.

References

1 Bauer, P. (1968). *Batteries for Space Power Systems*. National Aeronautics and Space Administration, NASA SP-172.

2 Gruntman, M. (2004). *Blazing the Trail – The Early History of Spacecraft and Rocketry*. Reston, VA: American Institute of Aeronautics and Astronautics.

3 Trout, J.B. (1972). *Apollo Experience Report – Battery Subsystem*. National Aeronautics Space Administration, NASA TN D-6976.

4 Cortwright, E.M. (1970). *Report of the Apollo 13 Review Board*. National Aeronautics Space Administration.

5 Hennigan, T.J. (1978). *Sealed Silver Oxide-Cadmium Batteries for Spaceflight: 1960–1977*, NASA CR-156838. Greenbelt, Maryland: Goddard Space Flight Center.

6 Ford, F.E., Rao, G.M., and Yi, T.Y. (1994). *Handbook for Handling and Storage of Nickel-Cadmium Batteries: Lessons Learned*. National Aeronautics Space Administration, NASA RP 1326.

7 *The NASA Standard 20Ah Nickel-Cadmium Battery Manual* (1980). National Aeronautics Space Administration, NASA CR 166700, McDonnell Douglas Corporation, St. Louis, MO.

8 *The NASA Standard 50Ah Nickel-Cadmium Battery Manual* (1981). National Aeronautics Space Administration, NASA CR 166701, McDonnell Douglas Corporation, St. Louis, MO.

9 Halpert, G. and Surampudi, R. (1997). An historical summary and prospects for the future of spacecraft batteries. In: *NASA Aerospace Battery Workshop*. Huntsville, AL.

10 Stockel, J.F., Van Ommering, G., Swette, L., and Gaines, L. (1972). A nickel-hydrogen secondary cell for synchronous satellite applications. In: *Proceedings of the 7th Intersociety Energy Conversion Engineering Conference*, 87–94. San Diego, CA.

11 Dunlop, J.D., Giner, J., Van Ommering, G., and Stockel, J.F. (1975). Nickel hydrogen cell. US Patent 3,867,199. Filed June 5, 1972 and Issued February 18, 1975.
12 Zimmerman, A.H. (2009). *Nickel-Hydrogen Batteries: Principles and Practice*. The Aerospace Press.
13 Dunlop, J.D., Gopalakrishna, R.M., and Yi, T.Y. (1993). *NASA Handbook for Nickel-Hydrogen Batteries*. NASA Reference Publication 1314.
14 Stockel, J.F., Dunlop, J.D., and Betz, F. (1978). NTS-2 nickel-hydrogen battery performance. In: *7th Communications Satellite Systems Conference*, 24–27 April 1978. San Diego, CA.
15 Hollandworth, R., Armantrout, J., Whitt, T., and Rao, G.M. (2004). Hubble space telescope 2004 update. In: *NASA Aerospace Battery Workshop*. Huntsville, AL.
16 Borthomieu, Y. and Bernard, P. (2009). Secondary batteries – nickel systems|nickel–hydrogen. In: *Encyclopedia of Electrochemical Power Sources* (ed. J. Garche), 482–493. Elsevier.
17 Wasz, M.L., Carter, B.J., Donet, C.M., and Baldwin, R.S. (1998). Destructive physical analysis of flight and ground tested sodium sulfur cells. In: *NASA Aerospace Battery Workshop*. Huntsville, AL.
18 Kumar, D., Rajouria, S.K., Kuhar, S.B., and Kanchan, D.K. (2017). Progress and prospects of sodium-sulfur batteries: A review. In: *Solid State Ionics*, 312.
19 Wasz, M. and Zimmerman, A. (2008). *The Effect of Teflon Emulsion on Hydrogen Electrode Properties and Performance in Nickel-Hydrogen Cells*. The Aerospace Corporation, TR-2008(8555)-2.
20 Wasz, M. (2010). Coping with T30 obsolescence in 2013. In: *NASA Aerospace Battery Workshop*. Huntsville, AL.
21 Marsh, R.A., Vukson, S., Surampudi, S. et al. (2001). Li-ion batteries for aerospace applications. *J. Power Sources* 97–98.
22 Semerie, J.P. (2000). Lithium-ion batteries for geosynchronous satellites: Qualification test results of the STENTOR battery. In: *35th Intersociety Energy Conversion Engineering Conference and Exhibition*, vol. 1, 621–628. Las Vegas, NV. http://dx.doi.org/10.1109/IECEC.2000.870846.
23 Croft, H., Kawtenjna, P., and Staniewicz, B. (2002). Li ion cell performance for space application. In: *Proceedings of the 17th Annual Battery Conference on Applications and Advances*. Long Beach, CA.
24 Smart, M.C., Ratnakumar, B.V., Chin, K.B. et al. (2002). Lithium-ion cell technology demonstration for future NASA applications. In: *37th Intersociety Energy Conversion Engineering Conference*, 297–304. Washington, DC.
25 Bruce, G.C. and Marcoux, L. (2001). Large lithium-ion batteries for aerospace and aircraft applications. In: *Proceedings of the 16th Annual Battery Conference on Applications and Advances*. Long Beach, CA.
26 Lurie, C. (1997). Evaluation of lithium ion cells for space applications. In: *IECEC-97 Proceedings of the 32nd Intersociety Energy Conversion Engineering Conference*. Honolulu, HI.
27 Smart, M.C., Ratnakumar, B.V., Whitcanack, L. et al. (2000). Performance characteristics of Yardney lithium-ion cells for the Mars 2001 Lander application. In: *35th Intersociety Energy Conversion Engineering Conference and Exhibit*. Las Vegas, NV.
28 Kelly, C.O., Friend, H.D., and Keen, C.A. (1998). Lithium-ion satellite cell development: past, present and future. *IEEE Aerosp. Electron. Syst. Mag.* 13 (6): 21–25.
29 Hossain, S., Tipton, A., Mayer, S. et al. (1997). Lithium-ion cells for aerospace applications. In: *IECEC-97 Proceedings of the 32nd Intersociety Energy Conversion Engineering Conference*. Honolulu, HI.
30 Gonai, T., Kiyokawa, T., Yamazaki, H., and Goto, M. (2003). Development of the lithium-ion battery system for space: Report on the result of development of the lithium ion battery system for space. In: *The 25th International Telecommunications Energy Conference*. Japan: Yokohama.

31 Jeevarajan, J.A. and Bragg, B.J. (1998). Evaluation of safety and performance of Sony lithium-ion cells, NASA CP-1999-209144, 27–29. In: *NASA Aerospace Battery Workshop*. Huntsville, AL.

32 Jeevarajan, J.A., Cook, J.S., Davies, F.J. et al. (2003). Comparison of performance and safety of two models of Panasonic Li-ion batteries. In: *The 203rd Electrochemical Society Meeting*. Paris, France.

33 Genc, D.Z and Thwaite, C. (2011). Proba-1 and Mars Express: An ABSL lithium-ion legacy. In: *Proceedings of the 9th European Space Power Conference*, ESA SP-690. Saint Raphael, France.

34 Mattesco, P., Thakur, V., and Tricot, H. (2005). Overview and preliminary in-orbit behaviour of the first lithium-ion batteries used onboard Eutelsat W3A GEO Telecommunications Satellite Mission. In: *Proceedings of the 7th European Space Power Conference*, ESA SP-589. Stresa, Italy.

35 Dalton, P.J., Schwanbeck, E., North, T., and Balcer, S. (2016). International space station lithium-ion battery. In: *NASA Aerospace Battery Workshop*. Huntsville, AL.

36 Dahn, J. and Ehrlich, G.M. (2019). Lithium-ion batteries, Chapter 26. In: *Linden's Handbook of Batteries*, 5e (ed. K.W. Beard and T.R. Reddy), pp. 26.1–26.79. New York: McGraw Hill.

37 *Energy Storage Technologies for Future Planetary Science Missions* (2017). National Aeronautics and Space Administration, Jet Propulsion Laboratory, California Institute of Technology, JPL D-101146.

38 Olivetti, E.A., Ceder, G., Gaustad, G.G., and Fu, X. (2017). Lithium-ion battery supply chain considerations: Analysis of potential bottlenecks in critical metals. *Joule* 1 (2): 229–243.

39 Igogo, T., Sandor, D., Mayyas, A., and Engel-Cox, J. (2019). Supply chain of raw materials used in the manufacturing of light-duty vehicle lithium-ion batteries, National Renewable Energy Laboratory, NREL/TP-6A20-73374.

40 Puglia, F., Gitzendanner, R., Gulbinska, M. et al. (2008). Lithium-ion cell materials for aircraft and aerospace – Next generation and domestic sources. In: *NASA Aerospace Battery Workshop*. Huntsville, AL.

41 Krause, F.C., Ruiz, J.P., Jones, S.C. et al. (2021). Performance of commercial Li-ion cells for future NASA missions and aerospace applications. *J. Electrochem. Soc.* 168.

42 Davis, L.A. and Filip, L. (2015). *How Long Does It Take to Develop and Launch Government Satellite Systems*. The Aerospace Corporation, ATR-2015-00535.

43 Barrera, T.P. and Wasz, M. (2018). Spacecraft Li-ion battery power system state-of-practice: A critical review. In: *International Energy Conversion Engineering Conference, AIAA Propulsion and Energy Forum*, AIAA 2018–4495.

44 Johnson-Roth, G.A. and Tosney, W.F. (2010). Mission risk planning and acquisition tailoring guidelines for national security space vehicles. The Aerospace Corporation, El Segundo, CA. TOR-2011(8591)-5.

2

Space Lithium-Ion Cells

Yannick Borthomieu, Marshall C. Smart, Sara Thwaite, Ratnakumar V. Bugga, and Thomas P. Barrera

2.1 Introduction

Since Sony's introduction of the first commercial secondary (rechargeable) lithium-ion battery (LIB) for portable electronics applications in the early 1990s, LIBs have rapidly become an enabling technology for many global energy storage applications [1]. In this revolutionary era of high specific energy LIBs for advanced electrified transportation, new consumer portable electronic devices, and expanding grid-energy storage capability, lithium-ion (Li-ion) cell technologies continue to enable new market applications. The increased performance characteristics of Li-ion cell technologies and emerging obsolescence of heritage nickel-hydrogen (Ni-H_2) cell technology in the mid-2000s led the space industry to a rapid transition to Li-ion cells for all spacecraft battery power systems. LIB-based electrical power systems (EPS) are now commonly used in spacecraft applications ranging from high-power Earth-orbiting satellites to long-duration planetary spacecraft missions.

This chapter describes the design, manufacturing, and testing of Li-ion cell technologies used in spacecraft applications. A special focus on both custom and commercial off-the-shelf (COTS) space Li-ion cell design features and their test and analysis methods used to demonstrate compliance to requirements is described. Additionally, primary (non-rechargeable) Li battery cell and thermal battery cell technologies are discussed to provide a comprehensive description of the various energy storage technologies used in space applications.

2.1.1 Types of Space Battery Cells

Many types of aqueous potassium hydroxide (KOH)-based, non-rechargeable (primary) Li battery, and rechargeable Li-ion cell technologies have been space-qualified for numerous spacecraft applications. Each space application has a unique set of performance, environmental, safety, quality, reliability, and cost requirements that dictates cell selection criteria. With few exceptions, all Earth-orbiting and Mars planetary mission spacecraft utilize an EPS that employs solar array power generation components coupled with a rechargeable battery energy storage system. Moreover, rechargeable LIBs provide an electrical energy storage capability for nearly every satellite, planetary, launch, crew transfer, and cargo vehicle used in spacecraft applications. Rechargeable LIBs

Spacecraft Lithium-Ion Battery Power Systems, First Edition. Edited by Thomas P. Barrera.

and non-rechargeable Li batteries are also used on the US astronaut extravehicular mobility unit (EMU) and portable astronaut crew electronics equipment used aboard the NASA International Space Station (ISS).

Energy storage with non-rechargeable Li batteries is commonplace for single-use applications requiring high specific energy, long storage (non-operational) life, or a short-duration discharge power demand. Due to their unique mission needs, Earth-orbiting satellite systems do not use non-rechargeable batteries as primary energy storage devices. However, launch vehicles (LVs) commonly use non-rechargeable batteries for short-duration avionics and flight termination system (FTS) power needs. Planetary orbiters, rovers, landers, probes, and sample return capsules are also commonly powered by non-rechargeable Li batteries. As further described in Chapter 6, many planetary mission spacecraft have used and continue to implement non-rechargeable Li batteries to meet mission-specific EPS requirements.

2.1.2 Rechargeable Space Cells

The various types of rechargeable cell technologies used in space applications are summarized in Table 2.1. Cell type selection is commonly based on mission-specific technical requirements largely dominated by orbit characteristics, mass, volume, and mission duration. Aqueous KOH electrolyte-based battery cell chemistries were favorable for space applications due to their tolerance to overcharge–overdischarge conditions, long cycle life, and compliance to on-orbit environmental operating conditions. Rechargeable Li-ion cell chemistry has a significant advantage in specific energy, high operating voltage, and low self-discharge rates when compared to heritage aqueous Ni-Cd and Ni-H_2 battery cell technologies. Due to their limited wet shelf and cycle life, rechargeable Ag-Zn battery cell technologies have been used in limited space applications, such as the Apollo spacecraft, Space Shuttle astronaut EMU, short mission duration LVs, and in early Mars Lander applications.

2.1.3 Non-Rechargeable Space Cells

Various types of electrochemical couples have been used as non-rechargeable cells and batteries in space applications. The performance characteristics of non-rechargeable Li and aqueous KOH based battery cells used in space applications are given in Table 2.2. Ag-Zn battery technologies have been used in numerous human-rated (crewed) and unmanned spacecraft missions. These

Table 2.1 Rechargeable space cell characteristics.

Type	Open circuit voltage (V)	Nominal voltage (V)	Specific energy (Wh/kg)	Operating temperature range (°C)	Self-discharge rate (% loss per month at 20 °C)	Overcharge tolerance
Li-ion	4.2	3.7	125–280	−10 to +50	2	No
Ni-H_2	1.5	1.3	60	−20 to +30	1% per h	Yes
Ni-Cd	1.3	1.2	45	0 to +40	20	Yes
Ni-MH	1.4	1.2	65	−20 to +65	30	Yes
Ag-Zn	2.0	1.5	105	−20 to +60	5	Yes
Ag-Cd	1.6	1.2	90	−25 to +70	5	Yes
Pb-SO_4	2.1	2.0	50	−10 to +40	8	Yes

Table 2.2 Non-rechargeable space cell characteristics.

Type	Specific energy (Wh/kg)	Energy density (Wh/l)	Operating temperature range (°C)	Nominal discharge voltage (V)	Capacity loss (% per year at 20 °C)
Ag-Zn	200	550	0 to +55	1.6	60
Li-CFx	524	905	−20 to +65	2.6	<1
Li-CFx-MnO_2	514	856	−40 to +85	2.6	<1
Li-SO_2	300	375	−60 to +71	2.9	<1
Li-$SOCl_2$	468	875	−60 to +85	3.2	<2.5
Li-MnO_2	270	445	−20 to +75	3.0	<1
Li-BCX	414	930	−40 to +70	3.4	<2

chemistries are commonly used for applications requiring one-time single-use for a limited period of time. The high specific energy, specific power, long calendar life, and wide operating temperature range characteristics are significant advantages of space-qualified non-rechargeable Li cell chemistries.

The first application of a spacecraft battery energy power system was achieved in 1957 when Sputnik I demonstrated that non-rechargeable Ag-Zn batteries could provide reliable short-term energy storage under stressing launch and on-orbit environments. Shortly afterwards, Vanguard I (launched March 17, 1958) was powered by six equally spaced solar cell clusters and a seven mercuric oxide-zinc (Zn-HgO) cell hermetically sealed non-rechargeable battery used for tracking transmitter power. Since these early applications, the most commonly used non-rechargeable battery cell chemistry used in space applications has been Ag-Zn. Despite its limited cycle life, Ag-Zn batteries were often used for limited rechargeable energy storage needs. However, most space applications have employed Ag-Zn batteries as a non-rechargeable energy storage device. Non-rechargeable Li-metal anode battery cell chemistries, such as Li-SO_2, Li-MnO_2, Li-$SOCl_2$, Li-BCX, and Li-CFx, have been widely used in various Earth-orbiting spacecraft (such as the NASA Space Shuttle and ISS), planetary probes, landers, rovers, Sample Return Capsules, and NASA astronaut equipment applications. Due to the high specific energy of Li non-rechargeable cell chemistries, compliance to safety requirements is critical for crewed spacecraft applications [2].

2.1.4 Specialty Reserve Space Cells

Although not used in Earth-orbiting satellite applications, thermal reserve batteries have been used in robotic planetary missions requiring reliable high power for pyrotechnic firing devices. For example, Li-FeS_2 thermal batteries (manufactured by EaglePicher Technologies) have been used on a number of NASA missions, including Mars Pathfinder, Mars 98, Mars Odyssey, 2003 Mars Exploration Rover, 2007 Mars Phoenix Lander, 2011 Mars Science Laboratory (MSL), and Mars 2020, because of their ability to provide high-power densities over a short time duration (minutes) upon activation. Their application was to fire NASA Standard Initiators (NSIs) for pyrotechnic events, such as those involved in the entry, descent, and landing (EDL) processes for rovers and landers [3]. In addition to firing NSIs, in the case of MSL and Mars 2020, thermal batteries were used to power the spacecraft power bus during the EDL process.

2.2 Definitions

To understand the design, performance, and safety characteristics of space LIBs, terms and definitions unique to space LIB industry practitioners are defined. In particular, capacity, energy, and depth-of-discharge (DOD) are essential to understanding Li-ion cell and battery characteristics. A more complete list of terms and definitions is given in Appendix A.

2.2.1 Capacity

Capacity is a quantity of electric charge expressed for batteries in coulombs or ampere-hours. The theoretical ampere-hour (Ah) capacity of an electrochemical cell is determined from the amount of the active material, either positive or negative, whichever is less [4]. In Li-ion cells, the cell capacity is proportional to the amount of the cathode material. The actual (measured) capacity is less than the theoretical capacity, depending on the discharge conditions, discharge rate, temperature, design, and age of the cell. Cell capacity (Ah) is calculated from the product of discharge duration (hours) and current (Amps) in a constant-current discharge. In discharges across a constant load or at a constant power, capacity is an integral current over time to the end-of discharge voltage (EODV) limit is reached [5]. Unlike most commercial cells and batteries, space-qualified cell capacities are often derated for performance margin, so the nameplate (rated) capacity will be lower than the actual or measured capacity. Figure 2.1 illustrates the various types of cell capacity

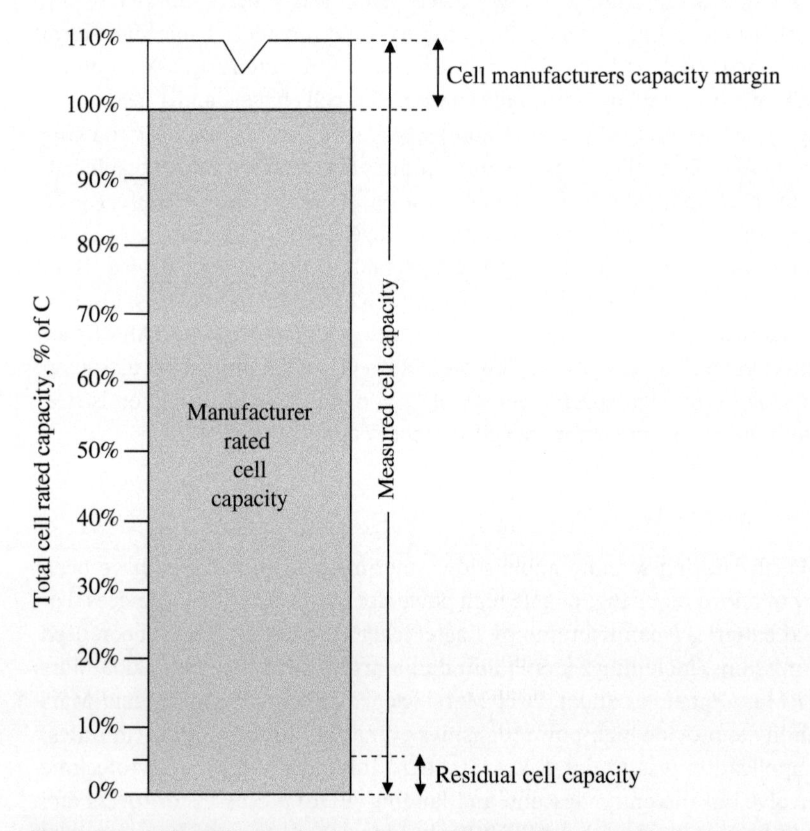

Figure 2.1 Li-ion cell capacity description. Cell operational capacity is always less than (or equal to) the measured cell capacity.

space industry practitioners utilize to demonstrate compliance to the cell, battery, and EPS-level requirements. The various types of cell capacity are as follows:

- Nameplate (rated) capacity – The minimum capacity of a cell specified by the manufacturer at the beginning-of-life (BOL) and at specified conditions such as the charge rate, end-of charge voltage (EOCV), discharge rate, EODV, and temperature. Cells usually have a measured capacity higher than the nameplate capacity (approximately 5–10%) to compensate for cell production lot-to-lot variability.
- Measured (actual) capacity – The capacity measured in a controlled test environment under various charge rates, EOCV, discharge rate, EODV, and temperature conditions.
- Operational capacity – The capacity measured in a controlled test environment corresponding to on-orbit and EPS operating conditions using EOCV and EODV control limits which are typically lower than the manufacturer recommended limits. Hence, the cell operational capacity is always less than or equal to the measured (actual) cell capacity.
- Residual capacity – Commonly called "low-rate reserve capacity," residual capacity is measured to quantify the EPS battery capacity reserves (margin) to support spacecraft-level contingencies requiring a low discharge rate capability.

2.2.2 Energy

Cell energy (Wh) is equal to the integral of the product of discharge current and voltage as a function of the time from the start of discharge to either the minimum cell voltage, minimum power subsystem battery voltage limits, the time at which the first battery cell reaches the lower cell EODV limit, or the end of a pre-determined fixed time period. Similar to capacity, this is measured at a defined charge voltage–current profile, discharge load profile, and temperature range. The cell discharge energy is commonly calculated from test data measured in a controlled test environment under constant-current discharge conditions. However, determining cell energy under constant-power discharge conditions is more representative, since most spacecraft EPS loads operate under constant-power conditions. Under constant-power discharge, current is lowest at the beginning of discharge and increases as the battery voltage decreases in order to maintain a constant-power input to spacecraft loads. Cell and battery energy requirements to support launch, transfer orbit, on-orbit energy, and reserve energy are allocated from the top-level parent EPS specification to the battery-level specification.

2.2.3 Depth-of-Discharge

The DOD of a cell or battery is the degree of discharge expressed as a ratio in a percentage of the capacity removed from the cell (or battery) for a defined discharge load profile and temperature to the cell nameplate (rated) capacity or the initial state-of-charge (SOC). In other words, the DOD is the inverse of the SOC. For example, 0% DOD corresponds to 100% SOC and vice versa. An advantage of using DOD values based on the rated cell capacity is that it is fixed for the lifetime of the cell. For example, if a 20% DOD is specified, the ampere-hours removed on discharge at the BOL are the same as the ampere-hours removed on discharge at the cell end-of-life (EOL). In practice, the spacecraft battery user is responsible for adjusting the cell (or battery) DOD target based on battery EOL performance data [6].

Cell DOD can also be expressed in terms of energy instead of capacity. Since satellite EPS user loads operate at constant power (and rarely at constant current), DOD based on energy is often used versus DOD based on capacity for satellite battery sizing and EPS power budget analyses.

2.3 Cell Components

Space LIB cells are available from various cell manufacturers who specialize in a diverse set of custom or COTS-type cell designs. While there is a limited set of custom space cell manufacturers, a wide variety of COTS Li-ion cell manufacturers exist in today's global marketplace. Li-ion space cell designs characteristically have different geometries (formats), chemistries, safety, and performance characteristics [7]. In general, space-qualified Li-ion cells have similar cell components when compared to commercial non-space qualified Li-ion cells. However, in space Li-ion cell designs, Li-ion cell chemistry may be tailored to demonstrate compliance with stringent mission-specific LIB performance, environmental, safety, and reliability requirements.

2.3.1 Positive Electrode

Commercial Li-ion cells utilize different positive electrode or cathode materials, depending on the type of applications, but only a select few are used preferentially in space applications [8]. Table 2.3 compares key performance characteristics of cathode materials used in space Li-ion cells. Cathode materials in space Li-ion cells include lithium cobalt oxide ($LiCoO_2$, or LCO), lithium nickel cobalt oxide ($LiNiCoO_2$, or NCO), lithium nickel cobalt aluminum oxide ($LiNi_{0.80}Co_{0.15}Al_{0.05}O_2$, or NCA), lithium

Table 2.3 Space li-ion cell cathode material comparison.

Cathode material	Formula	Specific capacity (mAh/g)	Mid-discharge voltage, V vs. Li	Comments
Lithium cobalt oxide (LCO)	$LiCO_2$	140–155	3.75	Good cycle/calendar life, moderate specific energy, flatter voltage profile and moderate safety, widely used in space missions
Lithium nickel cobalt oxide (NCO)	$LiNi_{0.8}Co_{0.2}O_2$	160–180	3.7	Good cycle/calendar life, moderate specific energy, sloping voltage profile and moderate safety, used in planetary missions
Lithium nickel cobalt aluminum oxide (NCA)	$LiNi_{0.8}Co_{0.15}Al_{0.05}O_2$	160–200	3.65	Excellent cycle/calendar life, high specific energy, sloping voltage profile and improved safety and widely used in space
Lithium manganese nickel cobalt oxide (NMC)	$LiNi_{1-x-y}Mn_xCo_yO_2$	150–200	3.75	Good cycle/calendar life, higher specific energy, sloping voltage profile and moderate safety, different Ni/Mn/Co ratios and widely used
Lithium manganese oxide (LMO)	$LiMn_2O_4$	110–120	3.95	Good cycle/calendar life, lower specific energy, flatter voltage profile, used with NCA and improved safety
Lithium iron phosphate (LFP)	$LiFePO_4$	150–160	3.30	Excellent cycle/calendar life, lower specific energy, flatter discharge profile and excellent safety, rarely used in space

nickel manganese cobalt oxide ($LiNi_{1-x-y}Mn_xCo_yO_2$, or NMC), lithium manganese oxide ($LiMn_2O_4$, or LMO) and rarely lithium iron phosphate ($LiFePO_4$, or LFP). NMC materials exist in a wide range of compositions for nickel, manganese, and cobalt, such as NMC111 (with a mole ratio of Ni, Mn, and Co being 1:1:1), NMC532 (Ni, Mn, Co ratio of 5:3:2), NMC622 and recently NMC811, or with an even higher Ni content. These materials are either used in their pristine form, with surface coatings, or with blends of these different cathode materials to optimize performance. The nickel-based cathodes show the highest capacity and NMC materials with a high nickel content are now the baseline for COTS 18650 cells. On-orbit spacecraft experience has demonstrated that LCO, NCO, NCA, and NMC811 cathode cell materials meet cell requirements for electrical, mechanical, and thermal performance. Cathodes are fabricated using conductive diluents and binders in a multi-step manufacturing process. The active material is coated on a thin aluminum foil substrate with a thickness ranging from 12 to 15 μm.

2.3.1.1 Lithium Cobalt Oxide

Developed in the 1980s for Li metal cells, LCO was the cathode material used in the first Li-ion cell for commercial applications [9–12]. Since then, LCO has been used extensively in commercial Li-ion cells due to its high voltage characteristics, flat discharge curve, simple synthesis, and a good cycle life that outweighs its high cost [13, 14]. Because of its lower specific capacity (140 mAh/g for 0.5 Li/Co), its specific energy is lower, which prompted the development of nickel-based cathodes. Li-ion cells with LCO cathodes have been widely used in numerous Earth-orbiting and planetary mission LIB applications.

2.3.1.2 Lithium Nickel Cobalt Aluminum Oxide

Lithium nickel oxide, $LiNiO_2$, is isostructural with $LiCoO_2$ with layers of monovalent lithium cations (Li^+) that lie between extended sheets of nickel and anionic oxygen atoms. Despite its higher specific capacity and energy, $LiNiO_2$ cannot be used due to its poor cycle life and high thermal instability. Substitution of 15–20% of Ni with Co and Al has alleviated these problems and resulted in an improved cycle life. Co increases the stability of the lattice and the Al aids in interfacial stability. $LiNi_{0.80}Co_{0.15}Al_{0.05}O_2$ (NCA) exhibits a high capacity (up to 200 mAh/g) de-intercalating 0.65 Li/Ni reversibly. The NCA cathode has demonstrated an excellent cycle and calendar life in numerous Earth-orbiting spacecraft LIB applications.

2.3.1.3 Lithium Nickel Manganese Cobalt Oxide

Mixed metal oxides with a layered structure also exist with manganese, cobalt, and nickel. $LiMnO_2$ is difficult to activate electrochemically, but the substitution of Mn with Ni and Co result in reversible cathodes. A popular composition of this family of mixed metal oxides is NMC 111 ($LiNi_{0.33}Mn_{0.33}Co_{0.33}O_2$), which has a specific capacity of ≥150 mAh/g over 2.5–4.2 V, excellent power capability, a good cycle life, and improved safety. The Ni content has been further increased in these compositions, such as NMC 532 and NMC 811, mainly to increase the specific capacity to ~200 mAh/g and thus achieve higher cell-level specific energy. In NMC cathodes, nickel undergoes redox reactions in two redox steps: Ni^{2+}/Ni^{3+} and Ni^{3+}/Ni^{4+}, while Co transitions from Co^{3+}/Co^{4+} [15]. However, manganese has been found to not participate in the cathode reaction but helps to keep the structure together, also providing thermal stability and lowering the cost [16].

2.3.1.4 Lithium Manganese Oxide

Lithium manganese oxide ($LiMn_2O_4$; LMO) has a cubic spinel structure and has the advantages of stability even at high charge voltages, and with a low material cost. It has a lower specific capacity compared to layered oxides and is not used in space Li-ion cells, except in blends with layered oxide cathodes in small proportions [17]. Its use is limited due to its lower specific capacity and also the risk of manganese dissolution into an electrolyte, which affects the stability of the anode SEI.

2.3.1.5 Lithium Iron Phosphate

Lithium iron phosphate ($LiFePO_4$; LFP) is another cathode material with an olivine structure used almost entirely in commercial applications because of its lower cost and better cyclic stability and safety without oxygen release on overcharge [18]. Even with good specific capacity of 160 mAh/g, it has low specific energy because of its lower but flat cell voltage of 3.3 V [19]. Similar to oxide cathodes, LFP exists in the delithiated form and is perfectly stable with an olivine structure that maintains its stability even after complete delithiation with a strong oxygen atom bonding within the PO_4 groups, and provides the overcharge capability [20, 21]. Despite its favorable safety characteristics, power capability, and cyclic stability, LFP cathodes are not typically used in space Li-ion cells, mainly because of its lower specific energy compared to other available cathode materials. Furthermore, its flat discharge curve is a risk to proper battery SOC determination, charge management, and cell balancing.

2.3.2 Negative Electrode

The desired characteristics for Li-ion cell negative electrode materials include a high specific capacity (mAh/g), low potential (vs. Li), adequate electrochemical stability, and good reversibility. Lithium metal, which has the highest possible specific capacity of all metal anodes, was studied extensively before the advent of Li-ion batteries, but without much success, due to the problems of poor reversibility and dendritic deposition causing serious safety issues. The anode materials widely used in commercial and especially space Li-ion cells are different carbonaceous materials and include graphite (both natural and synthetic), meso-carbon micro-beads (MCMB), or graphite with small amounts of Si for higher cell specific energy and energy density. The essential reason for the successful use of these materials is that they undergo single-phase intercalation reactions with lithium ions occupying the galleries of carbon hosts, similar to the cathode materials. This ensures excellent cyclability, an important requirement for space LIBs. The chemical and electrochemical stability of the carbon anode is provided by the surface film, which determines the durability of the anode. Anode manufacturing processes are similar to that of cathodes, but with aqueous binders in an aqueous process. The active material is coated on a thin copper foil substrate with a thickness ranging from 8 to 15 μm.

2.3.2.1 Solid Electrolyte Interphase

There is an innate thermodynamic instability of the electrolyte at the (carbon) anode potentials, which is kinetically stabilized by the anode surface film or solid electrolyte interphase (SEI). The SEI is formed from the products of the electrolyte reduction during the early stages of cell cycling or formation [22–26]. This is manifested as a first cycle irreversible capacity (typically 10%) and a low coulombic efficiency (approximately 90%), which quickly increases to >99% due to the stabilization of the SEI. In a carbonate electrolyte with $LiPF_6$ salt, the SEI is composed of a mix of Li-alkyl-carbonates, polymeric carbonates, and lithium fluoride. Electrolyte additives, such as vinylene carbonate (VC), may be added to facilitate the formation of a more robust SEI by preferential reduction [27]. The SEI layer on the anode's surface is the key for cell performance characteristics, such as rate, stability at extreme temperatures, cycle life, and safety [28].

2.3.2.2 Coke

Li-ion chemistry began with a disordered carbon, coke, as the anode material. Coke-type materials are formed when the heat-treatment temperature is limited to around 1000 °C. These materials have relatively small graphene sheets of lateral extent between 20 and 50 Å. The small graphene sheets are stacked in a parallel fashion, but with random rotations and translations between every pair of layers, characteristic of "turbostratic disorder." The reversible capacity of a typical commercial coke material varies between 160–220 mAh/g, which is significantly lower than that for graphite. Furthermore, amorphous carbon materials usually have a sloping lithium insertion voltage between 0 and 1 V. However, coke-type materials do have advantages over crystalline graphite, such as in the rate capability and with propylene carbonate (PC)-based electrolytes, which exhibit many attractive properties compared to ethylene carbonate (EC)-based solutions.

2.3.2.3 Hard Carbon

Non-graphitizable ("hard") carbon is another type of carbon anode material used in first-generation commercial 18650 Li-ion cell designs. It has the advantages of a higher reversible capacity and better cycling performance than graphite, but suffers from some deficiencies such as large irreversible capacity in the first cycle and high potential hysteresis between charge and discharge, and low electrode density. Because of the higher anode potential, the risk of Li plating is minimized. Early versions of EnerSys/ABSL space batteries utilized Sony 18650 Li-ion cell designs with hard carbon anode material.

2.3.2.4 Graphite

As carbon materials are continuously heated to 2000 °C, the lateral extent of the graphene sheets grows and the stacking becomes more parallel, the crystallinity increases, and the turbostratic disorder is relieved above 2200 °C so that, by around 3000 °C, synthetic graphite is produced. Graphite is favored primarily due to its high lithium intercalation capacity up to 372 mAh/g (corresponding to a composition of LiC_6) and its low and flat lithium intercalation voltage curve. The low and flat lithium intercalation potential of graphite (50–100 mV vs. Li) is a desirable factor for the maintenance of a flat cell voltage. Graphite, either natural or synthetic, is the most common anode material in space Li-ion cells.

Generally, Li-ion cells utilize graphite or other forms of carbons with a relatively low surface area to minimize the irreversible capacity loss. Even though the potentials of carbon anodes after lithiation are fairly close to Li (approximately 85 mV), the cells are designed to operate at specified voltages, temperatures, and cyclic conditions such that lithium plating on the anode is strictly avoided during charge.

2.3.2.5 Mesocarbon Microbead

MCMB carbon is another popular anode material for Li-ion cells due to its excellent performance in cycling and calendar life [29–31]. MCMB carbon is composed of spherical graphite beads of 6–25 μm average diameter with a stacked, sheeted structure. Due to its spherical shape, MCMB has a high packing density that results in high energy density and its small surface area decreases the irreversible capacity arising from the SEI formation. The SEI on the carbon or graphite anode is crucial in determining both the Li intercalation kinetics and its life characteristics. Most surfaces of spheres are edge plane surfaces, allowing for easy Li+ intercalation. Its low specific capacity (≤320 mAh/g) is the only reason for favoring natural graphite and surface-treated natural graphite that exhibit a reversible capacity higher that 350 mAh/g [7].

2.3.2.6 Si-C Composites

With the same objective to improve anode capacity, Si has emerged as one of the promising anode materials, at least as an additive to the carbon anode. Si offers a high theoretical specific capacity of approximately 4200 mAh/g based on the formation of the $Li_{22}Si_5$ alloy, which is about 10 times higher than that of conventional carbon-based anodes. However, Si is extremely challenging to implement in cells because of its large expansion up to 400% upon full lithiation, causing fragmentation of the particles, which exacerbate its reaction with the electrolyte to form the SEI. This fragmentation process also causes a loss of electronic continuity of the particles, which leads to electrochemically isolated particles. Another issue with Si is its low ionic and electronic conductivity. Nano-sized Si particles combined with graphite and/or conductive carbon (CC) are being investigated to obtain a viable silicon-carbon composite anode. A small amount (<10%) of Si-based material is blended with a graphite anode in new COTS 18650 cell designs [32].

2.3.2.7 Low-Voltage Resilience

The use of copper anode substrates results in poor tolerance of Li-ion cells to overdischarge conditions. At zero voltage for the cell, the anode potential may be >3.2 V vs. Li/Li^+ and there is a possibility for Cu to oxidize, though not near its oxidation potential. The dissolved copper can redeposit on the separator or cathode creating a risk to cell-level internal short circuit (ISC) conditions [33]. To mitigate this, other metals are being explored as anode substrates to achieve low-voltage (zero-volt) tolerance [34]. The main issues of using alternative current collector materials are their high cost and lower power due to a lower electronic conductivity.

2.3.3 Electrolytes

The primary function of the electrolyte is to effectively transfer the ionically conductive lithium ions back and forth between the anode and the cathode. The electrolyte is in contact with both electrodes and influences how quickly the energy can be released, along with the mass transport in the bulk of the electrodes and the charge transfer resistance, since it is dependent upon the mass transport of ions. Unlike the highly ionic conductive KOH-based electrolyte used in Ni-based rechargeable batteries (>50 mS/cm), the organic electrolytes that are used in Li-ion cells are much less conductive (approximately 10 mS/cm at ambient temperatures), but they enable the use of high-voltage electrodes and support a much wider temperature range of operation.

2.3.3.1 Room Temperature Electrolytes

To provide the required performance, organic electrolytes should demonstrate the following properties: (i) good ionic conductivity, (ii) a wide liquid temperature range, (iii) good electrochemical stability, (iv) good chemical stability, (v) good compatibility with the chosen electrode couple (such as desirable film-forming characteristics), and (vi) good thermal stability. Traditional Li-ion cell electrolytes consist of lithium hexafluorophosphate ($LiPF_6$) salt dissolved in a mixture of cyclic and linear organic carbonates [35]. The most common cyclic carbonate solvent employed is ethylene carbonate (EC), which has a high dielectric constant allowing for good salt solubility and upon electrochemical reduction produces a robust and protective SEI layer on the surface of the anode. The SEI layer prevents further electrolyte decomposition on the anode surface. The properties of the SEI layer, being ionically conductive but electronically insulative, allows the facile transport of lithium ions but prevents solvent molecules from co-intercalating into the graphite structure, leading to its exfoliation. In some cell designs, cyclic PC is used as a co-solvent in conjunction with EC,

Table 2.4 Properties of commonly used electrolyte solvents used in space Li-ion cells.

Type	Melting point (°C)	Boiling point (°C)	Dielectric constant (ε)	Density (d)	Viscosity (γ)
Ethylene carbonate	38	243	89.3 (40 °C)	1.321	1.85 cP (40 °C)
Propylene carbonate	−48.8	242	64.9 (25 °C)	1.205	2.4 cP (25 °C)
Dimethyl carbonate	3	90.5	3.12	1.069	0.585 cP
Diethyl carbonate	−43	126	2.82	0.975	0.748 cP
Ethyl methyl carbonate	−55	101.85	2.96	1.012	0.65 cP

but it is not typically used as the sole cyclic carbonate solvent due to the poor film-forming characteristics leading to co-intercalation and resulting exfoliation of the graphite particles. To improve the physical and electrochemical properties of the electrolyte, non-cyclic linear organic carbonates are also incorporated, including one or more of the following co-solvents: dimethyl carbonate (DMC), diethyl carbonate (DEC), and ethyl methyl carbonate (EMC) (Table 2.4). Electrolytes based on these formulations are also used in space-qualified Li-ion cells. Although many cell manufacturers do not typically disclose the exact nature of the formulations used, the reports that are available are usually EC-based binary and ternary mixtures of carbonates with the $LiPF_6$ salt in roughly 1.0 M concentrations. In some cases, additives such as vinyl carbonate (VC), are added to base electrolytes to meet supplier specific design requirements.

In some COTS 18650 Li-ion cells, a few other electrolyte salts have reportedly been used in conjunction with $LiPF_6$ to either improve safety, thermal stability, electrode filming behavior, or high-temperature resilience. These salts include lithium bis(oxalato) borate (LiBOB), lithium bis(trifluoromethanesulfon)imide (LiTFSI), lithium difluoro (oxalato) borate (LiDFOB), and lithium bis(fluorosulfonyl)imide (LiFSI). However, these salts have not been reportedly used in space Li-ion cell designs. In addition, numerous commercial electrolyte additives have been developed primarily to improve the interfacial filming at the electrodes, but they have not been adopted for space applications. The one exception is the addition of VC which results in much more robust SEI layers on carbon electrodes and provides good stability throughout the life of the cell [36]. Selected electrolytes have also been chemically tailored specifically for long-life low Earth orbit (LEO) space applications with resilience to the required elevated charge rates.

2.3.3.2 Low-Temperature Electrolytes

Although $LiPF_6$ in binary mixtures of organic carbonates, such EC + DMC or EC + DEC, can provide good life characteristics at ambient temperatures, the performance at low temperatures (<−10 °C) is generally very poor. By blending these linear and cyclic organic carbonates into ternary or quaternary compositions, suitable electrolytes can be formulated to ensure stability over a wide temperature range as needed in specific applications. For example, since a number of NASA missions require operational capability at low temperatures, an electrolyte consisting of 1.0 M $LiPF_6$ dissolved in EC + DEC + DMC (1 : 1 : 1 vol%) was developed at the Jet Propulsion Laboratory (JPL) and was subsequently used in the 2003 Mars Exploration Rovers (MER) and the 2011 Mars Science Laboratory Curiosity Rover, which required the capability of both charging and discharging the battery at temperatures as low as −20 °C [37, 38]. This same electrolyte formulation has also been used on other NASA missions, including the 2007 Mars Phoenix, Juno, Grail, Maven, and Mars 2020. Carbonate-based electrolytes have also been further optimized to provide improved performance at temperatures below −20 °C in space Li-ion cells, which involves lowering the EC-content in quaternary solvent blends, such as 1.0 M $LiPF_6$ dissolved in EC + DEC + DMC + EMC (1 : 1 : 1 : 2 vol%) [39]. The use of low-viscosity, ester-based solvents has also been employed to achieve improved low-temperature performance. For example, the NASA Mars InSight mission utilizes an electrolyte consisting of 1.0 M $LiPF_6$ in EC + EMC + methyl propionate (MP) (20 : 60 : 20 vol%) to support the requirements of both charging and discharging from −30 to +35 °C [38]. Higher salt concentrations can also be employed to improve the low-temperature performance of the cells, ensuring a good charge and discharge rate capability even at −20 or −30 °C. EaglePicher Technologies fabricated the cells and batteries for the aforementioned Mars missions.

2.3.4 Separators

A Li-ion cell separator functions to physically separate the anode from the cathode to prevent electronic contact, which would result in an ISC. It must be sufficiently porous to hold an electrolyte and allow for Li-ion transport during operation. To be effective, the separator material must possess the following physical and chemical characteristics: (i) sufficient electronic insulation, (ii) minimal ionic resistance, (iii) good electrolyte wettability, (iv) good mechanical and dimensional stability, and (iv) good chemical resistance to degradation being inert to the cell electrochemistry [40]. Furthermore, separators should have good uniformity, be resistant to tearing during the cell manufacturing process, display sufficient compressibility, and provide good mechanical strength at high temperatures. Other key properties that are commonly measured to characterize the membranes include (i) thickness (μm), (ii) Gurley number (s), or the time for a specified amount of air to pass through a specified area, (iii) ionic resistivity (Ω/cm), (iv) melt temperature (°C), (v) pore size, and (vi) porosity [41]. The Gurley number is a measurement of the air permeability, which can be used to indirectly determine the ionic resistance of the membrane [42].

Nearly all separators used in COTS Li-ion cells and custom space Li-ion cells consist of microporous polyolefin membranes consisting of polyethylene (PE), polypropylene (PP), or multi-layer composites, such as PP/PE/PP (also known as tri-layer separators) [43]. These membranes are fabricated by either a dry or wet extrusion process. In the case of the dry process, the polyolefin resin is melted and extruded into a film and then followed by a thermal annealing process and a stretching procedure. Manufacturers such as Celgard Inc. and Ube Industries use similar techniques to produce PE and PP membranes [44, 45]. The one disadvantage with respect to this process is that the mechanical properties are anisotropic since the stretching process produces slit-like pores in one direction. Alternatively, membranes can be fabricated by the wet process, which involves the use of hydrocarbon solvents or other low-molecular weight additives during fabrication, allowing the films to be extruded and then oriented biaxially. Examples of manufacturers that employ this technique include Toray (previously Tonen) and Setela (previously marketed by Exxon Mobil), and Asahi Kasei, who produce PE membranes [46, 47]. The main advantage of utilizing polyolefins films is that they are simple to manufacture and process, heat-sealable, and cost-effective. PE and PP have melting temperatures of 135 and 165 °C, respectively. The separators typically have high porosities (>40%) and good wettability with solvents, which enable good ionic transport through these thin membranes. The PE and PP separators have a reasonably high bulk puncture strength. Polyethylene terephthalate (PET) and polyvinylidene fluoride (PVDF) have also been used as separators in some commercial cells, but not for space applications [48, 49].

In typical space Li-ion cell designs, one layer of separator material is present between the two electrodes, with one of the electrode pairs being fully contained in a sealed bag. Separator thickness for these cells varies between 20 and 25 μm. However, even thinner membranes are available

(8–15 μm) that have been used in some commercially available Li-ion cells with the intent of increasing the specific energy, energy density, and power capability. Although thinner separator membranes result in a lower cell internal resistance, there is an increased risk of an ISC developing due to a foreign object debris (FOD), native object debris (NOD), or a metallic dendrite that could cause electrical contact between the two electrodes. A low impedance ISC caused by separator failure may lead to a high internal temperature that could create a thermal runaway (TR) safety hazard condition. Tri-layer separators, such the PP/PE/PE manufactured by Celgard (trade names being Celgard 2320 and 2325, with thicknesses of 20 and 25 μm, respectively), can provide a TR hazard control by providing an inherent "shutdown" feature designed to mitigate adverse cell reactions in the event of abuse conditions. In such a scenario, the inner PE layer melts before the two PP layers, which closes the pores and blocks the current flow that acts as a cell safety hazard control.

Ceramic coatings on the separators can enhance the overall safety of Li-ion cells. However, the effectiveness of various separator designs depends on both the size and geometry of the cell and battery configuration. Extensive modeling and multiple tests typically need to be done to evaluate the usefulness of separator technology for a given application. New separator materials, such as separators coated with aluminum oxides, can provide significant advantages with respect to the thermal, mechanical, and chemical stability under hazardous cell safety conditions [49, 50].

2.3.5 Safety Devices

A safe LIB power system space application is characterized by integrating the safety features of the Li-ion cell, module, LIB, EPS, and spacecraft together in order to meet spacecraft safety requirements. This layered approach to LIB safety begins with considering the safety devices designed into the cell. Cell-level safety devices are incorporated into Li-ion cell designs to provide verifiable controls against safety hazards such as overcharge, overdischarge, external short circuits, ISCs, and overtemperature abuse conditions. Cell-level protection to mitigate or prevent TR has also been incorporated into space-qualified Li-ion cell designs through electrode materials selection, tuning of safety devices, and other methods [51].

Since Li-ion cells combine highly energetic materials with flammable organic electrolytes in low-volume hermetically sealed containers, cell safety cannot be completely addressed at the cell chemistry level. Hence, internal cell-level safety devices or features are required in space-qualified Li-ion cell designs. For example, the NASA-ISS LIB orbital replacement unit battery cell (GS Yuasa model LSE134) contains a cell-level rupture plate, internal fusible link, and shutdown separators between electrode windings [52]. Internal safety devices such as pressure relief vents, current interrupt devices (CID), positive temperature coefficient (PTC) devices, and shutdown separators are common in small-format COTS 18650 space Li-ion cell designs. Safety devices used in custom large format Li-ion cells include pressure relief vents, leak-before-burst (LBB) cell cans, internal fusible links, and/or shutdown separator safety devices [53, 54].

2.3.5.1 Pressure Vents

Gaseous products are generated under electrical, mechanical, or thermal abuse conditions. A cell-level safety vent device can provide a safe means of releasing internal pressure in a controlled and directional manner before the cell reaches high pressures that can rupture the can. Relieving excessive internal gas pressure can also mitigate a rapid and uncontrolled increase in cell temperature, which may lead to a TR condition. The most common means to safely relieve excessive internal gas pressure in space-qualified Li-ion cells is by incorporating vents, burst disks, or rupture plates into the cell mechanical design. These pressure-relief devices are rated by the cell manufacturer to crack open and relieve internal gas pressure within a specified pressure range.

Cell venting with a rated safety vent device enables high-pressure gas to be relieved from the cell in a controlled and directional manner. Cell ejecta, such as decomposed electrode byproducts, may exit through the cell vent or other cell case rupture locations. Cell vents are commonly directed away from critical battery components to enable the pressure-relief function to operate and thus avoid unintentional battery damage. This safety design feature is a critical consideration when packaging and integrating Li-ion cells into the battery mechanical design.

Depending on cell packaging requirements at the battery level, high-capacity large format cell designs may have the cell vent device at the top, side, or bottom of the cell case exterior. In order to meet Range Safety requirements, custom and COTS 18650 cell cases must meet leak-before-burst (LBB) safety factor requirements with respect to the cell vent activation pressure and ultimate strength of the cell case [55]. Cell case construction compliance to Range Safety requirements verifies that the cell will leak through the vent, or other cell locations, before bursting in an uncontrolled manner. For example, Li-ion cells without built-in vent safety devices meet Range Safety requirements by allowing failure at weld closures sized to operate at safe venting pressures.

Vent disks are used in COTS 18650 cell designs to safely relieve pressure in a controlled manner to prevent sidewall rupture conditions [56]. In COTS 18650 cylindrical cell designs, the vent disk is integrated into the cell cap subassembly. The vent disk with specifically placed score marks is used as part of the sealing mechanism. Under high pressure (typically between 300 and 400 psi), the score marks, which are the weakest point of the disk, will break, allowing the cell to vent into the area below the cathode cap. To allow release to the atmosphere, the cathode cap incorporates vent holes. COTS 18650 cell designs also incorporate an LBB design to mitigate the risk of an uncontrolled cell burst event during an overpressure abuse condition.

2.3.5.2 Current Interrupt Devices

In addition to the internal vent disk, most COTS 18650 Li-ion cells contain non-resettable CIDs, which are located in the cathode region of the cell cap subassembly. The CID is a safety device that uses the internal pressure generated during the early stages of overcharge to activate a non-reversible mechanical switch designed to break the cell internal circuit. As shown in Figure 2.2, the CID is located in the current path between the cathode tab and the cell positive terminal.

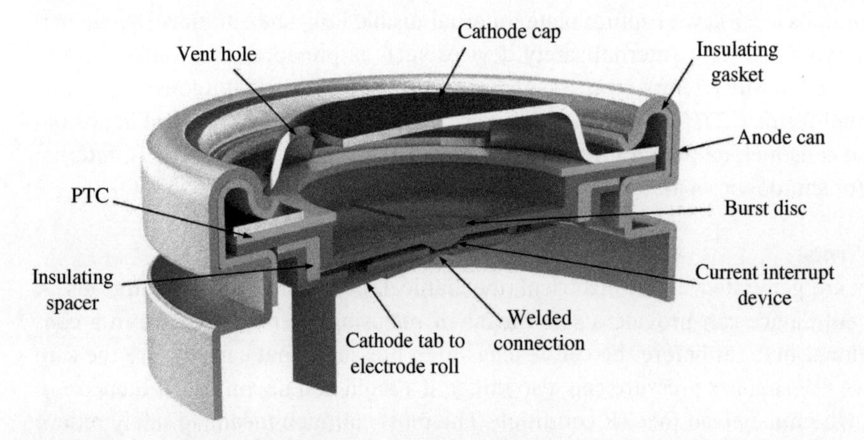

Figure 2.2 Cross-section of COTS 18650 Li-ion cell cap.

Increasing pressure within the cell, generated by gassing of the cell chemistry during overcharge, will cause the concave upper disk to flip upwards, breaking the spot-welded connection. This opens the cell internal circuit and prevents further charging. In the event of a continuing pressure rise within the cell, the cell vent disk serves as a safety device to release pressure in a controlled manner. Although the CID device is intended to provide overcharge protection, the CID will activate under other hazardous conditions, which causes an internal cell over-pressurization [57]. The CID is rated to operate at a lower pressure (approximately 175–200 psi) compared to the vent (300–400 psi).

2.3.5.3 Positive Temperature Coefficient

To protect against an overcurrent abuse condition (such as an external short circuit event), many COTS 18650 cells contain an internal PTC safety device, or simply a thermal fuse in the header cap. The PTC is a resettable thin polymeric disk with conductive elements embedded in this matrix. This disk is part of the cell current path within the cathode assembly. In normal operation, the PTC has a low resistance, which allows the cell to operate within the manufacturer's performance specifications. In the event of an overcurrent condition, the polymeric matrix heats up and expands, which reduces the contact between the conductive elements in the matrix. This results in a significant (many orders of magnitude) resistance increase within the PTC, which causes a rapid decrease of the current flow from the cell. The PTC operation is resettable and if the external short is removed, the PTC will cool, contract, and return to its original state.

In order to lower the internal resistance of some high-power density COTS 18650 cell designs, PTCs are not included as an internal safety device. Removal of the internal PTC lowers the cell-level internal resistance, thus increasing the cell power density. There are also limitations in using the device as protection for high-voltage batteries, since the PTC material can exhibit breakdown under high voltages and may fail with sparking and material breakdown [58]. PTC safety devices are not implemented in large format prismatic and elliptical–cylindrical cell designs.

2.3.5.4 Shutdown Separator

Li-ion cell separators commonly utilize separators (see Section 2.3.4), which possess a shutdown function capable of serving as an internal cell safety feature [59]. The shutdown separator is activated when the cells reach a temperature range from 120 to 140 °C. This high temperature causes a meltdown of the middle polyethylene-layer of the three-layer separator between the electrodes, preventing current flow within the cell stopping the electrochemical reaction and mitigating TR conditions. In addition, a ceramic-coated separator is used in selected COTS 18650 cells to minimize separator shrinking and the resultant internal short, which aids in mitigating the risk of the onset of a TR condition.

2.4 Cell Geometry

Custom and COTS 18650 Li-ion space cell manufacturers produce cell designs that can be classified by design features such as cell chemistry, safety device types, and geometrical shape (form factors). The most common form factors used in space-qualified Li-ion cells are cylindrical, prismatic, and elliptical–cylindrical geometries. Although commonplace in many terrestrial power system applications, Li-ion pouch cell prismatic formats have been used on a very limited basis for astronaut crew equipment and spacecraft EPS applications. Cell form factor dimensions have a significant impact on module and battery-level electrical, mechanical, thermal, and safety design features. For example, the cell form factor impacts packaging design, which impacts battery specific energy, volume, thermal dissipation characteristics, and safety [60, 61]. Process steps for electrode manufacturing and assembly are also a differentiator between various cell form factor designs.

2.4.1 Standardization

With the exception of rechargeable COTS Li-ion cylindrical cell designs, there is no space industry-wide standard that defines a common set of electrochemical systems, dimensions, terminals, nomenclature, or markings for custom space Li-ion cells. As such, Li-ion custom space cell manufacturers of cylindrical, prismatic, and elliptical–cylindrical shaped cells commonly use non-IEC designations for their products. Manufacturer-unique part number designations for custom space Li-ion cells include an electrochemical system, nameplate capacity, cell type (energy or power), identifiers for LEO or geosynchronous orbit (GEO) versions, and other cell-unique design features. Due to the widespread commercial use of COTS 18650 cylindrical Li-ion cells, the 18650-cell design designation has become a global standard. An 18650 cylindrical Li-ion cell has an approximate diameter of 18 mm and approximate length of 65 mm [62]. For example, electrochemistry designations such as LCR (L = lithium, C = cobalt, and R = rechargeable) indicate a Li-CoO_2 cathode chemistry cell, while NCR (N = nickel, C = cobalt, and R = rechargeable) indicates a Li-$NiCoO_2$ cathode chemistry.

2.4.2 Cylindrical

The cylindrical geometry is the most commonly used form factor for space-qualified Li-ion cells. The cylindrical form factor is also used exclusively by non-rechargeable Li cells used in various one-time use space applications. Space-qualified Li-ion cylindrical cells are hard-shell case designs manufactured using standard dimensions for diameters of 18, 21, 33, and 53 mm and lengths ranging between 65 and 252 mm. Above 53 mm in diameter, there are challenging manufacturing and thermal management issues with obtaining uniform and well-aligned electrode "jelly rolls". Thus, the largest diameter cylindrical Li-ion cells are used primarily for terrestrial applications with low cycle life requirements. These cell dimensions correspond to cell nameplate capacities ranging from 1.5 to 52 Ah.

During cell manufacturing, cylindrical cell electrodes and separator materials are wound together to form a jelly roll structure sized for capacity, energy, and/or power characteristics. The jelly roll structure contains dimensional allowances for electrode expansion–contraction caused by Li intercalation from the positive and negative electrodes during cycling [63]. As the cell electrodes expand and contract during cycling, cylindrical cells have the inherent ability to evenly distribute internal stress caused by electrode swelling. As a result, space cylindrical cells do not require unique packaging-compression restraints to maintain internal mechanical stability and a uniform geometrical shape. Stainless steel or nickel-plated steel is a common hard-shell case material used in COTS 18650 cell designs, while anodized aluminum is used as a mass-savings design feature in space-qualified custom cylindrical and elliptical–cylindrical cell cases [64–66]. Large format custom cylindrical space cell header assemblies are equipped with two terminals; one is a hermetic terminal feed-through using plastic, polymer, or ceramic insulator and the second is a connector to the cell case polarity. The COTS 18650 cell employs a cell design where the case, which is the negative terminal, is crimped around the cover, which is the positive terminal, to form a hermetic seal for the cell.

2.4.3 Prismatic

Prismatic (also known as "true prismatic") Ni-Cd and Ag-Zn cell designs were used successfully for decades in various spacecraft applications. Prismatic Li-ion cell designs differ greatly from cylindrical Li-ion cell geometries in aspect ratio, thermal dissipation characteristics, packing efficiency, and specific energy. In today's space cell marketplace, only custom prismatic space cell designs can be procured for spacecraft applications. Similar to cylindrical cell formats, the lower portion of a prismatic cell is the hard-shell case, while the top portion is referred to as the cell cover, cover assembly, or header. The cell header subassembly contains the cell terminals, safety vent (in some designs), and electrolyte fill tube. Cell cases and header subassemblies (not including the terminals) are manufactured using 304L stainless steel or aluminum [67]. An advantage of a prismatic Li-ion cell design is the conformal shape that can be tailored to fit into confined volumes. These characteristics yield a higher packing efficiency when compared to cylindrical cell geometries. Unlike cylindrical cell jelly-roll electrodes, prismatic cell electrodes are individual flat plates with a surface area and thickness proportional to the desired electrode pair capacity. The flat plate electrodes are manufactured, stacked, and inserted into the prismatic cell case in electrode pair quantities corresponding to the overall cell nameplate capacity. Prismatic and elliptical–cylindrical Li-ion cell form factors are highly susceptible to case wall deformation during extended charge–discharge cycling [68]. As a result, cell compression restraints are built into space LIB structures using prismatic or elliptical–cylindrical cell geometries to provide long-term mechanical stability. Cell nameplate capacities for space-qualified Li-ion prismatic cells range from 6 to 75 Ah.

2.4.4 Elliptical–Cylindrical

Custom elliptical–cylindrical Li-ion cells are a hybrid geometrical shape with physical characteristics of both cylindrical and prismatic cell form factors. Also available as COTS Li-ion cells for terrestrial applications, custom elliptical–cylindrical space Li-ion battery cells have an extensive on-orbit heritage on numerous types of Earth-orbiting satellite systems [69]. Similar to cylindrical cells, elliptical–cylindrical cells contain a wound jelly-roll (with an empty center core) electrode assembly inside an aluminum or stainless-steel hard-shell case. However, prior to a jelly-roll insertion, the jelly roll is flattened in a controlled manner to fit into the elliptical–cylindrical shaped case. The flat section of the pressed jelly-roll resembles the flat plate electrode geometry of a prismatic Li-ion cell. Similar to a prismatic form factor geometry, the flat surface of an elliptical–cylindrical cell is susceptible to deformation during charge–discharge cycling. As such, compression restraints are commonly built into the battery structures to provide long-term mechanical stability. Cell nameplate capacities for space-qualified elliptical–cylindrical Li-ion cells range from 6 to 190 Ah.

2.4.5 Pouch

Li-ion pouch cells have a prismatic geometry with a characteristic multi-layer laminate packaging material. Pouch cells are manufactured in various formats, which is advantageous for many commercial consumer products and applications. Similar to hard-case prismatic cells, pouch cells have a high packing efficiency that facilitates high module and battery-level specific energy. Although Li-ion pouch cells have been space-qualified for low-power astronaut crew equipment (such as laptop devices), NASA has identified a number of technical risks associated with pouch cell technology use in spacecraft battery power system applications [70]. For example, when exposed to simulated space environments, pouch cell performance limitations include poor hermeticity of cell seals, edge corrosion after long-term storage, and pressure growth during long-term charge–discharge cycling. Today, pouch Li-ion cells are only used for short-term LV and LEO nano-microsatellite missions [71].

2.5 Cell Requirements

Over the past 25 years, space Li-ion cell design solutions have evolved to meet a wide variety of mission-relevant performance and environmental requirements. As such, space-qualified LIB cell designs are required to comply with the spacecraft EPS battery mission-specific set of physical, mechanical, thermal, electrical, life, and safety requirements. Mission-relevant operational requirements corresponding to ground processing, launch, and on-station orbital environments commonly have the most significant impact on cell design features. Critical cell-level design features include cell-unique electrical (cell electrode and electrolyte type), mechanical (cell terminals and geometric cell case form factor), and thermal (electrode configuration and cell case material) design characteristics. Cell design features are developed to meet critical cell specification requirements such as capacity, operating voltage and temperature, cycle life, mass, environments, reliability, and safety. Detailed technical consideration to cell-level design compliance to LIB-level requirements is a space industry best practice applied early during spacecraft EPS preliminary design activities.

2.5.1 Specification

Spacecraft applications, such as short-mission expendable LVs, high-power commercial GEO satellites, or long-duration deep-space planetary missions, each have their own diverse set of cell-level performance and environmental requirements. Despite the unique characteristics of each spacecraft application, the specified Li-ion cell design, when assembled into a LIB, must be capable of providing reliable power to the EPS user loads for the entire mission duration. A space Li-ion cell specification defines the requirements for the design, performance, manufacturing, testing, and quality of a Li-ion electrochemical battery cell design intended for use in spacecraft LIB applications. Space Li-ion cell requirements are commonly documented in the cell manufacturer's product specification. Cell product specifications are defined, developed, and configuration managed according to each manufacturer's own internal proprietary processes. Various cell specification formats, such as single comprehensive documents, source control drawings, or other formats, are not uncommon in the space Li-ion cell manufacturer marketplace. Cell-level design and manufacturing details, such as proprietary cell chemistry formulations, electrode coating thicknesses, and electrolyte fill volume, are typically documented in a space Li-ion cell specification. Manufacturer proprietary cell acceptance and qualification test procedures and pass/fail requirements may also be included in the cell specification. This may include the cell formation test protocol requirements. Finally, proprietary raw cell materials and other component sub-tier supplier requirements procurement information may be contained in the manufacturer's cell specification.

2.5.2 Capacity and Energy

As discussed in Chapter 3, LIB-level capacity (or energy) requirements are derived from spacecraft EPS power requirements, power bus voltage limits, mission type, and LIB-level DOD. The LIB-level capacity requirements are decomposed and allocated to the cell specification. The BOL cell-level capacity specification requirements are specified using standard test conditions, such as charge–discharge rates, EOCV, and EODV. Since cell capacity is also a function of temperature, BOL cell capacity requirements are commonly specified over the intended LIB operating temperature range. Cell capacity may also be specified as a function of EOCV. These data are relevant to satellite EPS architectures that employ a ground-commandable battery management system (BMS) capability to vary cell EOCV limits. For large flight cell production lots, minimum guaranteed single-cell capacity, cell production lot average, and lot standard deviation-level capacity may be specified. Depending on the cell manufacturer, the difference between the nameplate capacity requirements and measured capacity may be significant. Quantifying these BOL cell capacity differences are relevant to LIB sizing and EPS energy balance analyses.

2.5.3 Operating Voltage

In general, cell manufacturers specify cell operating voltage limits based on cell chemistry characteristics, cycle life requirements, and safe operation thresholds. The cell manufacturer operating cell voltage range is specified at the maximum and minimum cell SOC corresponding to the EOCV and EODV limits, respectively. Cell manufacturer maximum EOCV limits may vary between 4.0 and 4.2 V, corresponding to a specified charge rate and temperature range. Manufacturer minimum cell EODV limits may vary between 2.5 and 3.0 V for a specified discharge rate and temperature range. The maximum cell operating voltage range specification requirements may be decreased from the nominal 4.2 to 4.0 V EOCV limit range to meet mission-unique cell cycle life requirements. The average BOL cell operating EODV requirements may also be specified in order to support mission-unique DOD or EPS power bus cut-off voltage characteristics.

2.5.4 Mass and Volume

Spacecraft industry experience has demonstrated that Li-ion cell mass is the largest contributor to total LIB mass properties. Cell mass properties are directly proportional to cell-level capacity and cell-level components such as the cell container material type and terminal design features. Although not typically specified in space Li-ion cell requirements specifications, cell-level specific energy targets can be derived from cell mass, capacity, and average voltage requirements. Cell dimensions, terminal height, and center of gravity data used for LIB design compliance is commonly specified in the cell specification mechanical interface control document (ICD).

2.5.5 DC Resistance

Cell-level DC resistance is a significant contributor to battery-level electrical and thermal performance characteristics. Since EPS power bus stability and dynamic performance depends on battery-level resistance, cell-level resistance is specified to enable battery design compliance to EPS requirements. BOL cell-level resistance is typically specified in the cell specification as a function of temperature, discharge rate, and SOC. Cell-level impedance may also be specified as a function of applied voltage and frequency range as required to meet certain battery and EPS-level design requirements.

2.5.6 Self-Discharge Rate

Li-ion cell self-discharge rates induce capacity loss during non-operational and storage time periods. To quantify the capacity loss due to self-discharge, capacity retention is commonly measured as part of a self-discharge test. Cell self-discharge rate data are primarily used to analyze cells for the presence of ISCs resulting from cell manufacturing defects, NOD, FOD, or other sources. Li-ion cell self-discharge rate data are also used to screen, select, and match flight cells prior to battery assembly. Self-discharge rate requirements are commonly specified as a function of cell SOC, temperature, and open-circuit voltage (OCV) stand-time (commonly between 7 and 30 days) durations.

2.5.7 Environments

Space Li-ion cells and batteries are designed, analyzed, and tested to comply with both induced and natural space environments expected during ground processing, launch activities, and on-orbit operation. Early in the preliminary satellite design phase, LIB specification environmental requirements are decomposed and allocated to the cell requirements specification. Induced and naturally occurring environments, which have a significant impact on space cell designs, include hot and cold temperature extremes, random vibration, shock, space vacuum, and radiation dosage levels.

2.5.7.1 Operating and Storage Temperature

Cell operating temperature requirements vary with the spacecraft mission-specific operating modes and expected on-orbit environments. Spacecraft and battery-level thermal control design features are specified to comply with manufacturer recommended cell operating temperatures. In general, space Li-ion cell manufacturers recommend cell operating temperature ranges as a function of cell charge and discharge rates. The relationship between cell operating temperature and charge rates is governed by meeting cell performance requirements while mitigating the risk of lithium plating at the cell anode. Lithium plating at a Li-ion cell anode can occur when charging at low cell operating temperatures and/or at high-charge rates. Lithium plating has been shown to irreversibly increase cell resistance, which reduces usable cell capacity [72, 73]. At elevated cell operating temperatures (>40 °C), Li-ion cells exhibit increases in cell resistance and reduced cell capacity, which may result in decreased cell cycle life. Typical space Li-ion cell operating temperature ranges during charge and discharge are 0 to +40 °C and −10 to +40 °C, respectively. Li-ion cells used in planetary exploration mission LIBs are required to operate at temperatures as low as −30 °C.

Controlled storage cell temperature environments are commonly documented in the cell manufacturer cell requirements specification or separate documentation. Controlled storage conditions included requirements for cell temperature, SOC, humidity, and packaging. In order to mitigate the effects of capacity fade during non-operating uncontrolled storage time periods, cell manufacturers commonly recommend controlled cell storage for non-operating time periods exceeding 30 days.

2.5.7.2 Vibration, Shock, and Acceleration

Applicable cell-level mechanical environmental requirements are derived from numerous sources, including transportation conditions, LV dynamics, and spacecraft appendage deployment events. For most applications, the most severe mechanical environments occur during spacecraft launch. From LV liftoff to orbital insertion, LVs induce significant levels of random vibration, shock, acceleration, and acoustic noise into the spacecraft environment. Encapsulated into the LV fairing, the satellite payload is exposed to severe mechanical load environments during launch. Therefore, the satellite EPS components are required to survive the worst-case launch environments unique to the dynamic performance characteristics of the relevant launch vehicle type. In terms of analysis and test protocols, the worst-case launch environments are incorporated into the EPS battery and cell qualification test programs. This includes a mechanical analysis and testing bounded by worst-case LV shock, vibration, and on-orbit spacecraft deployment environments. Battery location and mounting in the spacecraft bus structure may induce amplification and resonance factors into the battery structure, which may create different mechanical test environments for the cell subassemblies. Although not as severe as launch environments, ground transportation and shock requirements are also included in the cell requirements specification.

2.5.7.3 Thermal Vacuum

Operational spacecraft near-Earth environments are dominated by the solar intensity of the Sun, which creates unique radiation and thermal environments as a function of satellite orbit apogee and perigee. Operating outside the near-Earth environment, planetary mission spacecraft environments vary with distance from the Sun, atmospheric pressure, and thermal environment of the intended planetary body destination. Both Earth-orbiting and planetary spacecraft mission environments share similar thermal vacuum (TVAC) environments. The combination of mission-specific space vacuum pressures and thermal variations during satellite eclipse and solstice periods dictates TVAC requirements for all spacecraft subsystems. For example, the satellite external temperature, which varies with orientation to the Sun and orbital exposure time in solstice, can range between −170 and +140 °C. LIB cell requirements for surviving mission-relevant TVAC environments are derived from the spacecraft bus environmental specification, battery specification, or other applicable requirement source. In contrast to terrestrial Li-ion cell applications, it is challenging to maintain space Li-ion cell hermeticity under space vacuum environmental conditions.

2.5.7.4 Radiation

Depending on Earth-orbiting satellite apogee and perigee, spacecraft components may be exposed to significant on-orbit levels of naturally occurring space ionizing radiation. Industry experience has demonstrated that total ionizing radiation dosages may affect the physical characteristics of certain Li-ion cell internal components, such as electrode materials, separators, and electrolytes. Consequently, cell qualification testing is required to include analyses and/or testing for the susceptibility of Li-ion cell components to space ionizing radiation environments. In practice, the impact of radiation at the cell level is expected to be limited, based on the reported tolerance of Li-ion cells to high cumulative space radiation dosage levels [74, 75].

2.5.8 Lifetime

Space Li-ion cell lifetime, defined as the combination of ground (pre-launch) processing and on-orbit operating life, is a key performance requirement for all space LIB power systems. In general, Li-ion space cell life begins subsequent to electrolyte activation and completion of cell formation. Pre-launch cell life then includes all ground processing events such as cell and battery manufacturing (typically at least one year), spacecraft assembly, integration, and test (AI&T) (8–12 months), storage time (up to five years), and pre-launch operations (up to 12 months). On-orbit operating life includes launch and transfer (up to 30 days), GEO transfer orbit (up to 90–180 days for electric propulsion satellite systems), on-orbit testing (typically up to 30 days), and on-station service life. On-station service life can vary between 15 and 20 years (GEO satellite missions) to less than five years for LEO small satellite systems. Thus, for Earth-orbiting satellite applications, the total LIB cell lifetime requirements can be as high as 25 years or more.

2.5.9 Cycle Life

Cell cycle life specification requirements are derived from the mission lifetime requirements specific to the relevant Earth-orbiting or planetary mission operating regime. Cycle life requirements are based on orbit type, environments, mission lifetime, and EPS power system requirements. Orbit type defines GEO solstice, LEO sunlight, and eclipse period durations. For example, a spacecraft in a 95-minute LEO orbit experiences approximately 60 minutes of sunlight and 35 minutes of battery discharge in eclipse. These orbital conditions correspond to 15 battery charge–discharge cycles per day or 5475 battery cycles per calendar year. However, a spacecraft in a GEO orbit experiences two solstice periods lasting up to 137 days, each of which corresponds to two GEO equinox periods of 45 days each. These orbital conditions correspond to 90 battery charge–discharge cycles per calendar year. For example, a representative commercial GEO satellite LIB may be qualified for 15 years of on-orbit mission service life, which corresponds to 1350 charge–discharge cycles. Therefore, a satellite mission lifetime provides the minimum time period requirements for battery cell cycling according to the orbital regime and satellite user load demand. An EPS energy balance analysis is used to determine battery-level DOD requirements. Finally, as described in Chapter 3, cell-level DOD requirements may be dependent on LIB cell topology (p-s or s-p) design and battery-level DOD analysis.

2.5.10 Safety and Reliability

Li-ion cell safety requirements are derived from various government and non-government sources depending on whether the intended on-orbit spacecraft application is unmanned or crewed. However, regardless of mission-type, Li-ion cell ground processing activities such as transportation, handling, storage, and testing generally have a common set of safety requirement sources. Large format cell safety requirements are allocated from the LIB specification or cell manufacturer unique design requirements. Due to the large commercial market application of small-format COTS cell designs, cell safety requirements are typically derived by the cell manufacturer to meet industry-relevant safety standards [76–78].

The LIB product specification reliability requirements are decomposed and allocated to the cell specification. A space-qualified LIB will generally have its EOL reliability prediction governed by the cell-level reliability estimates. Cell-level reliability is typically estimated from a combination of analysis and mission-relevant reliability testing. Chapter 8 discusses real-time and accelerated life-cycle test strategies that may be used to estimate and predict cell-level reliability.

2.6 Cell Performance Characteristics

In selecting the proper Li-ion chemistry, cell design, and battery size for a particular space application, it is critical to fully test and analyze the baseline cell performance characteristics. There are many performance attributes that are common to nearly all space-rated LIBs, such as high coulombic efficiency, long life, and low self-discharge rates. However, since the Li-ion cell technology trade space comprises various chemistries and cell designs, there can be distinct differences in a number of performance characteristics. Key Li-ion cell performance characteristics include, but are not limited to, power capability, delivered energy, charge acceptance behavior, effect of DOD on cycle life, and charge–discharge rate behavior.

2.6.1 Charge and Discharge Voltage

In contrast to heritage aqueous KOH-based rechargeable space battery cells, Li-ion cells operate over a much wider voltage range since they possess organic solvent-based electrolytes that are stable to much higher voltages. Since Li-ion cell technologies encompasses a wide range of anode and cathode combinations, the voltage ranges of these systems vary accordingly, depending upon the chemistry selected. However, for chemistries relevant to space applications, which typically involve the use of graphitic anodes coupled with lithium metal oxide cathodes (such as LCO, NCA, NCO, and NMC), a common voltage range that can be employed is 2.70–4.20 V. In practice, however, in order to preserve the battery service life and to provide needed safety margins, narrower operating voltage ranges between 3.0 and 4.10 V are often baselined. The voltage profile during charging and discharging is determined by the choice of the cell anode and cathode chemistry. Since carbon anodes, with the exception of hard carbon, display relatively flat charge and discharge curves, the overall cell voltage is primarily governed by the behavior of the cathode chemistry.

To obtain fully charged cells, the most commonly used test charge protocol involves using a constant current-constant voltage (CC-CV) regime. Using a CC-CV test approach, the cell is charged at a constant current to the desired maximum charge voltage, termed cut-off voltage, and then is held at this constant voltage as the charge current is allowed to taper to a minimum value (approximately a C/50 rate). If high rates are used during the constant current regime a proportionately larger amount of charge capacity is put into the constant voltage mode. Furthermore, due to the higher cell resistance observed at lower temperatures, a higher proportion of the charge capacity is generally obtained in the constant voltage mode.

2.6.2 Capacity

As previously described, battery capacity is defined as the product of the discharge current and the time it takes to reach a minimum EODV. Discharge capacity of a Li-ion cell decreases with an increasing discharge rate, due to the increasing overpotentials associated with the electrochemical processes within the cell. Several factors influence the observed overpotential, including the cell ohmic resistance, the lithium intercalation/deintercalation kinetics, mass transfer kinetics, charge transfer kinetics, and solid-state diffusion at both the anode and the cathode. All these processes become more sluggish at low temperatures, resulting in a lower cell capacity, as illustrated by the data from a space-qualified Yardney 43 Ah prismatic cell (Figure 2.3). The discharge capacity obtained at −20 °C using a C/5 rate is 80% of that obtained at +20 °C [79]. The capacity will be lower if the preceding charge is at a low temperature, due to poor charge acceptance. Furthermore, unlike the Yardney 43 Ah cell, many Li-ion cell designs are not designed for charging at low temperatures, since lithium plating can occur rather than the desired lithium intercalation at the anode. However, for many space applications, batteries are controlled within a narrow operating temperature range (such as 0 to +30 °C), using active or passive thermal management control methods (such as heaters, radiators, or cold plates), which mitigates the risk of lithium plating. As described in Chapter 6, LIBs supporting planetary surface missions (such as rovers and landers) are often exposed to low temperature environments. These operating conditions necessitate more complex thermal management and Li-ion cell designs that are optimized for low temperature charge–discharge cycling. For non-optimized Li-ion cell designs, LIB operations are limited by cell performance at low temperatures.

These rate-limiting processes either at low temperatures or high-rate discharges can be partly mitigated in high-power cell designs, which utilize thinner electrode coatings, higher amounts of conductive diluents, and smaller particle sizes of electroactive materials. For space applications, however, the most commonly used cell designs tend to be optimized to provide high specific energy, while still providing reliable power capability to meet mission-specific requirements.

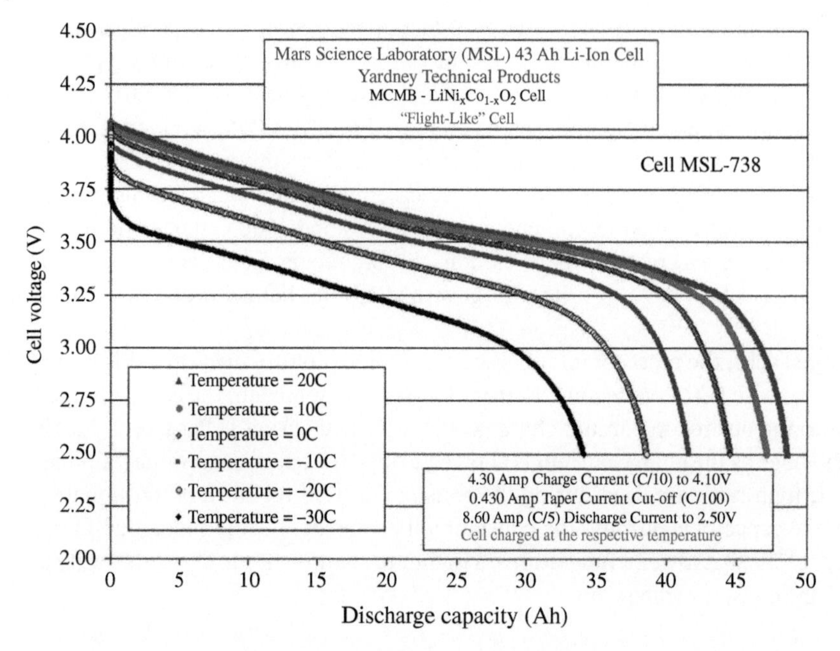

Figure 2.3 The discharge capacity of a Yardney 43 Ah Li-ion cell at various temperatures, using a C/10 charge to 4.10 V and a C/5 discharge rate to 2.50 V. *Source:* Courtesy of NASA-JPL.

2.6.3 Energy

For spacecraft operations, the energy contained in the battery is often more relevant than the capacity, since most spacecraft loads operate at constant power. The delivered energy of a Li-ion cell depends upon several factors, including the discharge current, operating temperature, and the conditions of the preceding charge history. Even though the coulombic efficiency of Li-ion cells is very high at BOL (>99.8%), the energy (Wh) efficiency may be lower due to polarization losses resulting from charge and discharge rates, operating temperature, and cell aging. Lower energy efficiency is obtained at high rates and low temperatures where the discharge voltages are lower and the charge voltages are higher. Lower energy efficiency is also due to higher polarization losses, which cause heat generation within the cell.

2.6.4 Internal Resistance

The internal DC resistance of a Li-ion cell is an important parameter to characterize since it can be predictive of cell performance and lifetime. The DC resistance of a battery is measured by the voltage drop obtained with the application of a current pulse. The observed voltage drop is governed by various processes that have different timescales. The initial, or instantaneous, voltage drop is due to ohmic contributions, including the ionic resistance of the electrolyte and the electronic resistances associated with cell-to-cell interconnects, cell internal tab connections, and current collector properties. At longer timescales, charge-transfer and mass-transfer kinetics in the electrolyte and at the intefaces, as well as ionic solid-state diffusion in the electroactive materials, contribute significantly to internal cell resistance characteristics. Therefore, the pulse amplitude and pulse duration can influence the measured cell DC resistance value, depending on the various polarization contributions. To further complicate DC resistance measurements, Li-ion cell resistance is a function of cell

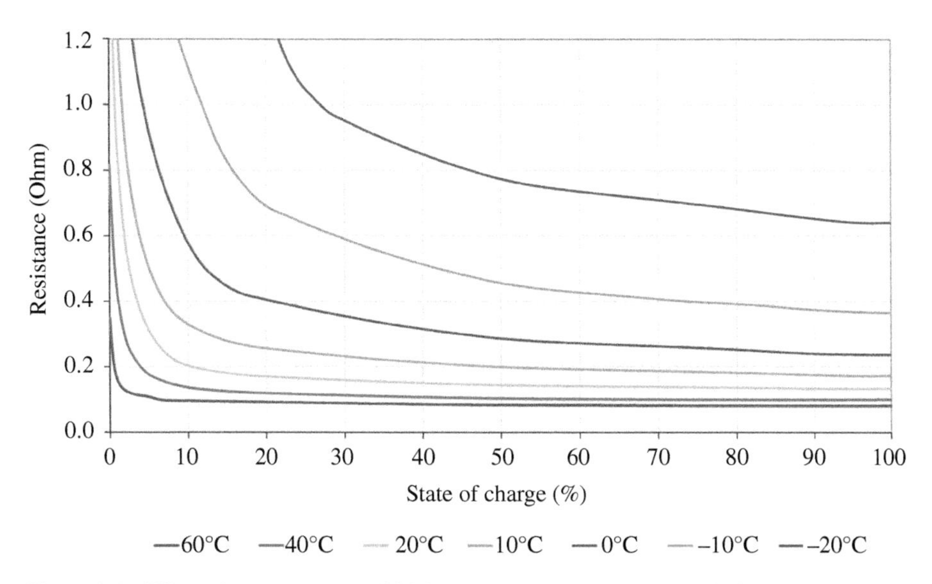

Figure 2.4 Effect of temperature and SOC on the internal resistance of COTS 18650 HCM cells. *Source:* Courtesy of EnerSys/ABSL.

SOC, temperature, and cell age. Figure 2.4 illustrates general trends for LCO-based Li-ion 18650 cells as follows: (i) the resistance increases at lower temperatures and decreases at high temperatures, (ii) the resistance increases with cell aging, and (iii) the resistance is generally higher at very low SOCs for an LCO-based cell design. Furthermore, large capacity cells have a lower cell resistance than smaller capacity cells and high-power cell designs generally have a lower internal DC resistance compared to high energy cell designs. The measured cell impedance is a combination of both the internal resistance and reactance of the cell to an AC stimulus. In practice, cell impedance is commonly measured at 1 kHz as a cell manufacturing quality control check.

2.6.5 Depth of Discharge

Cell DOD is a critical parameter that must be considered when determining if a particular cell design can meet battery mission service life requirements. Li-ion cell lifetime improves exponentially with a decrease in cell DOD. At high DODs, there are mechanical stresses associated with the expansion and contraction of the cell electrodes and thermal stresses from increased heat generation, which accelerates cell degradation mechanisms.

For LEO applications, the short orbital time (90–95 minutes), implies over 5000 charge–discharge cycles per year. Since most LEO satellite applications have a minimum life requirement of five or six years, the expected cycle life of the LIB can exceed over 30 000 cycles. To meet long cycle life requirements, LIBs operating in LEO orbital environments are commonly limited to DODs ranging from 10 to 30%. Also, due to the short duration of the LEO cycle, both the charge and discharge currents increase proportionately with DOD. A higher discharge current leads to increased heat dissipation, while a higher charge rate increases the risk of lithium plating. Lithium plating has been shown to greatly reduce battery lifetime since it leads to increased electrolyte decomposition and growth rates of the SEI film on the anode. Lithium plating is most likely when high cell charge currents are coupled with high charge voltages and lower operating temperatures. Thus, to mitigate these performance degradation modes, cell designs are used with an increased anode reserve

($C_{anode}/C_{cathode}$ ratio) or on-orbit operational protocols to lower cell EOCVs or maintain elevated charge temperatures (>10 °C) as needed.

For GEO and medium Earth orbit (MEO) applications, the cycle life requirement is significantly lower compared to LEO missions, with only a few thousand cycles needed to support a 10–15-year mission lifetime. However, the battery calendar life is longer and DOD is much deeper (60–80% depending on the cell design) for GEO and MEO satellite orbits. Under these cycling conditions, the cycle life degradation is primarily attributed to cell cathode aging. Real-time cell-level life-cycle testing is commonly conducted to space qualify a cell design to the relevant mission operating conditions. This approach verifies that the selected battery cell design can provide the desired operational on-load voltage at the specified DOD to support all spacecraft operations including safe mode contingencies and possible mission extension plans.

2.6.6 Life Cycle

Li-ion cell cycle life is dependent on, but not limited to, DOD, charge–discharge rates, operating temperature, EOCV, and calendar life. Relative contributions of all these operational parameters are strongly dependent upon cell design features. In general, to obtain a high cell cycle life it is preferred to limit the DOD, maintain low charge rates (<*C*/3 rates), reduce the EOCV, and operate in a temperature range of approximately 10–25 °C. For certain Earth-orbiting or planetary mission EPS architectures, operational parameters can be adjusted by ground mission operators to ensure that the battery cells provide the needed on-orbit cycle life.

Figure 2.5 shows the performance of NCO-based EaglePicher 7 and 10 Ah cells with varying DOD LEO cycling (14, 24, and 30%) at 23 °C with a 3.95 V charge voltage [80]. It is apparent from the observed decay in the EODV that the performance degradation is more dramatic with a deeper DOD. The trend of a shallower DOD providing superior life characteristics during LEO cycling is well documented. Excellent performance was also obtained with GS Yuasa's Generation III high-capacity LCO-based 110 and 51 Ah Li-ion cells subjected to a 25% DOD real-time LEO cycling regime at 15 and 20 °C, respectively. As shown in Figure 2.6, the Gen III LSE110 (110 Ah) and LSE51 (51 Ah) cells exhibit over 45,000 and 47,000 cycles, respectively when charging to 4.10 V [79].

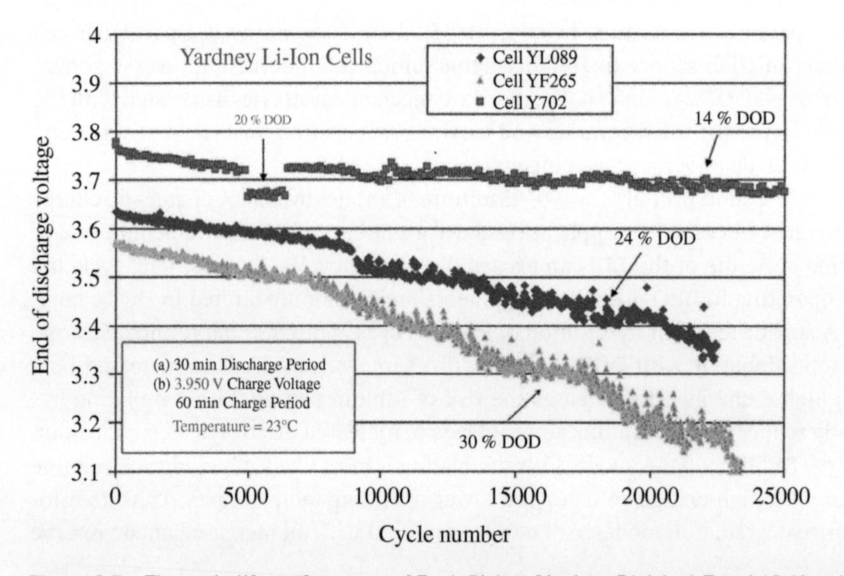

Figure 2.5 The cycle life performance of EaglePicher (Yardney Division) 7 and 10 Ah cells with varying DOD (14%, 24%, and 30%) LEO cycling profiles at +23 °C. *Source:* Smart et al. [80]/with permission of John Wiley & Sons.

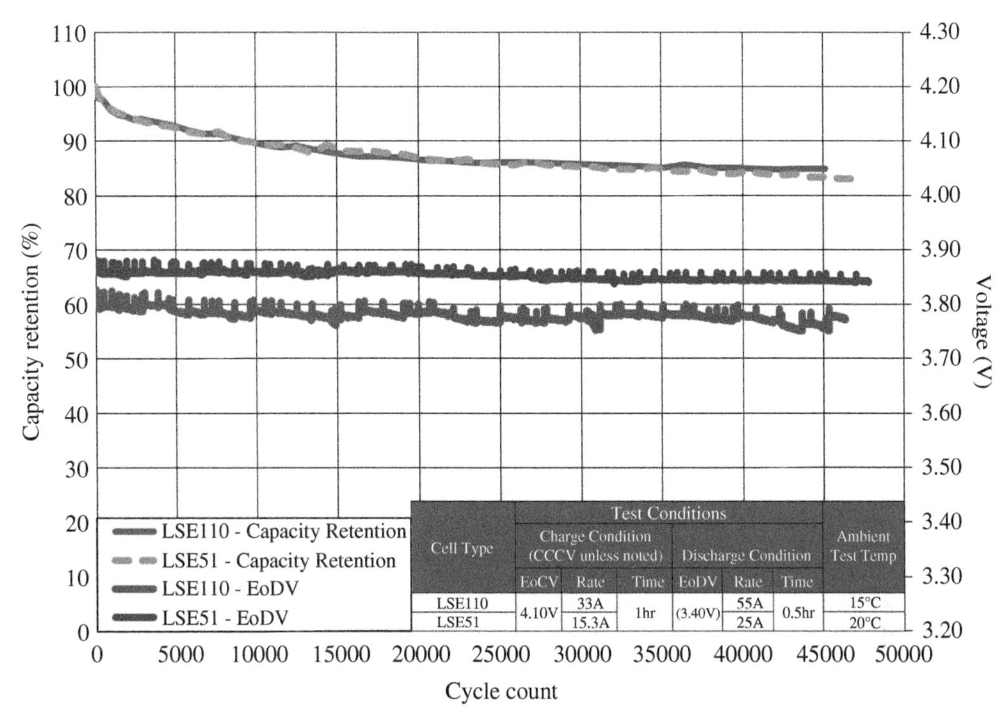

Figure 2.6 The cycle life performance of GS Yuasa Gen II 100Ah Li-ion cells subjected to 25% DOD LEO cycling at +25 °C.

Figure 2.7 shows the performance of an NCA-based Saft VES16 Li-ion cell during a semi-accelerated 80% DOD GEO cycling regime with a charge voltage of 4.05 V over 45 GEO seasons [81]. These data demonstrate that the Li-ion cells can support several GEO seasons over several years. The methodology employed to accelerate the life test consisted of removing the non-solstice periods of the mission when the battery was not charging or discharging.

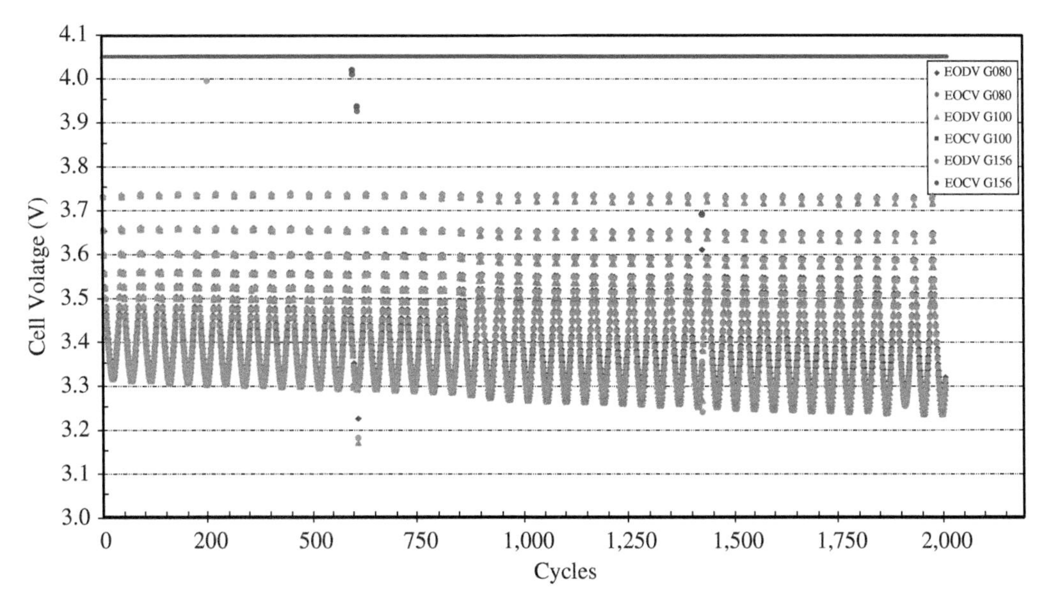

Figure 2.7 Cycle life performance of Saft VES16 D-size Li-ion cells subjected to a semi-accelerated 80% DOD GEO cycling profile at +20 °C.

2.7 Cell Qualification Testing

Li-ion cell qualification requirements for space LIB applications can vary widely between the commercial space industry and government agency customer stakeholders. Qualification test and analysis requirements are derived from numerous sources, including space industry guidelines, consensus standards, and cell manufacturer specifications. Li-ion cells are typically qualified using a combination of cell-level acceptance testing (AT), lot acceptance testing (LAT), and qualification testing at the cell and/or battery level of integration. As with other spacecraft component qualifications, the Li-ion cell qualification test and analysis scope is dependent on program technical, schedule, cost, and risk management constraints.

In general, custom and COTS Li-ion space cell designs are qualified and screened using two different methodologies:

- A space custom-manufactured Li-ion cell qualification test program includes cell-level AT followed by cell qualification at the battery level of integration, and
- COTS Li-ion cells are qualified by cell type, which includes LAT of large cell quantity lots followed by battery-level qualification and AT.

Despite these different qualification test sequence approaches, there are many common qualification verification test and analysis methods used to space-qualify custom and COTS Li-ion cell designs. This section summarizes the general flow and scope of test and analysis used by the space industry to space-qualify custom-manufactured and COTS Li-ion cells.

2.7.1 Test Descriptions

Figure 2.8 provides an example of a typical COTS Li-ion cell LAT and qualification test plan [82]. Depending on the batch type (acceptance or qualification), different tests are applicable for full justification. The health check characterization test sequences (capacity, energy, internal resistance) are performed after each critical test, such as vibration, shock, thermal vacuum test, radiation exposure, cycle life evaluation, and safety tests. Additional safety tests may be performed after environmental testing to ensure cell requirement compliance. Also, cells may be subjected to the anticipated mission radiation dose testing prior to life-cycle testing. A typical cell qualification campaign can be divided into electrical characterization, environmental loads, safety tests, and life-cycle testing.

2.7.1.1 Electrical

The objective of electrical characterization testing is to establish the baseline performance of the cells and the operational capability of the selected cell under application-relevant load conditions. These data may be used to develop cell and battery analysis to facilitate predictions of cell cycle life. Typically, cell capacity, energy, and internal resistance is measured for a given operating voltage range defined by the cell supplier or user. This allows one to establish the baseline performance of the cell to be used as a reference for the life test programs. In addition, the OCV vs. SOC relationship is established since it is a fundamental element of any electrical cell analysis.

Rate capability testing involves cell capacity measurements at different charge and discharge rates as a function of temperature. This provides data on the effect of both temperature and charge and discharge rates on cell performance. Test cells are charged and discharged at a given rate to EOCV and EODVs specified by the cell manufacturer. The effect of discharge rate and temperature on cell performance is also recorded and documented.

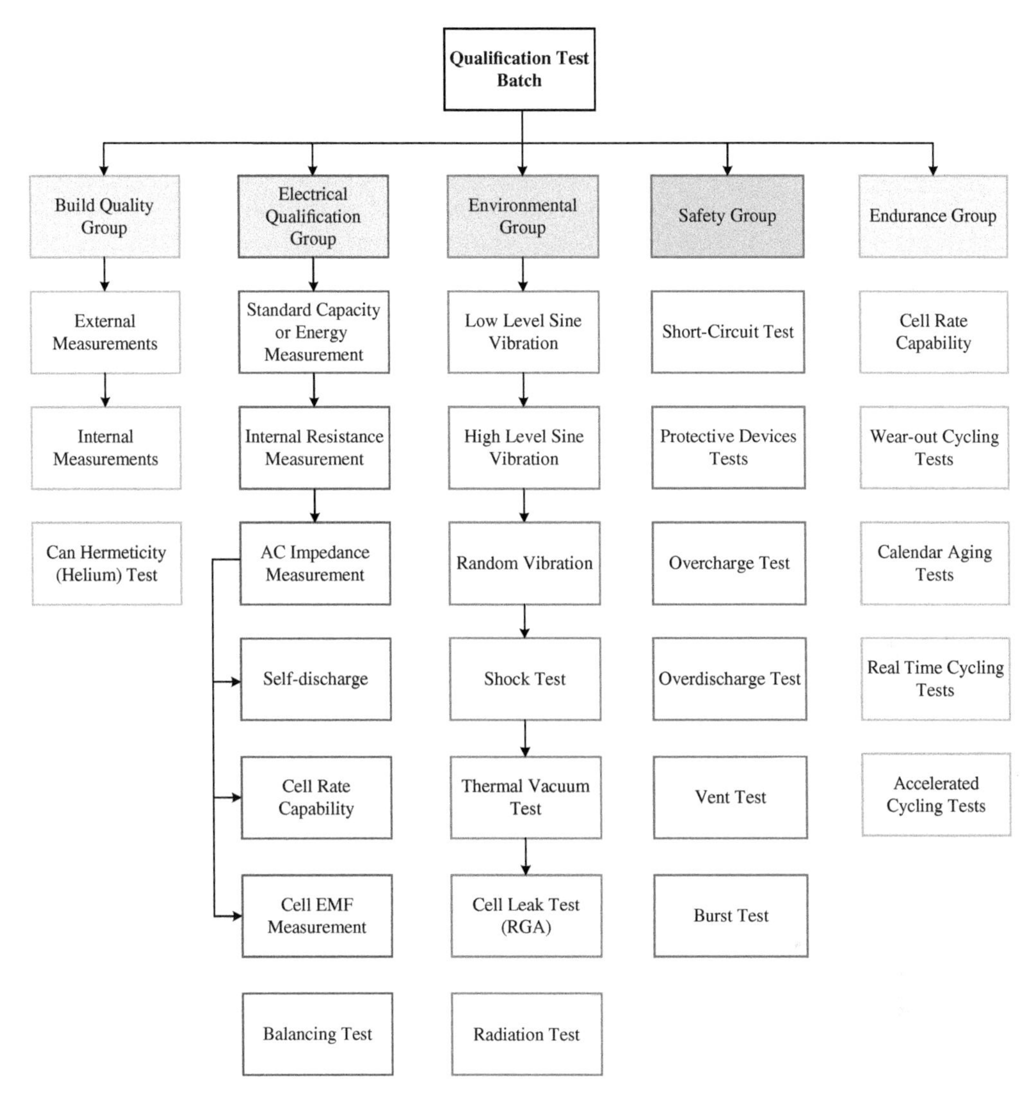

Figure 2.8 General Li-ion cell qualification test flow derived from ECSS-E-HB-20-02A.

2.7.1.2 Environmental

The environmental testing aims to subject the cells to the environmental stresses expected during the target mission and may be expanded to include other identified applications. These tests cover the mechanical, thermal, and radiation environments and are conducted to standard space qualification practices described in more detail in Chapter 3. During mechanical testing, the cell is subjected to the sine, random vibration, and shock loads in the mission-simulated environment. Prior to each load type, a low-level sine is performed to measure the Eigen frequency and identify any performance degradation occurring during the test. The cells are mounted in a mechanical jig, which is representative of the anticipated battery-mounting configuration. The test set-up and protocols are similar to those performed at the battery level discussed in Chapter 3. After the test sequence, a sample of cells may be subjected to a destructive physical analysis (DPA), which includes chemical and internal mechanical component inspection.

Cells are also subjected to TVAC testing to confirm their suitability under vacuum conditions and survival at low and high temperatures. The number of cycles (from 3 to 8 cycles) and the temperature limit range for TVAC testing can vary depending on the user. The test temperature range is defined based on the selected application. Radiation testing is also a critical test for space cell qualification. The cell is subjected to a total ionizing dose of radiation, the magnitude of which varies depending on the mission-specific orbit. Typically, the highest radiation levels are encountered for GEO applications and planetary exploration. To cover these cases, cells are qualified to withstand several MRads of radiation. During these tests, cell capacity measurements are performed at regular intervals in order to monitor the impact on the cell state of health and establish cell radiation hardness levels. The sensitive components in the cell are the cell electrolyte polymeric seal insulators, electrode bonders, and the separator [74, 75].

2.7.1.3 Safety

Cells equipped with integrated safety devices have a sequence of safety tests as part of the qualification campaign to characterize these safety devices, as well as verifying the compliance of the battery to transportation guidelines and regulations. For example, cells are tested for an external short circuit by applying a conducting path of sufficiently low resistance between the positive and the negative terminals to simulate an inadvertent short circuit (such as accidental contact made with metallic tools). The short circuit is maintained until the cell or battery can be demonstrated to be safe according to an accepted pass–fail criterion. Test measurements include cell temperature, voltage, and current.

Overcharge tests are conducted to characterize the cell behavior under abusive overcharge conditions. Typically conducted under ambient temperature conditions, test cells are charged at a specified current to an EOCV greater than that specified by the cell manufacturer. The cell is maintained at the maximum EOCV until a non-hazardous failure mode occurs (either an internal disconnection or an open-circuit failure), such that further overcharge is either impossible or will not lead to hazardous events such venting, fire, or explosion.

Cell overdischarge testing is performed by discharging at a specified current to an EODV lower than the recommended cell manufacturer lower voltage limit. As part of this test, the cell can be forced into reversal in order to assess the cell behavior under such conditions. During an overdischarge reversal, copper dendrites may cause formation of an ISC, which may be characterized as a resistance. The resistance depends on the discharge current used during the cell reversal. A number of charge and discharge cycles may be performed in order to characterize how the cell reversal stabilizes.

Cell venting and burst capabilities are characterized to establish the mechanical safety factor. A safety factor of three is required for the cell burst protection versus the worst-case operating pressure. During the venting and burst tests, the activation pressure for both features is measured. Similar to other safety abuse tests, cell burst testing is destructive and is therefore performed on dedicated non-flight cell samples. The tests are performed on an empty cell can with and without the cell vent feature installed or inhibited.

2.7.1.4 Life-Cycle Testing

The cell qualification plan includes cell or module-level life testing that can be conducted using real-time and accelerated methodologies. Life-cycle test data are used for battery-level service life estimates, cell reliability predictions, and EPS specialty analyses. Cell and battery-level life-cycle testing is further discussed in Chapter 8.

2.8 Cell Screening and Acceptance Testing

The COTS cell LAT is closely related to COTS cell qualification testing. The objective of the cell LAT is to establish homogeneity within cell production lots and ensure the qualification test results remain valid for the cells proposed for flight battery integration. Additionally, data are measured and recorded for each flight cell design that allows optimization of battery level performance and ensures that mission requirements are met. Differences in cell screening protocols depend on the program unique requirements and battery cell procurement strategy. For LIB manufacturers procuring custom cells or having an internal cell production line, cell screening is performed as part of cell LAT. Additionally, tailored lot acceptance procedures are performed to guarantee performance consistency between flight cell production lots. Battery manufacturers who procure COTS Li-ion cells for space applications typically procure large quantities (lots) of cells, which in some cases may be thousands of cell units, where a single production lot is defined as all of the cells being manufactured from a single electrode coating run and batch of materials. In the initial run of a new cell variant, a subset of the procured production lot is subjected to a comprehensive cell qualification procedure. The qualified production lot of cells would typically be used across several flight battery projects over a finite period of time. In the event of a further procurement of the same cell variant for other missions, a subsequent production LAT is performed to screen the cells and confirm consistency between different lots [83, 84].

2.8.1 Screening

The main objective of the battery cell screening process is to identify outliers (in terms of statistical distribution) and reject cells that could lead to battery-level cell-to-cell divergence, battery underperformance, anomalies, or failures during mission use. Battery failures can be induced by internal cell anomalies such as ISCs, which lead to a slow cell voltage decrease over time and create a cell imbalance within the battery. This is particularly an issue if there is no cell voltage monitoring and balancing circuitry or if the balancing capability of such circuitry is exceeded by the soft short current. This phenomenon can then lead to a decrease in the battery performance and a risk of cells being overcharged or overdischarged. Therefore, cells exhibiting evidence of internal shorting need to be identified and excluded during the cell screening process.

A cell ISC could be induced due to the presence of FOD or NOD particles introduced during the cell manufacturing process or due to any conducting species dislodged from the electrode. Manufacturing defects such as metallic burrs due to electrode slitting, non-uniform active material thickness along the electrode length, or separator damage are also possible sources [85]. To identify the presence of soft shorts, several approaches can be implemented during the cell screening process, such as OCV stand tests of different durations being the most common method. The ultimate objective is to achieve a cell screening process that can identify 100% of manufacturing workmanship mishaps or design defects. Although some of the cell failures may evolve during the course of battery usage, especially when operated outside the manufacturer's recommended usage conditions, this scenario is extremely rare when employing a rigorous and controlled cell screening process.

An important aspect of the cell screening process is defining the pass/fail criteria. Specification requirements are based upon measurable performance characteristics such as cell capacity, internal resistance, dimensionality, and mass. To account for flight cell lot-to-lot variability, statistical variations of $\pm 3\sigma$ are added after a reasonable number of production lots have been evaluated. For other key features, such as the cell self-discharge rate, seal leakage, cell internal construction, CID activation or vent pressure, the pass/fail criteria depend on the cell design. The screening criteria generally are based on the cell supplier's recommendation, especially with respect to the use of custom cell formats rather than COTS cells. A select number of the more extensive screening tests are performed on a limited set of flight lot samples. During these screening tests, statistical analysis is used to determine the allowed quantity of cell failures depending on the lot size and the required quality level. Finally, depending on the rejection rate of a dedicated parameter for a given lot, the complete lot could be identified as not suitable for flight. This would depend on the cell suppliers' rejection rates database on all manufactured lots versus the particular one being screened. Once cell screening is completed, flight cells are selected based on a cell matching criterion such that the overall battery performance is as homogeneous as possible. The cell selection is performed based on the main performance characteristics measured during the cell screening process.

2.8.2 Lot Definition

An important selection criterion is whether all the batteries required for the mission must be built from one cell manufacturing lot or if they can be built with cells manufactured from different production lots. This topic is mainly relevant for large-format cells since for COTS cells the lot size is much larger and a single production lot can be used for the flight LIB. A cell lot is defined as having a single lot of materials, process campaigns (such as slurry mixing, electrode coating run, and welding), and a common formation cycle for all cells within a lot [86]. Depending on the cell supplier's manufacturing processes, it can become challenging to fulfill the single lot requirement across all battery units required for a mission. The lot size could be limited by features of the manufacturing processes. For example, the size of the slurry mixer, length of the electrode foils, level of automation of the internal tab and electrode roll fabrication and capacity of the electrolyte filling facility are some of the features that can be limiting factors depending on the required number of cells within a battery. In addition, it is desirable to have the qualification battery, the flight battery, and the flight spare batteries all fabricated from the same cell lot. However, it is possible to relax the single lot requirement and limit it to a few key aspects, such as only requiring the positive and negative electrodes and electrolyte to all come from a single manufacturing lot. With respect to the electrode coating, it can be a matter of discussion as to whether the raw materials themselves should define a lot or whether the coating run should define the lot. Basically, to define the lot mixing strategy, a number of factors are considered, including cell heritage, the manufacturing process maturity, the mission-unique application, and the battery size requirements. It is useful to consider that qualification data for cell-life performance typically use only a limited number of lots, which are then used to justify all future missions even though the cells are not from the flight cell production lot. The cell screening process, including LAT, provides justification for the lot mixing by conducting comparator tests on every lot. Therefore, a relaxation of the single lot requirement is possible, but needs to be assessed on a case-by-case basis. It is preferable to keep a single production lot of cells within one flight battery, which is qualified to use the cells from the same production lot.

2.8.3 Acceptance Testing

The screening process separates cells that are outliers from a performance perspective, since use of outlier cells can degrade the overall battery performance or result in battery failure. This requires both an acceptance testing of the cell lot to ensure that the performance is to the required specification and a confirmation of a consistent performance of cells within the lot. Figure 2.9 shows an example of a cell LAT flow commonly used to screen COTS Li-ion cells for use in satellite applications.

The first and functionally most important selection criterion is the capacity and/or energy measurement performed on all cells from the cell lot. This guarantees that the cell performance is to

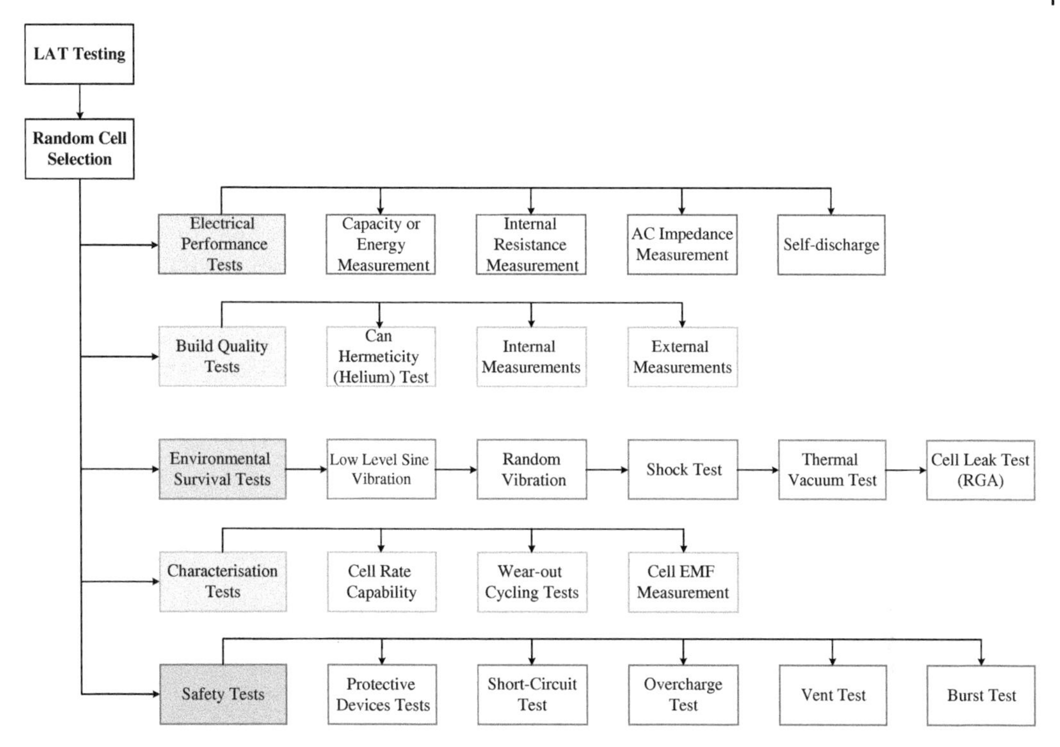

Figure 2.9 General lot acceptance test flow derived from ECSS-E-HB-20-02A.

specification and is also a key parameter for verifying the overall cell assembly process. Cell internal resistance is measured on all cells, but the methodology employed varies from supplier to supplier. Some measure at different DODs while others measure at the end of charge, 50% SOC, and the end of discharge. Additionally, some suppliers use the cell AC impedance measured at 1 kHz as a cell selection criterion.

As described previously, soft shorts internal to the cell are a particular concern for long-term battery level reliability and safety. These internal soft shorts can be induced by any FOD or NOD introduced into the cell during the manufacturing process or by any conducting species dislodged from the electrode. To identify potential soft shorts, an OCV stand test or self-discharge test is performed by monitoring OCV as a function of time in a fully or partially charged state. The OCV drop per unit time and remaining cell capacity at the end of the test are measured to screen cells for internal soft shorting.

Dimensional and mass measurements are performed on all cells from the lot, or on a smaller sample size for large cell lots with narrow manufacturing tolerances. These measurements are compared against the battery supplier specifications to verify conformity. Non-conformities could be due to active material thickness variations, missing components, an incorrect amount of electrolyte filling, or improper component selection.

Cell internal construction can be assessed for build quality using X-ray inspection and a DPA. X-ray inspection suitability needs to be assessed for each cell type as it provides more information when performed on a large-format cell than on a small-format cell. For the smaller cells, X-ray tomography can be performed to provide inspection images of cell internal components. DPAs may be conducted on a small quantity of cells taken randomly across the production lot before and after selected test events.

Cell hermeticity is another feature of the utmost importance since electrolyte leakage cannot be tolerated for safety and performance reasons. Typically, a helium leak test is performed on all cell lots with helium spread around the cell seals, with a detector measuring the level of helium on the other side. Some space LIB standards require a proof pressure test on 100% of the cells prior to the helium leak test, even for sealed cells that operate at a lower pressure, which helps to verify weld integrity [87].

All cell designs include a pressure-releasing vent safety feature whose functionality is verified for every cell lot. This verification cannot be performed on all cells as it is a destructive test and venting is irreversible. Depending on the cell design, the feature is either checked at the cell level or at the cell cover level only. Samples can be collected at the start and end of each manufacturing campaign or at random within the campaign. The definition of a manufacturing campaign depends on the implementation process of the venting and is discussed with the cell suppliers. The collected samples are then tested and the venting pressure is recorded.

Burst pressure of the cell is an important design feature from a safety perspective and therefore this parameter is also recorded to establish the burst-to-vent ratio (using data from the vent testing discussed previously). As with the vent test, the burst test is a destructive test. Samples are collected in the same way and submitted to a dedicated test set-up. In order to perform this test, samples may be collected where the cell electrode roll is not included so as to measure the burst pressure. The influence of the electrode roll during the test itself depends on the cell design and should be discussed with the cell suppliers.

2.9 Summary

Since the launch of PROBA-1, space Li-ion cell manufacturers have incrementally improved Li-ion cell performance and safety characteristics to enable advanced space LIB technologies. Space Li-ion cell technologies have successfully supported LIB-based EPS architectures aboard Earth-orbiting, cargo transport, crew transfer, LVs, and planetary mission applications in all types of orbital environments. In meeting the evolving demands of space-based LIB applications, Li-ion cell manufacturers have traditionally used a conservative approach to adopting next-generation improvements to cell-level designs intended to increase specific energy, extend cycle life, mitigate materials obsolescence risk, and reduce cost. More recently, the increase in spacecraft applications employing space-qualified COTS 18650 battery cells has caused large-format Li-ion cell manufacturers to include small-format Li-ion cell design options in their product line portfolios. In addition, future planetary exploration missions will require advancements in energy storage technologies to enable long duration deep space missions operating in extreme environments. In some cases, the ability to operate in these extreme environments with enhanced specific energy and life will enable new missions that were previously considered unattainable with conventional energy storage technologies.

Acknowledgments

Some of the work described here was carried out at the Jet Propulsion Laboratory, California Institute of Technology, under contract with the National Aeronautics and Space Administration (NASA) (80NM0018D0004).

References

1 Nagura, T. and Tozawa, K. (1991). *Progress in Batteries and Solar Cells*, vol. 10, 218. Brunswick, OH: JEC Press Inc. and IBA Inc.

2 Bragg, B.J., Casey, J.E., and Trout, B.J. (1994). *Primary Battery Design and Safety Guidelines Handbook*. NASA RP 1353. National Aeronautics and Space Administration (NASA).

3 Ratnakumar, B.V., Smart, M.C., Kindler, A. et al. (2003). Lithium batteries for aerospace applications: 2003 Mars exploration rover. *J. Power Sources* 119–121: 906–910. https://doi.org/10.1016/S0378-7753(03)00220-9.

4 Linden, D. and Reddy, T.B. (2019). Basic concepts. In: *Linden's Handbook of Batteries*, Chapter 1 (ed. K.W. Beard and T.R. Reddy), 5. New York: McGraw Hill.

5 Air Force Space Command (2008). Lithium-ion battery for launch vehicle applications. Space and Missile Systems Center Standard, SMC Standard SMC-S-018.

6 Salim, A. and Isaacson, M. (2007). Alternate methodologies for sizing lithium-ion batteries. In: *Space Power Workshop*, The Aerospace Corporation, Los Angeles, CA.

7 Gulbinska, M.K. (ed.) (2014). *Lithium-ion Battery Materials and Engineering: Current Topics and Problems from the Manufacturing Perspective*. Springer.

8 Chikkannanavar, S.B., Bernardi, D.M., and Liu, L. (2014). A review of blended cathode materials for use in Li-ion batteries. *J. Power Sources* 248: 91–100.

9 Goodenough, J.B. and Mizushima, K. (1981). Electrochemical cell with new fast ion conductors. US Patent 4,302,518.

10 Goodenough, J.B. and Mizushima, K. (1982). Fast ion conductors. US Patent 4,357,215.

11 Mizushima, K., Jones, P.C., Wiseman, P.J., and Goodenough, J.B. (1980). $LixCoO_2$ ($0 < x$ 1): A new cathode material for batteries of high energy density. *Mater. Res. Bull.* 15: 783–789. https://doi.org/10.1016/0025-5408(80)90012-4.

12 Yamahira, T., Kato, H., and Anzai, M. (1991). Nonaqueous electrolyte secondary battery. US Patent 5,053,297.

13 Scrosati, B. (2000). Recent advances in lithium-ion battery materials. *Electrochim. Acta* 45: 2461–2466. https://doi.org/10.1016/S0013-4686(00)00333-9.

14 Lu, C.H. and Yeh, P.Y. (2002). Microstructural development and electrochemical characteristics of lithium cobalt oxide powders prepared by the water-in-oil emulsion process. *J. Eur. Ceram. Soc.* 22: 673–679. https://doi.org/10.1016/S0955-2219(01)00366-1.

15 Kim, J.M. and Chung, H.T. (2004). The first cycle characteristics of $Li[Ni_{1/3}Co_{1/3}Mn_{1/3}]O_2$ charged up to 4.7 V. *Electrochim. Acta* 49: 937–944. https://doi.org/10.1016/j.electacta.2003.10.005.

16 Dahbi, M., Saadoune, I., Gustafsson, T., and Edström, K. (2011). Effect of manganese on the structural and thermal stability of $Li_{0.3}Ni_{0.7\text{-}}yCo_{0.3\text{-}y}Mn_{2y}O_2$ electrode materials (y 0 and 0.05). *Solid State Ion.* 203: 37–41. https://doi.org/10.1016/j.ssi.2011.09.022.

17 Feng, J., Song, B., Lai, M.O. et al. (2012). Electrochemical property of $LiMn_2O_4$ in overdischarged conditions. *Funct. Mater. Lett.* 5: 1250028. https://doi.org/10.1142/S1793604712500282.

18 Padhi, A.K., Nanjundaswamy, K.S., and Goodenough, J.S. (1997). Phospho-olivines as positive-electrode materials for rechargeable lithium batteries. *J. Electrochem. Soc.* 144: 1188–1194. https://doi.org/10.1149/1.1837571.

19 Dahn, J. and Ehrlich, G. (2019). Lithium ion batteries. In: *Linden's Handbook of Batteries*, Chapter 26, 5e (ed. K.W. Beard and T.R. Reddy). McGraw Hill.

20 Li, G., Yang, Z., and Yang, W. (2008). Effect of $FePO_4$ coating on electrochemical and safety performance of $LiCoO_2$ as cathode material for Li-ion batteries. *J. Power Sources* 183: 741–748. https://doi.org/10.1016/j.jpowsour.2008.05.047.

21 Xu, F., He, H., Liu, Y.D. et al. (2012). Failure investigation of $LiFePO_4$ cells under overcharge conditions. *J. Electrochem. Soc.* 159 (5): A678–A687. https://doi.org/10.1149/2.024206jes.

22 Aurbach, D., Markovsky, B., Shechter, A. et al. (1996). A comparative study of synthetic graphite and Li electrodes in electrolyte solutions based on ethylene carbonate dimethyl carbonate mixtures. *J. Electrochem. Soc.* 143: 3809–3820. https://doi.org/10.1149/1.1837300.

23 Peled, E., Golodnitski, D., Menachem, C., and Bar-Tow, D. (1998). An advanced tool for the selection of electrolyte components for rechargeable lithium batteries. *J. Electrochem. Soc.* 141: 3482–3486. https://doi.org/10.1149/1.1838831.

24 Imhof, R. and Novak, P. (1998). In situ investigation of the electrochemical reduction of carbonate electrolyte solutions at graphite electrodes. *J. Electrochem. Soc.* 145: 1081–1087. https://doi.org/10.1149/1.1838420.

25 Chagnes, A. and Swiatowska, J. (2012). Electrolyte and solid-electrolyte interphase layer in lithium-ion batteries. In: *Lithium-ion Batteries—New Developments* (ed. I. Belharouak). Rijeka, Croatia: InTech. https://doi.org/10.5772/1358.

26 Yang, H., Bang, H., Amine, K., and Prakash, J. (2005). Investigations of the exothermic reactions of natural graphite anode for Li-ion batteries during thermal runaway. *J. Electrochem. Soc.* 152: A73–A79. https://doi.org/10.1149/1.1836126.

27 Jehoulet, C., Biensan, P., Bodet, J.M. et al. (1997, 1997). *Proceedings of the Electrochemical Society 97-18 (Batteries for Portable Electric Vehicles)*, 974–985. Pennington, NJ: The Electrochemical Society Inc.

28 Dahn, J.R., Sleigh, A.K., Hang, S. et al. (1993). Carbons and graphites as substitutes for the lithium anode. In: *Lithium Batteries; New Materials, New Materials, New Perspectives* (ed. G. Pistoia), 1–47. North Holland: Elsevier.

29 Beguin, F. and Frackowiak, E. (ed.) (2009). *Carbons for Electrochemical Energy Storage and Conversion Systems*, 1e. Boca Raton: CRC Press.

30 Barsukov, I.V., Johnson, C.S., Doninger, J.E., and Barsukov, V.Z. (ed.) (2006). *New Carbon-Based Materials for Electrochemical Energy Storage Systems: Batteries, Supercapacitors and Fuel Cells, NATO Science Series II: Mathematics, Physics and Chemistry*. Dordrecht, Netherlands: Springer. https://doi.org/10.1007/1-4020-4812-2.

31 Hirota, N., Itabashi, T., and Maki, S. (2011). Composition for battery. US Patent 20110159360A1.

32 Heenan, T.M.M., Jnawali, A., Kok, M.D.R. et al. (2020). An advanced microstructural and electrochemical datasheet on 18650 Li-ion batteries with nickel-rich NMC811 cathodes and graphite-silicon anodes. *J. Electrochem. Soc.* 167: 140530.

33 Guo, R., Ouyang, M., Lu, L., and Feng, X. (2016). Mechanism of the entire over-discharge process and over-discharge-induced internal short circuit in lithium-ion batteries. *Sci. Rep.* 6 (1): 30248. https://doi.org/10.1038/srep30248.

34 Tsukamoto, H., Kishiyama, C., Nagata, M. et al. (2003). Lithium ion battery capable of being discharged to zero volts. US Patent 6,596,439B1.

35 Xu, K. (2004). Nonaqueous liquid electrolytes for lithium-based rechargeable batteries. *Chem. Rev.* 104: 4307–4417.

36 Simon, B. and Boeuve, J.P. (1997). Rechargeable Lithium Electrochemical Cell. US Patent 5,626,981.

37 Smart, M.C., Ratnakumar, B.V., and Surampudi, S. (1999). Electrolytes for low temperature lithium-ion batteries based on mixtures of aliphatic carbonates. *J. Electrochem. Soc.* 146: 86.

38 Smart, M.C., Ratnakumar, B.V., Ewell, R.C. et al. (2018). The use of lithium-ion batteries for JPL's Mars missions. *Electrochim. Acta* 268: 27–40. https://doi.org/10.1016/j.electacta.2018.02.020.

39 Smart, M.C., Ratnakumar, B.V., Whitcanack, L. et al. (2003). Improved low temperature performance of lithium-ion cells with quaternary carbonate-based electrolytes. *J. Power Sources* 119–121: 349–358.

40 Santhanagopalan, S. and Zhang, Z. (2011). Separators for lithium-ion batteries. In: *Lithium-Ion Batteries: Advanced Materials and Technologies*, 1e (ed. X. Yuan, H. Liu and J. Zhang), 197–251. CRC Press.

41 Vengopal, G., Moore, J., Howard, J., and Pendalwar, S. (1999). Characterization of microporous separators for lithium-ion batteries. *J. Power Sources* 77 (1): 34–41.

42 Arora, P. and Zhang, Z. (2004). Battery separators. *Chem. Rev.* 104: 4419–4462.

43 Zhang, S.S. (2007). A review on the separators of liquid electrolyte Li-ion batteries. *J. Power Sources* 167: 351–364.

44 Celgard. www.celgard.com (accessed 27 November 2021).

45 Ube Industries, Ltd. www.ube.com (accessed 27 November 2021).

46 Toray Battery Separator Film Korea Limited. http://www.toray-bsf.com/en (accessed 29 November 2021).

47 Asahi Kasei. http://www.asahi-kasei.com. (accessed 27 November 2021).

48 Lee, H., Yanilmaz, M., Toprakci, O. et al. (2014). A review of recent developments in membrane separators for rechargeable lithium-ion batteries. *Energ. Environ. Sci.* 7: 3857.

49 Orendorff, C.J. (2020). The role of separators in lithium-ion cell safety. *Electrochem. Soc. Interface* 21 (Summer): 61–65.

50 Luiso, S. and Fedkiw, P. (2020). Lithium-ion battery separators: recent developments and state of art. *Curr. Opin. Electrochem.* 20: 99–107. https://doi.org/10.1016/j.coelec. 2020.05.011.

51 Liu, K., Liu, Y., Lin, D. et al. (2018). Materials for lithium-ion battery safety. *Sci. Adv.* 4: 6. https://doi.org/10.1126/sciadv.aas9820.

52 Dalton, P.J., Schwanbeck, E., North, T., and Balcer, S. (2016). International space station lithium-ion battery. In: *NASA Aerospace Battery Workshop*, Huntsville, AL.

53 Doughty, D. and Roth, P.E. (2012). A general discussion of Li ion battery safety. *Electrochem. Soc. Interface* 21: 37.

54 Balakrishnan, P.G., Ramesh, R., and Kumar, T.P. (2006). Safety mechanisms in lithium-ion batteries. *J. Power Sources* 155: 2. https://doi.org/10.1016/j.jpowsour.2005.12.002.

55 United States Space Force Command Manual (2020). *Range Safety User Requirements. AFSPCMAN* 91–710.

56 Xu, B., Kong, L., Wen, G., and Pecht, M. (2021). Protection devices in commercial 18650 lithium-ion batteries. *IEEE Access* 9: 66687–66695. https://doi.org/10.1109/ACCESS.2021.3075972.

57 Mikolajczak, C., Kahn, M., White, K., and Long, R.T. (2011). *Lithium-Ion Batteries Hazard and Use Assessment.* Fire Protection and Research Foundation, Springer.

58 Darcy, E., Davies, F., Jeevarajan, J. et al. (2010). Li-ion cell PTC device withstanding thresholds. Appendix G. In: *NASA Aerospace Flight Battery Program*, NASA TM-2010-216727, vol. II (ed. M.A. Manzo, J.C. Brewer, R.V. Bugga, et al.). NESC-RP-08-75, pp. 118–148. National Aeronautics and Space Administration.

59 Roth, E.P., Doughty, D.H., and Pile, D.L. (2007). Effects of separator breakdown on abuse response of 18650 Li-ion cells. *J. Power Sources* 174: 2. https://doi.org/10.1016/j.jpowsour. 2007.06.163.

60 Maiser, E. (2014). Battery packaging – technology review. *AIP Conf. Proc.* 1597: 204. https://doi.org/10.1063/1.4878489.

61 Schröder, R., Aydemir, M., and Seliger, G. (2017). Comparatively assessing different shapes of lithium-ion battery cells. *Procedia Manuf.* 8: 104–111. https://doi.org/10.1016/j.promfg.2017.02.013.

62 International Electrotechnical Commission (2017). Secondary cells and batteries containing alkaline or other non-acid electrolytes – Secondary lithium cells and batteries for portable applications– Part 3: Prismatic and cylindrical lithium secondary cells, and batteries made from them. *International Standard*, IEC-61960-3.

63 Oh, K., Epureanu, B.I., Siegel, J.B., and Stefanopoulou, A.G. (2016). Phenomenological force and swelling models for rechargeable lithium-ion battery cells. *J. Power Sources* 310: 118–129. https://doi.org/10.1016/j.jpowsour.2016.01.103.

64 Saft. www.saftbatteries.com (accessed 26 November 2021).

65 GS Yuasa Lithium Power. http://www.gsyuasa-lp.com (accessed 26 November 2021).

66 EnerSys. www.EnerSys.com (accessed 26 November 2021).

67 EaglePicher Technologies. www.eaglepicher.com (accessed 26 November 2021).

68 Bang, H.J., Thillaiyan, R., and Quee, E. (2013). Mechanical force and pressure growth in Li-ion cells during LEO cycling. In: *NASA Aerospace Battery Workshop*, Huntsville, AL (5–7 November 2013)

69 Pusateri, T. and Aldrich, C. (2014). State of GS Yuasa lithium-ion space products and introduction to life and thermal modeling capabilities. In: *Space Power Workshop*, The Aerospace Corporation, Manhattan Beach, CA.

70 Manzo, M.A., Brewer, J.C., Bugga, R.V. et al. (2010). NASA Aerospace Flight Battery Program. *NASA/TM-2010-216727/*, Volume I, NESC-RP-08-75.

71 Mendoza, O., Sone, Y., Bolay, L.J. et al. (2018). The REIMEI Satellite Li-ion batteries after more than 13 years of operation. In: *NASA Aerospace Battery Workshop*, Hunstville, AL.

72 Ratnakumar, B.V. and Smart, M.C. (2010). Lithium plating in a Li-ion cell – A correlation with Li intercalation kinetics into graphite. *ECS Trans.* 25 (36): 241.

73 Smart, M.C. and Ratnakumar, B.V. (2011). Effect of electrolyte composition on lithium plating in lithium-ion cells. *J. Electrochem.* 158 (4): A379–A389.

74 Ratnakumar, B.V., Smart, M.C., Whitcanack, L.D. et al. (2004). Behavior of Li-ion cells in high-intensity radiation environments. *J. Electrochem.* 151 (4): A652–A659.

75 Ratnakumar, B.V., Smart, M.C., Whitcanack, L.D. et al. (2005). Behavior of Li-ion cells in high-intensity radiation environments: II. Sony/AEA/ComDEV cells. *J. Electrochem.* 152 (2): A357–A363.

76 Manual of Tests and Criteria (2019). *United Nations Recommendations on the Transport of Dangerous Goods, Model Recommendations*. Part III, Section 38.3, 7th ed.

77 UL Standard 1642 (2020). *Lithium Batteries*, 6e. Underwriters Laboratory.

78 International Electrotechnical Commission, IEC 62133-2 (2017). Secondary cells and batteries containing alkaline or other non-acid electrolytes – Safety requirements for portable sealed secondary cells, and for batteries made from them, for use in portable applications – Part 2: *Lithium Systems*.

79 Aldrich, C., Bergmark G., Pusateri, P. et al. (2019). Performance review of GS Yuasa Generation 3 space cells and introduction of Generation 4. In: *Space Power Workshop*, The Aerospace Corporation, Manhattan Beach, CA.

80 Smart, M.C., Ratnakumar, B.V., Whitcanack, L.D. et al. (2010). Life verification of large capacity Yardney Li-ion cells and batteries in support of NASA missions. *Int. J. Energy Res.* 34: 116–132.

81 Borthomieu, Y. and Reulier, D. (2012). VES16 cell and battery design. In: *NASA Battery Workshop*, Huntsville, AL.

82 European Cooperation for Space Standardization (2015). Space Engineering – Li-Ion Battery Testing Handbook. *ECSS-E-HB-20-02A*. ESA Requirements and Standards Division, ESTEC.

83 Pearson, C., Thwaite, C., and Russel, N. (2005). Small cell lithium-ion batteries: The responsive solution for space energy storage. In: *AIAA 3rd Responsive Space Conference*, 1–13.

84 Pearson, C., Thwaite, C., Curzon, D., and Rao, G. (2004). The long-term performance of small-cell batteries without cell-balancing electronics. In: *2004 NASA Aerospace Battery Workshop*, 1–23.

85 Darcy, E. and Smith, K. (2010). Advanced mitigating measures for the cell internal short risk. *Electric Aircraft Symposium,* Rohnert Park, CA.

86 McKissock, B., Loyselle, P. and E. Vogel, E. (2009). Guidelines on lithium-ion battery used in space applications. *NASA/TM-2009-215751.*

87 United States Space Force Space Systems Command Standard (2022). *Lithium-ion Battery Standard for Spacecraft Applications.* SSC Standard SSC-S-017, 20 January 2022.

3

Space Lithium-Ion Batteries

Sara Thwaite, Marshall C. Smart, Eloi Klein, Ratnakumar V. Bugga, Aakesh Datta, Yannick Borthomieu, and Thomas P. Barrera

3.1 Introduction

For over 20 years, lithium-ion battery (LIB) technology has proven to be an enabling energy storage option for the commercial electronics, electrified transportation, grid energy storage, and space markets. More specifically, the high specific energy and high energetic efficiency characteristics of LIB-based electrical power system (EPS) architectures have enabled higher power satellite payload capability, reductions in thermal control subsystem complexity, and simplified spacecraft ground and mission operation requirements. As a result, spacecraft manufacturers have deployed higher performing and more reliable cost-effective on-orbit systems. To mitigate new technology readiness risks, space LIB manufacturers commonly leverage LIB technology advances associated with commercial portable electronics and electric vehicle EPS designs. This risk reduction approach includes innovations in LIB supply chain management, manufacturing and production processes, and design-for-safety improvements.

However, Earth-orbiting satellites and planetary mission spacecraft operate in stressing space environments and operating conditions not characteristic of terrestrial portable consumer electronics or electric vehicle applications. Launch and on-orbit environmental requirements dictate the majority of mechanical, thermal, and electrical design features associated with space-qualified LIBs. In addition, the spacecraft mission-specific duty cycle and service life has a significant impact on how LIBs are designed in terms of cell selection, energy, operating temperature, and charge–discharge management. This chapter discusses the performance, design, analysis, and testing of space-qualified LIB technologies based on the current space industry state-of-practice. A systems engineering approach to requirements definition, analysis, design, and testing objectives is employed. Based on space LIB industry experience, an emphasis on requirements, guidelines, and best practices is used to provide a framework for LIB design strategies.

3.2 Requirements

As discussed in Chapters 1 and 2, LIBs are the preferred energy storage technology for meeting spacecraft EPS requirements. LIB technologies provide a mission enabling design solution to key energy storage requirements such as mass, volume, energy, and power over a wide range of

Spacecraft Lithium-Ion Battery Power Systems, First Edition. Edited by Thomas P. Barrera.

spacecraft mission lifetimes. The primary LIB operational requirement is to provide power to the satellite bus and payload subsystems during on-orbit eclipse periods. When the satellite is illuminated by the sun, the satellite solar panels provide power to the user loads and on-board EPS electronics to recharge the LIBs. In the late 1990s, widespread adoption of Li-ion cell technology for space applications became feasible due to the rapidly growing market demand for LIB energy storage solutions in the commercial portable electronics market. Compliance to stringent LIB cell requirements for space applications was demonstrated by comprehensive testing campaigns using custom and commercially available Li-ion cell designs.

In order to be viable for spacecraft applications, space-qualified LIBs must safely and reliably demonstrate performance characteristics that meet demanding electrical, mechanical, and thermal spacecraft requirements. To date, extensive ground and on-orbit experience with all types of successful spacecraft LIB-based EPS have demonstrated the following [1]:

- The Li-ion cell type selected for the spacecraft LIB must comply with the performance, design, and verification requirements of the LIB specification.
- Since spacecraft LIBs cannot be replaced or serviced after launch, LIB performance, safety, and reliability characteristics must be verified prior to EPS and spacecraft integration to meet mission and product assurance requirements.
- High-reliability spacecraft LIB-based EPSs are commonly designed to be at least single-fault tolerant to failure mode scenarios, which is achieved through the battery design or at the EPS level.
- Spacecraft LIBs and the associated battery management system (BMS) must be compliant with the relevant worst-case space environmental requirements for all phases of the intended mission application. In particular, electronic hardware and software/firmware included in the BMS must comply with requirements specific for space applications.
- Human-rated (crewed) and robotic spacecraft LIB-based EPS designs must consider the relevant safety requirements applicable to the mission specific application to mitigate safety risks to personnel, flight hardware, and facilities.

The wide variety of spacecraft mission objectives creates a diverse set of program performance, safety, test, and quality assurance requirements for satellite bus and payload subsystems. These unique satellite design requirements cause the majority of satellite EPS LIBs to be manufactured from custom-designed components. Custom satellite LIBs commonly incorporate available custom or commercial off-the-shelf (COTS) Li-ion cells into their architectures. As advanced next-generation satellite EPS architectures penetrate the growing space market, satellite manufacturers have transitioned to more highly optimized modular bus platforms. To meet customer demands for lower satellite mass and lower non-recurring cost impacts, modular and scalable LIB building block designs have become commonplace in the satellite industry. Modular LIB designs can also be qualified using proto-flight test protocols tailored to meet the unique satellite system volume, mass, energy, and power requirements. Spacecraft LIB designs vary across battery and satellite manufacturers based on a diverse set of customer-defined and manufacturer-derived EPS specification requirements. As such, a common standard of LIB form, fit, and function performance requirements has not yet been developed for the spacecraft EPS applications.

3.2.1 Battery Requirements Specification

It is an engineering best practice to define a set of battery requirements based on parent EPS and battery manufacturer-derived requirements. Along with a statement-of-work (SOW), these requirements are documented in a configuration-managed battery requirement specification. The battery specification is contractually levied by the customer procurement authority onto the battery manufacturer to ensure the battery unit provided will meet the satellite EPS specification requirements. The specification must cover those topics specific to the battery, but general design and test guidelines may be allocated from satellite-level or other sources of parent requirements. LIB specification development is a collaborative activity between the battery manufacturer, spacecraft system engineering subject matter experts, program management, and other product stakeholders.

The scope of the battery specification is to establish the performance, lifetime, design, analysis, manufacturing, testing, and verification requirements for the flight battery. A battery specification documents the battery end-item definition to include the nominal operation as well as the worst-case satellite power user demands, environments, and contingency operating conditions. This allows the battery manufacturer to estimate the battery capacity needed to comply with the predicted end-of-life (EOL) mission performance in consideration with predicted on-orbit use and associated battery performance. Battery specific performance characteristics, such as reliability, charge/discharge, impedance, and magnetic moment, critical to EPS performance are also defined in the specification.

The battery specification will also define mechanical and electrical interface requirements such as mass and volume limits, thermal environment characteristics and dissipation limits, electrical pin-out restrictions, power, and telemetry (signal) requirements. Transportation, handling, and storage requirements will also be documented. Launch-site Range Safety requirements are also included, since safety-critical operations may occur during ground processing rather than on-orbit. Battery-specific test requirements not covered by the general environmental testing specifications will be provided. This can include LIB electrical performance testing when exposed to the associated mechanical and thermal environments. Finally, quality assurance requirements and provisions for the purpose of requirements verification during acceptance, proto-qualification, and qualification testing are defined. The battery specification is configuration-managed by the procuring authority (such as the battery or satellite manufacturer).

Recurring satellite programs benefit from a released battery specification with mature, stable, and verified requirements. However, new satellite programs with custom LIB designs commonly revise the baseline battery specification as relevant hardware component testing and analysis is re-baselined. Changing parent requirements from updates to spacecraft environmental operating conditions or EPS analysis may create requirement gaps that impact battery specification requirements stability. As such, battery design, manufacturing, and test risk can be mitigated by early program retirement of missing requirements or to-be-determined specification elements.

3.2.2 Statement of Work

In addition to the battery procurement specification, the battery SOW is a necessary procurement work product that specifies the task requirements of the battery manufacturer for the contracted work effort. Since battery technical requirements are documented in the battery specification, the SOW will generally not contain battery design technical requirements. Battery SOW task requirements may include, but are not limited to:

- Battery risk reduction units, engineering models (EMs), and special test equipment that supports flight battery development but is categorized as non-flight hardware,
- Procurement quantities for flight, qualification, EM, life cycle testing, and assembly, integration, and test (AIT) ground test battery hardware deliverables,
- Battery manufacturer shipping, handling, storage specification manual, and user handbook,

- Quality assurance, technical status reporting, formal program design reviews and schedule, period of performance, requirements, applicable (government and non-government) documents, and
- Supplier (contract) data requirements list, which lists and describes the data requirements such as content, format, delivery, and approval for the battery procurement.

Since LIBs have characteristics of quality degradation or drift with age and use, they are categorized as life-limited items. A battery SOW will typically levy requirements on to the battery manufacturer to report life-limited item data, such as battery identification (serial and model number), time since Li-ion cell formation, and accumulated battery charge–discharge cycles (or total hours of operation) at the time of delivery to the battery customer. Tracking cell and battery performance changes with ground storage, age, or use enables updates to pre-launch on-orbit EPS energy balance and power budget predictions. To begin the battery procurement process, the battery SOW and specification procurement data package is formally delivered by the battery customer to the battery manufacturer for bid consideration.

3.2.3 Voltage

The LIB voltage range is defined by the spacecraft EPS requirements. All platform and payload units of the spacecraft are typically connected to a power bus and can operate properly only if the power bus voltage stays within a given range. The LIB voltage is specified that for all mission-specific conditions, which are dependent upon the spacecraft EPS bus voltage regulation architecture, the minimum and maximum voltage requirements are met. The number of cells (or cell strings) in series is selected so that the allowable battery voltage range matches as closely as possible the required EPS bus voltage range.

3.2.4 Capacity

The battery capacity is also defined from EPS requirements, which are summarized in the spacecraft power budget document. The power budget defines, for each phase of the spacecraft mission, how much power must be delivered by the battery, and for how long. This allows the maximum energy required to be delivered by the LIB to be determined. Once the energy requirements are specified, the approximate battery capacity (sizing) can be calculated using the average battery voltage required to support the EPS power bus. After the Li-ion cell model has been selected, the battery configuration and topology can be assessed in terms of quantity of cells to be connected in parallel (to fit with the desired capacity). Other considerations that can affect the battery design include peak power load demands, charge rate, and temperature. The final LIB design and analyses are normally performed by the supplier to ensure that the most efficient configuration is employed to meet the mission requirements documented in the battery requirements specification.

3.2.5 Mass and Volume

Mass properties of spacecraft LIBs are defined by the sum total of electrical, mechanical, and thermal components built into the flight battery design. LIB major components such as Li-ion cells, supporting structure, wire harness, radiation shielding, power/signal/test connectors, passive and active thermal control devices, safety devices, voltage/temperature sensors, and electrical control circuitry, when needed, account for the total LIB mass. In space-qualified LIB designs, Li-ion cells contribute to approximately 60–85% of the total battery mass. The LIB cell count and form factor also have a significant impact on the mechanical and electrical packaging approach that defines the battery dimensions. LIB design trades may also include options to integrate or distribute the BMS electronics subassembly (when required) to meet mass, volume, safety, and reliability requirements. LIBs used in crewed spacecraft applications that have safety requirements to mitigate cell-to-cell thermal propagation hazards may incorporate additional electrical, mechanical, and thermal design features, which increase battery mass and volume [2].

3.2.6 Cycle Life

LIB cycle life includes charge–discharge cycling during ground processing, pre-launch, transfer orbit (if required), and on-orbit service life requirements. Cycle life is affected by many operational and non-operational conditions such as depth-of-discharge (DOD), operating voltage range, charge–discharge rates, and operating temperature [3–5]. The effect of an accumulated calendar life during ground storage and on-orbit performance also impacts the cycle life by reducing available LIB capacity. The on-orbit required service life for telecom geosynchronous orbit (GEO) spacecraft can vary between 15 and 18 years, while low Earth orbit (LEO) spacecraft lifetime varies from months to up to 12 years. During on-orbit satellite operations, battery discharge occurs during eclipse periods or station-keeping activities when user loads exceed solar array capability. In order to maintain the EPS energy balance, battery charging immediately follows battery discharge periods. The frequency of eclipses depends on the spacecraft orbit type. GEO spacecraft batteries are cycled once a day during equinox seasons (two 45-day seasons per year) totaling 90 charge–discharge cycles per calendar year. The required DOD varies based on the day of the season and can peak as high as 80%. Due to the lower orbital altitude of LEO spacecraft, the batteries experience a much larger number of cycles per year. Depending on orbital parameters, LEO spacecraft batteries undergo up to 16 eclipse periods per day, equating usually to around 5500 cycles per year at DODs ranging between 10 and 40% with these LIBs. For planetary exploration missions, including landers and rovers, the cycle life requirements are much less uniform and vary depending upon the spacecraft design, the power generation source, and the science operation objectives.

3.2.7 Environments

Space-qualified LIB designs must comply with stringent naturally occurring space environments to meet spacecraft mission performance and life requirements. As such, space battery designs commonly incorporate design solutions that meet unique space environmental conditions such as vacuum, extreme temperature, radiation, and microgravity. In addition, mission-specific conditions, such as ground transportation, storage, vibration, acoustic, and shock environments induced by launch vehicle (LV) and spacecraft operation, are commonly incorporated into space battery product specifications. For planetary mission spacecraft, LIBs must comply with extreme space environments, such as wide temperature ranges (−120 to +465 °C) and high radiation dosage. Compliance is often achieved by implementing active thermal management to maintain a much narrower temperature range and the use of shielding to reduce the radiation exposure of the battery.

3.3 Cell Selection and Matching

Space LIB manufacturers utilize various methodologies when selecting flight cells for flight LIB manufacturing. As discussed in Chapter 2, flight cells are acceptance tested and screened before becoming candidates for flight battery integration.

3.3.1 Selection Methodologies

Cell selection is a critical process step toward identifying cells as candidates for flight battery integration. The methodology relies on data collected during the cell screening process, which may be adjusted based on the LIB design characteristics and spacecraft mission-specific requirements. To ensure consistent performance, all the cells in flight LIBs are typically required to be from one production lot. For configuration-controlled cell designs, a single-cell production lot is defined as all cells manufactured in a single or multiple production run manufactured from the same anode, cathode, electrolyte, and separator material sub-lots with no change in processes, drawings, or tooling, and within a defined production run or time period [6]. The activation of the cells, or the "filling date," should also occur within a specified time period.

Cell screening data are also collected and analyzed for the cell selection process. For each of the performance parameters, the criteria for matching and tolerance band for each parameter within the cell strings and within the overall battery depend on the battery supplier heritage, battery design, and the end application. The typical methodology employed is for the measured performance metrics generated on each of the cells to be organized into bins of common characteristics. Ultimately, the objective is to have all cells within a string to come from a single bin. The performance difference within a single bin is defined by the battery supplier and is based on its institutional experience and heritage. These guidelines can sometimes be relaxed based on the mission-specific application, or in cases in which the requirements are not stringent. For LIBs without cell balancing electronics, bin requirements are more stringent, since a more homogeneous performance is required from each battery cell.

3.3.2 Matching Process

The primary performance cell characteristic used for the flight cell matching process is cell capacity measured during the flight cell lot acceptance test (LAT) process. For flight cells to be arranged in a cell string, a small divergence in cell capacity ensures that the total battery capacity meets the spacecraft mission requirements as defined by the battery sizing analysis. The cell self-discharge rate is also critical in the cell selection process, especially for LIBs without cell balancing electronics. If no cell voltage equalization capability is implemented, a single cell with a higher self-discharge rate could negatively impact the performance of the entire cell string for the ground and on-orbit life of the LIB.

Battery cell internal resistance may also be considered in the cell matching process. These data can be used to sequence the cells within a string and/or the strings within a battery to compensate for the internal terminal connection and harness resistances. This cell string arrangement then aids in equalizing the string voltage and capacity performance. Another use of the internal resistance data can be to reduce the thermal gradient within the battery. In order to achieve this, thermal analysis is required to identify the hot and cold locations within the battery. The arrangement then reduces the thermal gradient within the battery as the thermal analysis considers an identical string performance in the computation. The cell matching process is commonly based on a combination of LIB manufacturer experience and customer unique requirements.

3.4 Mission-Specific Characteristics

Space industry experience has demonstrated that a compliant spacecraft EPS design is based on mission-relevant operational environments and usage modes. A significant step in defining the EPS design architecture is sizing the solar array, battery, and power electronics. Solar array and battery sizing are based on an EPS power (energy) budget analysis of worst-case user load requirements, orbital operating characteristics, and other mission-specific constraints. A proper EOL battery sizing analysis is characterized by including sufficient battery energy to support the longest orbital eclipse times, transient or electric propulsion peak loads in solstice, and reserve energy to support worst-case fault recovery scenarios. To illustrate how mission-specific parameters can influence the battery design and sizing, including aspects of the power subsystem architecture, several different spacecraft applications are described and discussed.

3.4.1 LIB Sizing

Sizing LIBs for a satellite or planetary mission application requires a detailed understanding of the mission and how the LIB will be utilized. There is a direct relationship between the satellite application, the orbit where it is located, and the battery utilization for a number of cycles, lifetime, cycle duration (orbit), and DOD. Typically, Li-ion cell design selection and battery sizing are conducted concurrently to optimize key requirements such as performance, mass, volume, and cost. Traditional battery sizing guidelines can be used when sizing an LIB for any space-rated application. LIB sizing design input parameters are quantified based on a worst-case analysis of various mission critical requirements. Key design input parameters include, but are not limited to:

- Total mission service lifetime
- Worst-case payload power demand (eclipse and non-eclipse)
- Worst-case eclipse duration
- Maximum allowable DOD during the mission eclipse period
- Capacity degradation and cell resistance growth due to mission operational life (number of charge and discharge cycles)
- Number of failed cells or cell strings per battery
- Charge and discharge DC–DC converter efficiency
- Harness voltage drop between the battery interface and the power regulation control unit
- Reserve energy required for spacecraft fault recovery
- Operating and non-operating voltage and temperature

In general, however, battery sizing is largely dependent on mission requirements for EOL performance characteristics such as spacecraft operational life, minimum and maximum EPS bus voltage, allowed charge/discharge currents, and usable capacity/energy. Redundant cells, or cell strings, may be added to the battery topology to meet EOL performance and reliability requirements in the event of a cell or cell string failure. EPS-level characteristics such as power distribution and control electronic efficiencies and battery harness voltage drop are also critical inputs to the LIB sizing analysis. The orbital characteristics of various Earth-orbiting and Earth–Sun (Lagrange) satellites and their generalized battery requirements are listed in Table 3.1.

3.4.2 GEO Missions

GEO satellites are positioned in a circular orbit in the Earth's equatorial plane at an altitude of 35 786 km and an inclination of 0°. Such an orbit is geosynchronous, meaning that its orbital period matches the Earth's rotational period. A GEO satellite moving West to East in such an orbit remains stationary with respect to one Earth location, which permits point-to-point and point-to-multi-point communications with ground station operators. GEO satellites are primarily used for telecommunications by operators that have assembled fleets covering the Earth, including

Table 3.1 Earth-orbiting mission characteristics and satellite battery requirements.

Characteristics	Low earth orbit (LEO)	Medium earth orbit (MEO)	Geosynchronous orbit (GEO)	Highly elliptical orbit (HEO)	Lagrange points (L2)
Representative satellites	ISS; Globalstar, Iridium, Cubesats, OneWeb, Starlink	GPS, Galileo, GLONASS, O3b	GPM, TDRS, GOES, COMSATS, METEOSAT	Van Allen Probes, Tacsat 4, Molniya	JWST, Herschel, Planck, GAIA, Euclid, PLATO
Applications	Scientific, Earth Observation, Communication	Navigation	Communications, Weather	Scientific, Communications	Scientific
Altitude	< 2000 km	2000 km to GEO	35 856 km	Typical Perigee in LEO and Apogee Near GEO	1 500 000 km
Orbit Type	Circular	Circular	Circular	Elliptical	Lissajous Orbit around L2
Orbit Duration	Up to 127 min	127 min to 24 h.	24 h	From 36 to 100 h	Synchronous with Earth-Sun System
Mission Duration	1–12 yr	Up to 14 yr	15–18 yr	From 5 to 20 yr.	>10 yr
Eclipse period	12–16 per day	Up to 7 wk per year	2 Periods of 45 d per year	3–4 wk per year	None
Eclipse Duration (per cycle)	Up to 35 min	Up to 67 min	Up to 72 min	Up to 67 min	None
Sunlight Duration (per cycle)	Up to 92 min	Up to 200 d per year	2 Periods of 137 d per year	Up to 300 d per year	Continuous
Charge-Discharge Cycles	Up to 5840 per year	Up to 2500	1350 to 1620	<500	<200 (station keeping)
Depth of Discharge	10–40%	60–80%	60–80%	Up to 60%	25% during launch and ascent
Operating Temperature	0–40 °C	0–40 °C	0–40 °C	0–40 °C	0–40 °C

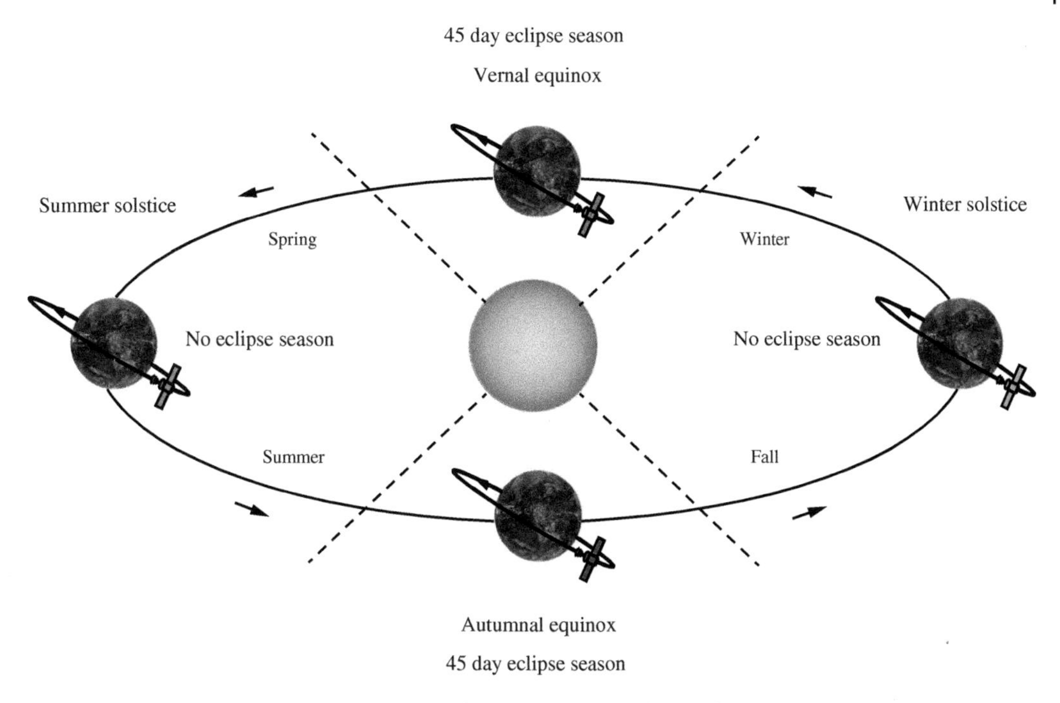

Figure 3.1 Representation of a satellite in a GEO orbit in a Sun-Centric Referential. Credit: Julianna Calin.

direct-home TV broadcasting, voice, and Internet. GEO satellites are also used for Earth operational meteorological observation and augmentation of global navigation satellite systems [7].

Due to the obliquity of the ecliptic plane and its position far from the Earth, a GEO satellite passes through the Earth's shadow cone once a day only during the equinox (seasonal) periods on each side of March 21st and September 23rd (two times, 45 days per year) each calendar year (Figure 3.1). GEO satellites experience 90 eclipses per year with an eclipse duration increasing during the first 22 days, up to a maximum duration of 72 minutes and decreasing for the last 22 days of the eclipse period (Figure 3.2).

Commercial GEO satellite LIBs may be cycled up to 1620 times during a representative 18-year mission. This relatively low number of charge–discharge cycles allows the battery to be used with a relatively high DOD, typically between 60 and 80%. GEO satellites utilizing electric ion-thrust propulsion subsystems to perform maneuvers during station-keeping operations may require high-power for several hours. As such, GEO satellites utilizing electric propulsion subsystems may impart up to 2 additional charge–discharge cycles per day on the EPS LIB while in solstice. In order to reduce the solar array and cost, many satellite manufacturers have chosen to support these high-power events using the EPS battery. The effect of electric ion-thrust propulsion engine power demands on the LIB design is commonly included in the EPS battery sizing, EPS power budget analysis, and qualification life cycle test plans.

3.4.3 LEO Missions

LEO satellite missions have a circular orbit at altitudes ranging from 160 to 2000 km above the Earth's surface. The LEO orbital period can thus vary between 90 and 120 minutes, which corresponds to approximately 12–16 orbital periods during a 24-hour period. Depending on the beta angle, which is the angle between the orbital plane of the satellite and the direction from which the sunlight is coming, each orbit can include an eclipse, whose duration may reach up to 35 minutes.

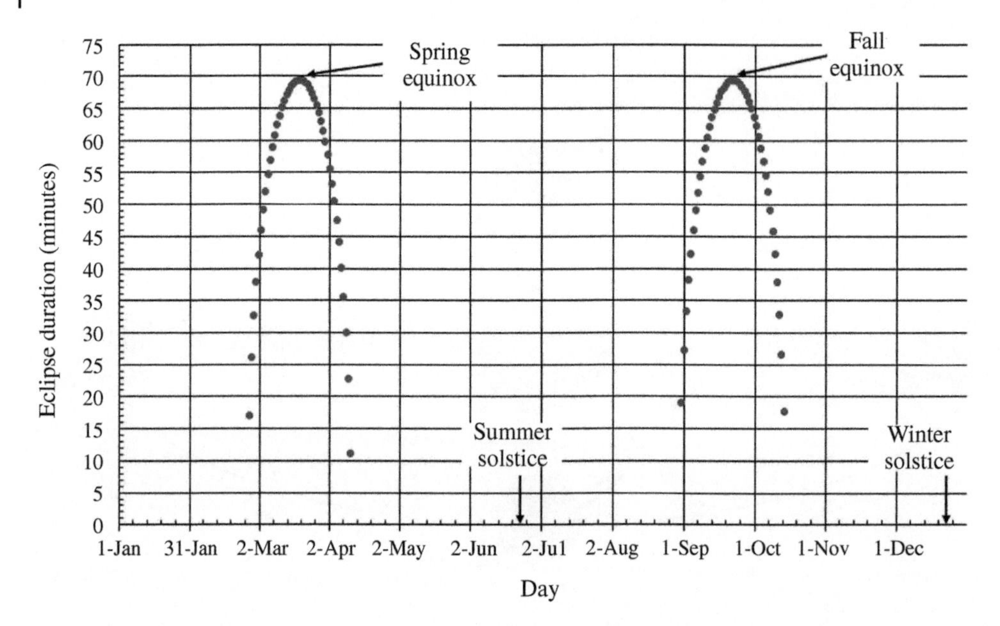

Figure 3.2 GEO eclipse season for a representative commercial communications satellite, located 98° E, in a slightly inclined orbit.

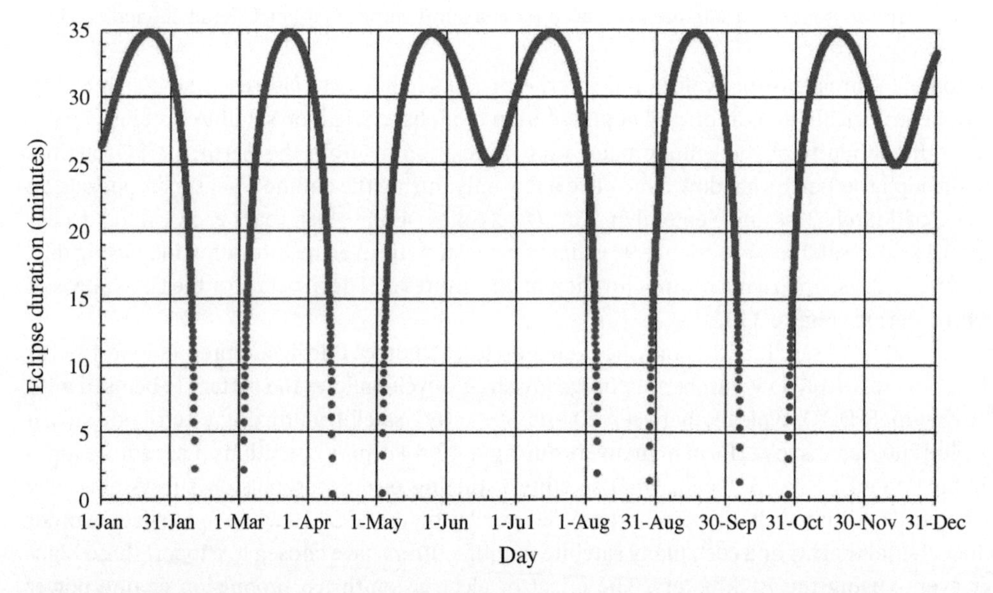

Figure 3.3 Variation of eclipse duration for a representative LEO satellite (altitude 1336 km, inclination = 66.038°).

LEO satellites can also have orbits that are inclined from 0 to 90° from the equatorial plane. There is an influence of the beta and inclination angle on the eclipse occurrence and duration during the year. The eclipse duration corresponds to the time when the satellite is running in the Earth's shadow at the opposite side of the Sun's position. Considering a 95-minute average LEO spacecraft orbit period, up to 15 battery charge–discharge cycles will be completed per day (Figure 3.3). For example, the NASA International Space Station (ISS) LIBs, when operated at specified power levels and temperatures, are qualified to a LEO design life of 10 years (approximately 60 000 charge–discharge cycles). To meet LEO mission performance and lifetime requirements, most LEO satellite battery DODs vary between 10 and 40%.

Most LEO satellites are used for Earth or space observation, science, and telecommunication missions. Some examples of such satellites are the NASA-ISS [8], Globalstar™ [9], Iridium™ [10], Starlink [11], and the Copernicus family of satellites [12].

3.4.4 MEO and HEO Missions

The medium Earth orbit (MEO) is generally circular, with altitudes extending from the upper altitudes of the LEO regime to altitudes lower than the GEO orbital region. As a consequence, spacecraft operating in MEO have orbital periods ranging from 2 to 24 hours. A common MEO application is the family of satellite applications providing global navigation system services. Examples of MEO-based global navigation satellite systems include the US Global Positioning System (GPS), the European Galileo, and the Russian Global Navigation Satellite System (GLONASS) applications.

Earth-orbiting satellites operating in a highly elliptical orbit (HEO) have locations of perigee in LEO and apogee near GEO. The retired Russian Molniya communications satellite system is an example of an HEO application that provided communications services to high northern latitudes. The Molniya series of satellites employed three satellites in three 12-hour orbits with an Earth apogee distance of 39 400 km and perigee at 1000 km. Other representative HEO missions include the NASA Van Allen Probes space weather laboratory [13], the Sirius Satellite Radio (Radiosat) digital radio communications system [14], and the Naval Research Laboratory TacSat 4 technology demonstrator [15].

The MEO and HEO satellite eclipse duration depends on the selected orbit and varies from 35 to 85 minutes. Moreover, due to their high apogee, the MEO and HEO satellites have full solstice periods and eclipse seasons. On a typical MEO orbit of 20 000 km, the eclipse seasons are longer than for GEO (75 days instead of 45) and include two eclipses per day instead of one. For battery sizing, MEO requirements are similar to GEO satellite EPS requirements including cycle life and eclipse time. However, the high radiation environment of the Van Allen belts presents a hazard to Earth-orbiting HEO satellites. As such, HEO satellite EPS components, including the battery, must be shielded to protect components from the effects of radiation on components degradation.

3.4.5 Lagrange Orbit Missions

Unlike typical Earth orbiting LEO or GEO spacecraft, there are five points in space where the gravitational forces of the Sun and the Earth are balanced with the centripetal forces of an orbiting spacecraft [16]. These five points were identified by the eighteenth century Italian–French mathematician Joseph-Louis Lagrange, who solved the General Three Body mathematical problem and are named in his honor. The L1, L2, and L3 points are all in line with each other, while L4 and L5 are at the points of equilateral triangles.

L4 and L5 are inherently stable, but as a result tend to contain natural occurring objects such as interplanetary dust and asteroids, making them unsuitable for most missions. L1–L3 are in "unstable equilibrium," meaning objects will slowly drift away into their own sun-centric orbit unless station-keeping thrusters are periodically utilized. Most Lagrangian missions will employ quasi-stable periodic orbits called "halo orbits" about L1, L2, or L3 points following a Lissajous curve trajectory. Although these halo orbits are not perfectly stable, a relatively small amount of station keeping can keep a spacecraft in its desired Lissajous orbit for an extended period of time. Satellites in L1–L3 orbits are in continuous sunlight, so the battery discharge occurs only during launch and an early orbit phase, or in a case of contingency station-keeping maneuvers or fault scenarios. As such, satellites in Lagrange point orbits are commonly used for Sun observations and deep space telescopes.

3.5 Interfaces

As described in Section 3.2.1, the LIB product specification establishes the performance, design, analysis, manufacturing, testing, and verification requirements for the end-item flight battery. A major element of the LIB specification is the electrical and mechanical interface control document (ICD) requirements. Sometimes referred to as the external interface requirements, the ICDs define the structural, power, thermal, and venting interface requirements between the battery and the spacecraft. LIB thermal control requirements are defined by the thermal interface between the battery and the spacecraft structure. In addition to the top-level parent EPS specification battery requirements, spacecraft bus subsystem ICD requirements are linked to the battery specification. This includes the harness, telemetry and control, propulsion, thermal control, communication, attitude determination and control, and structure bus subsystem specification ICDs.

3.5.1 Electrical

Electrical ICD requirements are typically documented within the LIB specification or as a separate engineering drawing document. Power and signal connector type and pin assignment (with wire gauge) requirements are specified in the battery electrical ICD. These requirements include but are not limited to:

- Power, signal, and test connector interface requirements for battery charge–discharge control, telemetry, and grounding.
- Signal connector interface requirements for battery, module, cell-level voltage, cell or module temperature sensor, heater control, BMS signal, telecommand, and telemetry.
- Test connector interface requirements for charge–discharge control and voltage monitoring.

3.5.2 Mechanical

Similarly, the mechanical ICD requirements are typically documented within the LIB specification or as an engineering drawing. Electrical connector locations, attachments for mechanical ground support equipment, battery baseplate mounting hole and attachment layout, mass limits, center-of-gravity and moment of inertia, external finish, stay-out zones, and dimensional requirements are commonly specified in the battery mechanical ICD.

3.5.3 Thermal

Thermal interface requirements ensure that the LIB design will have provisions for transferring heat loads from the battery to the spacecraft mechanical interfaces and then to space via a radiator element. The mechanical ICD may contain engineering data that specifies thermal interface requirements. On-orbit vacuum space environments limit transfer and dissipation of battery heat loads to radiative and conductive heat transfer modes. As such, various active and passive means to provide battery thermal control have been successfully used in various spacecraft applications. LIB thermal control design features may include integrated radiator panels, temperature sensors, heat pipes, cold plates, louvers, or heaters. To meet pre-launch LIB thermal management requirements, ground cooling supplied to the encapsulated spacecraft in the LV fairing provides a means to thermally manage battery temperature.

3.6 Battery Design

The LIB or pack design solution is governed by the selected Li-ion cell design. As described in Chapter 2, the cell design requirements are documented in the cell specification. Cell requirements are verified by test, analysis, or other program-approved compliance methods. The cell form, fit, and function has the greatest impact on the LIBs electrical, mechanical, thermal, and safety design features. As such, spacecraft LIBs with a modular design based on a common building block approach provides a variety of packaging solutions to the satellite customer.

LIB designs may be viewed as consisting of Li-ion cells electrically connected in series and/or parallel to meet voltage and capacity requirements. Custom and COTS space LIB components may vary widely depending on the battery supplier design solution to meet the program battery specification requirements. However, components such as Li-ion cells, harness and bus bar connectors, voltage and temperature sensing, switches and relays, heaters, BMS, electronic control boards, firmware, and safety devices are common to many designs. All space-qualified LIBs will also contain a battery chassis, baseplate, or other structural components necessary to meet battery requirements. Approved materials of construction such as adhesives, conformal coatings, and electrical and thermal conducting or insulating materials are documented in the LIB bill-of-materials (BOMs). Spacecraft customers may also require custom design features, such as implementation of current sense shunts, bypass switches to increase battery reliability, cell-to-cell thermal propagation-resistant design features, or added battery cell voltage or temperature telemetry.

A main characteristic that LIB suppliers use as a figure-of-merit in battery design is the ratio of cell-to-battery mass. This figure-of-merit quantifies the additional battery mass due to non-cell subassembly battery components. Therefore, throughout the design phase, the mass properties of the battery are continually assessed to optimize the cell-to-battery mass ratio. Since meeting all battery design environmental and performance requirements is extremely challenging, each battery manufacturer achieves battery design modularity based on their internal heritage and qualification testing. For specific needs, all space battery manufacturers are able to develop customized battery designs, although this may also lead to a higher development effort and risk.

3.6.1 Electrical

The spacecraft EPS requirements drive the key battery design features based on the operational bus voltage range and spacecraft power requirements. Other key electrical requirements are derived from the mission profile, which provides the nominal/peak battery charge/discharge power and/or current needs as well as the required number of cycles over the mission lifetime. This required battery capacity/energy during the mission lifetime is used to establish the required electrical configuration in order to have a battery DOD consistent with the available heritage data, based on the cell supplier life cycle tests. The life cycle tests usually include various DOD, end-of-charge voltage (EOCV), currents, and temperature conditions. This allows identification of the most appropriate operating conditions for the battery to meet mission-specific service lifetime requirements. The overall EPS architecture also influences the battery electrical design. Some functions required for the BMS (such as relays that allow disconnection of the battery from the spacecraft bus or the cell voltage balancing system) may either be implemented in the battery or in another component connected to the battery.

3.6.1.1 S-P and P-S Design

LIB cell packaging architecture, or topology, is typically defined in terms of the chosen cell electrical packaging configuration. As many space-qualified Li-ion cell designs are available with capacity ratings ranging from 1.5 to 190 Ah, space battery manufacturers have adopted two different cell electrical configurations to meet high-power, high-capacity spacecraft EPS requirements. These battery cell architectures are commonly known as series-parallel (s-p) and parallel-series (p-s) topologies [17].

The s-p topology is an electrical configuration in which individual cells are connected in series to produce strings, which in turn are connected electrically in parallel (Figure 3.4). The number of cells in series within each string determines the battery output voltage. The total capacity of the battery is then sized appropriately for the energy and power needs of the application by adjusting the number of parallel strings. This topology is typically used for batteries when the number of parallel strings (p) is larger than the number of cells connected in series (s). By using multiple strings, the battery can survive cell failures without any loss in voltage. As a consequence, the s-p battery design configuration is mainly used in conjunction with unregulated satellite buses, where the EPS bus voltage is determined by the battery voltage. In order to meet safety and reliability requirements of most applications, a protection device that mitigates against individual cell overcharge and short-circuit hazard events, such as a cell internal current interrupt device (CID) or an active electronic BMS, is mandatory for these applications. In the event of a cell failure, safety hazard controls electrically opens the string, thus reducing the overall battery capacity but not the battery voltage. S-p battery architectures also have an advantage in simplifying battery telemetry processing, since individual cell monitoring may be autonomous or not required, and only battery voltage needs to be processed. In s-p electrical configurations, harness resistance between each battery module and the spacecraft interface is minimized and cells are matched within cell strings so that the overall string internal resistances are matched. If there is any imbalance in the string resistances, the string currents will also be imbalanced. At the beginning of a discharge, the strings with lower internal resistances will initially be discharged at a higher current, resulting in

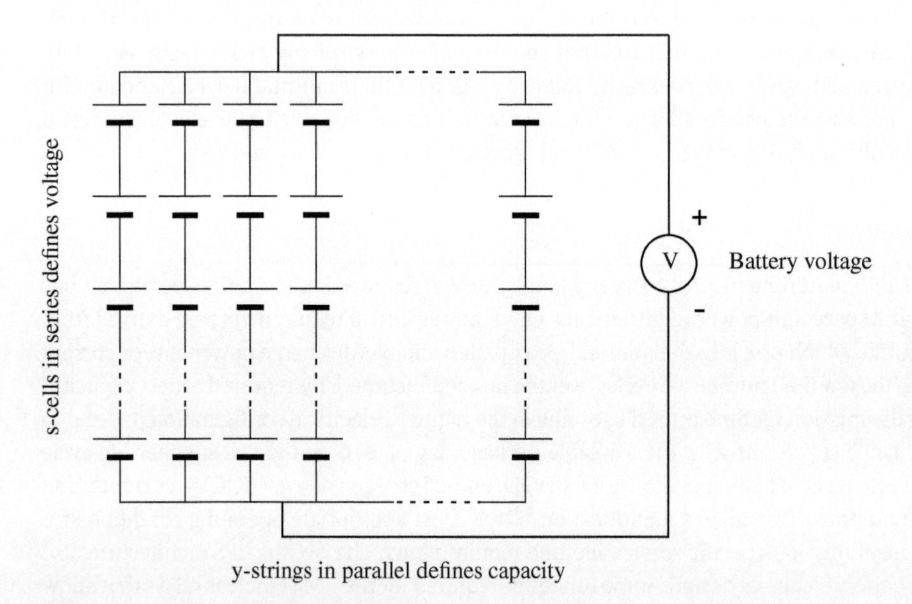

Figure 3.4 LIB electrical s-p topology.

a faster decrease of their state-of-charge (SOC). As a result, an open circuit voltage (OCV) difference between strings will be created, which will result in a decrease of the current in the low-resistance strings. This tends to cancel the current imbalance. A mirror effect will occur in the charge, and when the battery is at rest, the most charged strings will discharge into the least charged strings. In summary, all strings eventually remain balanced, but the inhomogeneous currents can result in unequal degradation between cells (due in particular to localized heating) that could affect the lifetime of the battery. To mitigate these effects, individual cell balancing systems can be used in s-p configurations to guarantee cell-to-cell voltage balancing during the entire lifetime of the mission. Conversely, for an s-p configuration with widely distributed battery modules in the physical bus architecture, there is a possibility to adjust the harness resistance for each module, thus equalizing the overall resistance for each battery module.

An LIB p-s topology is typically used when the number of parallel cells (p) is smaller than the number of cells connected in series (s) (Figure 3.5). The LIB p-s topology may require individual cell voltage monitoring and cell-to-cell voltage balancing. The cells that are connected in parallel constitute one parallel cell bank that defines battery capacity. The cell banks are electrically connected in series to meet battery voltage requirements. The number of cell banks must be compatible with the designed satellite voltage range. Moreover, in the event of the failure of one cell, which leads to the failure of an entire bank of cells (depending upon the design), a voltage decrease will occur at the battery level. This topology is well-suited for satellites employing fully regulated EPS power busses, such as those commonly used on GEO communication satellites.

Using this approach, the extent of the cell balancing circuitry required is reduced, since only s-number of units are required to perform the parallel cell bank balancing. p-s battery designs may not require cell balancing electronics which results in a reduction in battery complexity when compared to a comparable s-p configuration requiring cell balancing. As all the cells within the same cell

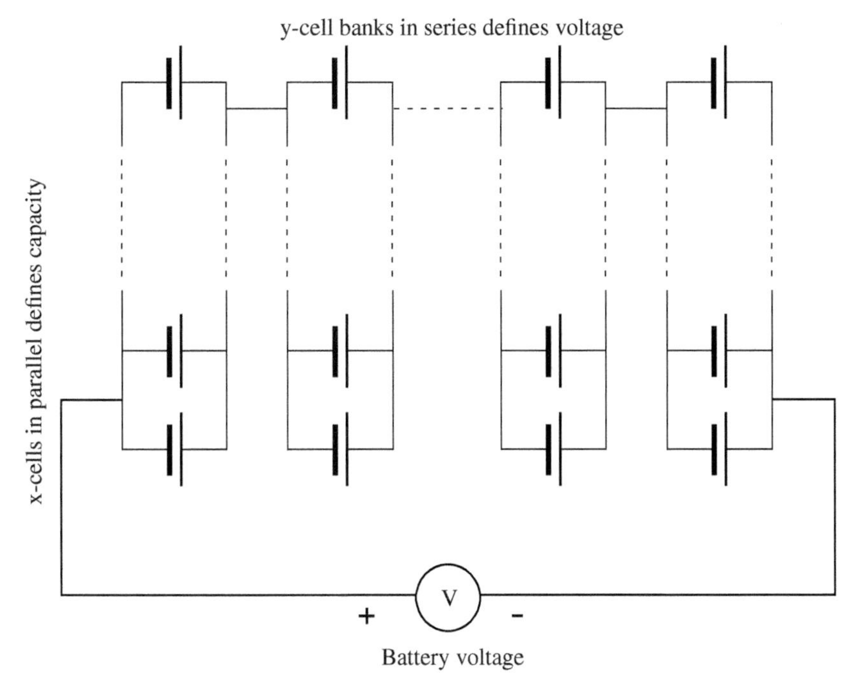

Figure 3.5 LIB electrical p-s topology

Table 3.2 LIB s-p and p-s topology comparison.

Characteristic	s-p topology	p-s topology
EPS bus compatibility	Optimum for low DOD; unregulated bus	May require partially or fully regulated bus
Typical satellite power	Low and medium power	Medium and high power
Cell capacity	<15 Ah	>15 Ah
Individual cell telemetry	String-level	Cell bank-level
End of charge management	Battery voltage control	Cell bank voltage control
Cell failure case consequence:	Potential loss of one string	Potential loss of one cell bank
• On battery voltage	Limited impact when p is high	–3.6V if the cell failure leads to bank failure
• On battery capacity	Battery capacity reduced by 1/p	Battery capacity unchanged

bank have the same voltage, the number of telemetry lines is also reduced. By using a telemetry system configured for the p-s topology, a detailed health assessment of the battery can be obtained, owing to all of the individual cell bank voltage measurements that are performed during the course of the entire mission. Each parallel cell bank can also be equipped with a cell bypass system in order to protect the battery from failure. In the event of a cell failure, the cell bank can be bypassed, resulting in the reduction of the battery voltage, while still maintaining the same capacity. In the event that a cell experiences an internal short circuit (ISC), it will subsequently become an external short to the adjacent parallel cells. This can worsen the outcome of such a fault and increase the probability of thermal runaway propagation. As a consequence, for safety-critical applications (such as human-rated missions), the interconnecting tabs are sized to fuse at a relatively low current. Table 3.2 lists design characteristics of s-p and p-s LIB configurations commonly used in space applications.

The battery electrical architecture is achieved with the establishment of the power circuitry made of the cells, busbars, and wiring for the cell's serialization and/or parallelization. The power interface to the spacecraft is customer dependent and can be achieved either by using connectors or busbars. Connector interface requirements define the connector type (rectangular, circular, push–pull, screw), number of pins, size of pins, and whether they are provided on a connector holder or as flying leads. Similarly, power busbar designs and interface connections are customer specific requirements.

Additionally, telemetry circuitry can be required to monitor battery voltage, battery current, cell voltages, and temperature. Battery telecommand lines may also be needed for heater, bypass, and cell balancing activation lines. The telemetry/telecommand circuitry varies for each LIB design as it is directly driven by the spacecraft electrical architecture and also depends on the design needs for bypass and balancing.

Cell-to-cell voltage balancing is another key aspect to be considered, as it aids battery performance over the service lifetime and serves an important role in terms of safety. Over time, if the battery cell voltages diverge from one another, this can lead to cell overcharge and/or overdischarge, which in turn may lead to inhomogeneous degradation and safety issues associated with overcharge of cells. This risk must be addressed, either by matching cell characteristics (with a focus on self-discharge) during the cell selection process or by implementing a battery cell balancing circuit either as an integrated design solution within the battery or as a separate EPS component balancing system. In practice, for a given electrical configuration, an s-p topology requires more channels for cell voltage monitoring and cell balancing than a p-s topology. On the other hand, in a p-s topology, each channel may need a higher balancing current capability.

A battery cell bypassing solution is an important aspect to be considered in an LIB design [18]. This is typically required in a battery with p-s configurations as a cell failure can negatively impact battery performance. In case of a cell open-circuit failure, the cell will not provide any power, leading the other cells in the parallel cell bank to provide more load and hence degrade faster. For a battery with a 1p configuration, a cell open-circuit failure will result in a battery open circuit. In this case a bypass switch is mandatory. A cell can also fail by having a leakage current higher than the balancing current capability of the BMS. In this case, it is also required to remove the failed cell bank by bypassing it, in order to preserve battery-level functionality. Relevant test and analysis methods have been used by space LIB manufacturers to demonstrate that, in certain p-s topologies, bypass switches can be removed [19–21].

Double electrical isolation is also a common requirement for space-qualified LIB designs. It is required as a mitigation measure against internal battery shorting hazards. The main design features are the battery cells, with the battery potential being isolated from the battery structure connected to the spacecraft ground. This can be achieved through several means such as insulating materials (Kapton tape, potting, heat-shrink sleeves) or gaps that are controlled by mechanical elements to ensure a constant physical gap. Compliance to double electrical isolation requirements is specific to each battery manufacturer design.

3.6.1.2 Analysis

The majority of space LIB electrical performance specification requirements are verified by test and analysis compliance methods. However, some LIB electrical performance and environmental requirements are verified by analysis-only due to flight hardware test limitations. Electrical engineering analysis methods, including worst-case circuit, survivability, voltage-drop, power interface, and battery cell life, are commonly used to demonstrate compliance with the LIB requirements. Survivability in space environments includes a worst-case analysis for total-dose radiation based on the satellite mission-specific orbit, internal electrostatic discharge (IESD) effects, single-event upsets on active circuitry and monitoring devices, dielectric materials and no-floating metal (or grounding scheme) verification, and electromagnetic interference (EMI) and electromagnetic compatibility (EMC) analyses to analyze the self-compatibility of the battery and its compatibility with other EPS components and subsystems. Other battery specialty analysis types include magnetic dipole analysis at worst-case battery discharge currents, micrometeoroid and orbital debris (MMOD) protection assessments [22–24], and materials contamination (outgassing) analysis for total mass loss and volatile condensing materials.

Radiation environments are an important aspect to be considered when establishing battery design requirements [25–27]. All elements within the battery have a given radiation hardness capability. The battery design needs to take into account the radiation environment during the mission and provide the required protection so that the most sensitive elements do not exceed their hardness, including project specific margins. For battery designs including electronics, radiation aspects also need to be taken into account such as the total non-ionizing dose, single-event effects, and single-event transients. Radiation environments can lead to the need for localized radiation shielding in the form of a reinforced aluminum plate or spot potting [28, 29].

3.6.2 Mechanical

LIB mechanical requirements are established to ensure that the battery design's physical characteristics meet mass, volume, and structural-mechanical environments throughout the ground, launch, and on-orbit lifetime of the spacecraft [30]. In addition to mechanical interface requirements, the main requirements are the battery natural frequency and mechanical environmental loads. During its lifetime, a battery is subjected to sine/random vibrations and shock loads. Sine vibrations are typically generated on-ground during battery transportation and during maneuvers of the LV. Random vibration is generated during the launch ascent phase and shock loads occur during launch stage separations and satellite payload, radiator, and solar array deployments. Besides the battery design itself being required to comply with these mechanical requirements, the internal components of the cells themselves need to inherently display the requisite robustness [31].

Acceleration and maneuvers of the LV result in quasi-static and sine accelerations that can typically reach 20 g, at frequencies lower than 100 Hz. The interaction of the LV fairing with the atmosphere results in acoustic noise that can be translated into a random vibration spectrum. Mechanical shock is generated once the satellite is separated from the LV and when deployment mechanisms are activated. It can also occur during pyro events during cruise and entry descent and landing (EDL) in a planetary environment.

A final environmental aspect is the risk of an MMOD strike, which can for some Earth-orbiting satellites be a critical requirement for battery mechanical design. Typically, batteries are mounted inside the spacecraft structure such that impact from an MMOD event is not a risk or battery design requirement. However, in some cases batteries are externally mounted, thus exposed directly to the space environment. MMOD protection can either be implemented in the battery design or at the satellite level, which carries a significant mass penalty to the entire spacecraft system.

3.6.2.1 Packaging

As described in Chapter 2, most Li-ion cells used in space applications are cylindrical, prismatic, or elliptical-cylindrical in geometry. The geometric shape of the cell has a direct impact on the requirements driving the LIB mechanical packaging design. The electrode assembly of a cylindrical cell is made by spirally winding electrodes and separator. Such an assembly, called a jelly roll, is then inserted in a cylindrical tube. This tube will tend to keep its shape uniform when the electrode volume changes. When designing the fixation of the cell to a structural part, care is taken not to exert inhomogeneous local pressure on the cylindrical tube, since this could lead to significant degradation of the cell performance [32]. Prismatic cells possess flat areas that can bulge due to electrode swelling in designs where the cells are unrestrained. To secure the cells, it may be necessary to apply a mechanical compression force on cell flat areas.

3.6.2.2 Structural Mechanical Analysis

Mechanical analysis needs to be performed to confirm that the battery design can meet the mechanical requirements. Tools such as finite element analysis are commonly used by battery manufacturers. Modal analysis allows the extraction of battery natural frequencies that can be correlated to battery vibration test results. Static, shock, random, and sine vibration analysis methods are also performed to confirm that the stress levels experienced by every element of the battery structure are well within the yield and ultimate limits of the structural material.

As discussed in Chapter 2, a Li-ion cell is an assembly of many components (electrodes, electrode substrates, electrolyte, separator, cell can, terminals, and safety devices) with homogeneous and non-homogeneous mechanical properties. Furthermore, these properties can interact during battery operation. For example, the volume of the electrodes depends on the cell SOC, which can therefore modify the stiffness and damping properties of the cell [33]. Also, mechanical stress can affect the electrical performance of the cell (such as internal resistance or self-discharge) without any visible mechanical damage. Hence, elementary tests are usually performed at the cell level and used to model the cell mechanical properties at a macroscopic scale [34]. Furthermore, life cycle tests normally incorporate a flight-like mechanical interface to verify the long-term performance of the cell design.

3.6.3 Thermal

Spacecraft thermal control subsystems are designed to the expected mission-specific on-orbit environment [35]. However, spacecraft environmental temperatures experienced during transportation, pre-launch, launch, and ascent must also be predicted to ensure battery temperature limits will not be exceeded during these ground operations and launch-site mission phases. The applicable satellite thermal control subsystem and environmental specification requirements are allocated to the LIB specification to derive the LIB thermal design solution. LIB thermal requirements are not derived in the same manner as other satellite bus equipment as space-qualified batteries typically operate with narrow operating temperature ranges to best meet service and performance life requirements. Additional LIB thermal requirements are derived from:

- Battery cycling during worst-case (hot and cold) satellite orbital beginning of life (BOL) and EOL eclipse and solstice duration characteristics,
- Thermal environments during interplanetary cruise trajectories,
- Battery pack exposure to the Earth, deep-space environment, and spacecraft structure,
- Satellite contamination environment on exposed battery radiator surfaces,
- Battery pack thermal radiator properties,
- Infrared backloading from the solar arrays and other appendages (such as antennas), and
- Thermal interface definition to ensure battery cells remain within their temperature limits during all mission phases.

In general, on-orbit LIBs are cycled within the unit acceptance operating temperature range that is recommended by the cell manufacturer. Most space cell manufacturers recommend cell operating temperatures between 0 and +40 °C. However, optimal operating temperature ranges may vary depending on cell chemistry and relevant life test performance results. For spacecraft contingency operations (such as safe mode or dead bus events), the temperature range can be extended to survival temperature limits based on worst-case battery thermal analysis. LIB survival temperatures may range from −10 to +60 °C depending on allowances for performance degradation, which may result in reduced on-orbit service life. These thermal environments are characterized in terms of their possible frequency of occurrence and cumulative duration to enable completion of the battery thermal design details.

In addition to operational temperature ranges, battery pack temperature gradient requirements need to be defined as they may have a significant impact on meeting battery electrical and mechanical performance requirements. Battery heat dissipation is another driving requirement from a system point of view in terms of how much heat is dissipated as well as how it is dissipated. The heat dissipation also increases with battery life, due to the increase of battery resistance with time. Typically, battery heat transfer is defined by conduction only with radiative exchanges occurring between the cells.

In order to keep the battery within the specified operating temperature range, thermal hardware requirements can also be implemented specifying the need for internal heaters, thermostats, thermal blankets, number of thermistors, and the location of the temperature reference point (TRP). This is mainly driven by the spacecraft thermal control subsystem strategy.

3.6.3.1 Design

Satellites are generally operated in a narrow temperature range and therefore thermal control systems for the LIB can be less complex. However, planetary landers and rovers are designed to operate over a wider range of temperatures and thus require more sophisticated thermal management of the battery. For example, the 2003 Mars Rovers batteries were designed to operate over a temperature range of −20 to +30 °C and the thermal management was achieved by utilizing a combination of resistive heaters and a thermal mechanical switch activating a heat pipe rejection system [36, 37]. Even though battery temperature range requirements are defined based on the cell manufacturer heritage, at a system level the battery is maintained in a narrower range where heater set-points are turned ON/OFF (cycled) to maintain the battery at a target temperature. For example, for a battery operational temperature range of 0 to +40 °C, the nominal heater cycles on and off and can be set between 12 °C (heater ON) and 15 °C (heater OFF). This provides a smaller temperature amplitude during spacecraft operations and simplifies battery in-flight performance trend analysis. Moreover, this can allow reducing battery calendar fading by reducing the temperature diurnal during mission phases when the LIB is not cycled during solstices phases for GEO missions. This thermal control strategy helps in reducing the heater power consumption required by the battery thermal control subsystem.

Minimizing thermal gradients within a battery is critical to ensuring that cells age at the same rate and thus perform uniformly for the duration of the battery service life. If large temperature gradients exist, the cells showing relatively high temperatures will degrade faster even though the overall battery temperature is maintained within the normal operating temperature range. Moreover, the battery cell temperature directly impacts the cell internal resistance and the colder cells (and cell strings) will have a lower discharge current compared to the hotter strings. Therefore, it is critical to minimize thermal gradients between cells in a single module and between battery modules contained in a multi-module battery architecture.

Battery heat dissipation characteristics depend on the on-orbit satellite operational mode, battery energy demand, and satellite environments. Consideration to optimizing heat transfer paths between the battery and space environment includes mounting the battery baseplate to a radiator panel, use of a sufficient number of mounting points, and minimizing thermal loads from satellite bus heat leak sources. Mounting the battery to minimize incident sunlight on battery heat transfer surfaces during on-orbit operations also aids battery heat dissipation characteristics. The heat dissipation is one of the major challenges at spacecraft level and drives the selection of the battery panel material (carbon-fiber-reinforced polymer honeycomb or aluminum honeycomb). For highly dissipative LIBs, heat pipes are also used in order to spread the battery heat dissipation over a larger area of the battery pack radiator.

A significant source of battery-level heat dissipation also includes ohmic heating due to electrical contact resistance at cell and battery pack terminal connections. Other heat and power dissipation sources include battery-level internal electronic interfaces, bypass switches, fuses, and internal battery harnesses. Heat is generated by the battery during discharge and can be dissipated to the spacecraft structure within the limits provided by the LIB specification heat flux requirements. During charge, the cell chemistry can exhibit endothermic behavior, seen as a temperature decrease at the battery level. However, depending on the battery charge rate and SOC, this behavior may be dominated by the ohmic heat generation, resulting in exothermic behavior leading to heat waste and thermal dissipation from the battery during the charging.

In order to remove cell-generated heat, battery designs incorporate thermally conductive materials placed in contact with the cell cases. This can be achieved in the form of a sleeve around the cell up to a certain length, a thermal control subsystem cold plate mated to the battery baseplate, or the cells being directly potted on an aluminum baseplate. This hardware is then connected through other thermally conductive mechanical parts to the spacecraft panel in order to create a heat dissipation path. Materials selection for these mechanical parts is critical in order to guarantee the thermal design of the battery. For spacecraft that allow radiative exchanges for battery heat dissipation, the battery design would then require its mechanical structure to be coated with space-qualified black paint in order to increase its emissivity and allow an efficient thermal exchange.

In terms of a battery-heater thermal control design, various design options are available that will keep the battery cells at or above their cold operating temperature limit. One option involves implementing the battery thermal hardware internally with the heaters and temperature sensors placed within the structure. The advantage of having internal heaters is optimization of the installed heater power to maintain battery thermal control within the operating temperature range. Another option does not include internal battery heaters and instead heaters are mounted external to the battery and directly on to the satellite-side of the battery mounting interface. This leads to a higher heater power and duty cycle, which can impact performance at spacecraft level. Additionally, thermostatically controlled survival heaters can be implemented. The goal is to have these survival heaters available as a back-up solution in case of an on-orbit spacecraft safe mode contingency. The survival heaters function to maintain the battery within its qualified temperature range.

Finally, battery thermal control is managed based on the location of the TRP, where temperature sensors are located to control battery heaters. This is defined by the spacecraft thermal control design. These sensors are located at the battery mounting interface or internally (as close as possible to or on the cells). The battery mounting interface solution allows the customer to design its spacecraft thermal control and fix its panel temperature range. The battery manufacturer accounts for the temperature offset between the battery mounting interface and the battery cells as the qualification data are usually based on the cell temperature. In addition to the TRP temperature sensors, additional sensors can be implemented in order to monitor the cold and/or hot cells depending on the TRP location. The number of temperature sensors for the TRP depends on the spacecraft thermal control system requirements.

3.6.3.2 Analysis

Thermal analysis is performed to verify that the temperature of the battery critical components, such as the cells, remain within the specified operating range over mission life. Thermal analysis also verifies that the temperature gradient between the hottest and coldest cells is compatible with the mission requirements and does not induce premature aging or significant imbalance of cell-to-cell electrical performance within the battery. This analysis may also account for the possibility of failed cell(s). As an input to this analysis, the battery manufacturer is provided with the spacecraft panel temperature profile as well as the thermal interface conductance value that provides analysis boundary conditions. The battery manufacturer and satellite manufacturer collaborate to conduct a battery thermal analysis using the nominal and worst-case battery charge–discharge duty cycles as well as worst-case environmental heat loading.

A battery thermal analysis is performed based on a detailed thermal mathematical model, which is built with a defined number of nodes depending on the battery design. The node locations are selected so that the coldest and hottest battery regions are analyzed, thus allowing assessment of the temperature gradient across the battery. Additionally, the temperature offset between the cell and the spacecraft panel as well as the rejected heat are assessed to verify compliance to the battery dissipation and components temperature requirements. Indeed, for each battery component, a rated temperature value is defined with a given derating rule to achieve a derated temperature. The output of the analysis is then compared to the derated temperature to assess the component compliance.

3.6.4 Materials, Parts, and Processes

All LIB materials of construction must be capable of meeting mission-specific requirements when exposed to the severe environment of space. As such, only space-qualified Li-ion cells, parts, materials, and processes may be used to design and manufacture spacecraft LIBs. The approved BOMs list the Li-ion cell part number and all remaining components, parts, and materials used to manufacture the LIB. Key characteristics of materials suitable for the space environment are maintained by various space agencies in the form of databases. The NASA Materials and Processes Technology Information System (MAPTIS) database or other approved LIB manufacturer parts and materials databases are common space industry accepted resources [38]. These design databases provide various material properties for use by LIB suppliers for developing compliant BOMs. In addition, material properties are used by LIB design analysts when considering the environments in which the materials will function. Candidate LIB materials of construction are commonly space-qualified based on satellite program-specific requirements including, but not limited to:

- Capability to survive space vacuum environments without degradation of material properties, especially any thermo-optical properties required for thermal control. This includes requirements for low materials outgassing of volatile components. LIB materials outgassing products may adversely affect satellite solar panels, radiators, or payload equipment sensitive to volatile condensable materials.
- Ability to survive both ionizing and non-ionizing radiation environments. LIB total dose-level requirements vary widely depending on orbit type and LIB location within the satellite bus structure. Extensive engineering analysis and testing has demonstrated that space-qualified Li-ion cells are tolerant to high levels of non-ionizing radiation. Associated LIB electronics are also analyzed, tested, and selected to ensure proper functionality after exposure to the required total dose levels, including the effect of highly charged particles on semi-conductor component lifetime. Significant radiation shielding may be necessary for LIB designs operating in the extreme radiation environments of HEO and deep space.
- Satellites operating in LEO will be exposed to atomic oxygen environments which are known to aggressively erode and degrade materials. LIB units mounted to the spacecraft exterior will be exposed to atomic oxygen fluences, which may adversely impact battery materials reliability and lifetime. This is generally limited to exposed LIB chassis materials, such as multilayer insulation (MLI)-based thermal blankets or Kapton tape used for battery thermal and electrical insulation. Space-qualified Li-ion cell performance, safety, and reliability are not impacted by atomic oxygen exposure.
- Extreme hot and cold stressing temperature environments may be experienced during satellite missions. As such, LIB construction materials should be selected so as to survive satellite non-operational and contingency-mode temperature extremes. In this case, the temperature constraint is likely to be governed by LIB cell operating survival temperature limits rather than battery chassis materials.

- All satellite systems are characterized by low mass designs that utilize lightweight materials of construction chosen to meet stringent electrical, mechanical, and thermal requirements. LIB mass savings approaches have traditionally included use of space-grade aluminum alloys and polymers and other mass-saving features as webbing and pocketing of housing elements.
- Identified prohibited and restricted materials are explicitly not permitted for space flight due to adverse performance in vacuum or zero gravity. The use of pure unalloyed tin is strictly controlled by spacecraft manufacturers due to the development of tin whiskers, which may result in electrical failure of satellite components and systems [39, 40]. This prohibition has a direct effect on the use of commercial lead-free solder, since pure tin solder must be avoided in preference for tin-lead or silver solder alloys. Materials that sublimate under vacuum (such as zinc and cadmium), are toxic (such as beryllium), or unstable in vacuum conditions (PVC) are also strictly controlled. Certain coatings, such as chromates, may also be restricted for use on exterior battery surfaces. LIB designs may also have corrosion protection and control requirements to protect against galvanic corrosion when dissimilar metals are used in intimate contact.

3.6.4.1 Parts

Space-qualified LIB designs commonly follow satellite program approved electrical, electronic, and electromechanical (EEE) parts plans. EEE parts plans include controls for selecting approved parts lists, screening, qualification, derating, and radiation analysis. Although LIB cells are not classified as EEE parts, all other electronic parts used in LIB designs are commonly included in the LIB EEE parts plan and inventory. Representative LIB EEE parts include switches, relays, connectors, resistors, thermistors, heaters, and any active or passive electronics piece-parts used on electronics boards.

3.6.4.2 Cleanliness

During manufacture and test of the battery hardware, it is important to maintain exacting standards of cleanliness. This ensures that dust, manufacturing foreign objects debris (FOD), native objects debris (NOD), and other loose particles and undesired chemical deposits are reduced to a minimum. Elimination of any conductive FOD or NOD particles is essential to avoid the risk of shorting within the battery unit and is rigorously monitored by test and inspection. Although non-conductive particulate or molecular (chemical) contamination may not represent a direct threat to the battery, it may readily spread to spacecraft locations that are adversely affected, for example cryogenic optics in the instrument payload [41, 42]. To reduce the risk of contamination, it is a general practice to build and test space hardware in controlled certified clean room environments.

Certain mission types have specific cleanliness requirements more restrictive than traditional space industry standards. For example, planetary lander missions must consider the possible contamination of the planetary environment with traces of living organisms from Earth [43]. Planetary protection plans are developed for such missions to include the most rigorous standards of cleanliness for all lander hardware, including battery units. Similarly, scientific missions may have strict requirements for low out-gassing materials and molecular cleanliness. Many industry techniques for management of this level of biological and chemical cleanliness involve the application of heat (dry heat sterilization and high temperature bake-out of materials). Finally, LIB exposure to out-of-specification temperature limits may permanently degrade cell performance characteristics.

3.6.5 Safety and Reliability

Industry experience has demonstrated that the high specific energy and volumetric energy density of LIBs create a safety risk for the hazardous release of energy under abusive (user-induced) or non-abusive (non-user-induced) conditions in the intended spacecraft application. The rapid release of LIB electrical and chemical energy may lead to potentially catastrophic outcomes for the impacted systems and surroundings. As such, battery manufacturers, users, and other stakeholders are required to ensure the safety of personnel, hardware, and facilities throughout the entire life cycle of a spacecraft battery. This includes all phases of battery design, development, manufacturing, assembly, testing, handling, storage, and transportation. Design requirements for safety features at the cell, module, and battery level are commonly used to mitigate the risk of safety incidents. At the EPS level, BMSs are employed to monitor the LIB state of health (SOH) and other operational conditions that may pose a safety risk to personnel and flight hardware. Understanding battery safety characteristics also includes identifying battery failure modes that may impact spacecraft mission assurance objectives.

In general, the reliability of the power management and distribution (PMAD) electronics, solar array, and battery energy storage component hardware dictates the overall EPS reliability. More specifically, predicting and measuring Li-ion battery reliability is critical to assessing the probability that the EPS will meet spacecraft performance requirements during the intended mission lifetime. On-orbit operating conditions such as orbit type, environments, and power duty cycle also have a significant effect on EPS component reliability throughout the design life of the spacecraft.

3.6.5.1 Human-Rated and Unmanned Missions

Spacecraft EPS operating requirements such as bus voltage, charge management, and operating temperature have a significant impact on battery safety and reliability. To mitigate the severity and consequences of a battery safety incident, applicable industry battery safety standards and guidelines are commonly part of a space battery specification. Compliance with space battery safety requirements ensures that the spacecraft battery power system will perform its intended operational functions safely under relevant configurations and environments. Human-rated (crewed) and unmanned spacecraft LIB specifications commonly include requirements to protect against hazardous abuse conditions such as overcharge, overdischarge, ISCs, external short circuits, and high temperatures.

Safety requirements for space LIB systems vary greatly between crewed and unmanned spacecraft applications. Battery safety requirements for crewed space applications generally differ from unmanned battery applications due to the presence of crew members, enclosed crewed environments, and mounting (or use) in unpressurized spaces adjacent to habitable volumes. Crewed applications utilizing LIB and other battery technologies typically utilize safety and hazard control requirements derived from the NASA-JSC 20793 standard or other sources [44]. In these applications, failure tolerance levels are defined by the number of mutually independent catastrophic hazard control measures (or features) capable of either maintaining safe operation, or enabling a safe shutdown, after the imposition of the battery failure. Fault tolerance for batteries also includes consideration to the toxicity as well as the energy content of the failure. Crewed space vehicles with habitable areas, such as the NASA ISS and crew transport vehicles, commonly utilize a combination of dual-fault tolerance requirements with a design for minimum risk (DFMR) approach to comply with battery fault tolerance requirements [45, 46].

Unmanned Earth-orbiting and planetary mission spacecraft battery safety requirements are not required to comply with JSC-20793. However, unmanned spacecraft may be required to comply with applicable NASA, US-Air Force, or US-Navy government safety certification standards. Industry regulatory organizations such as the Underwriters Laboratory (UL), International Electrotechnical Commission (IEC), and Radio Technical Commission for Aeronautics (RTCA) also publish consensus battery safety standards applicable to unmanned spacecraft batteries. Crewed and unmanned spacecraft Range users operating on the US Air Force East and West coast Ranges are subject to the Range Safety requirements found in the Air Force Space Command Manual (AFSPCMAN) 91-710 [47]. AFSPCMAN 91-710 provides battery safety requirements for ground and launch support personnel and equipment, systems, and material operations. These requirements include procedures for safe receiving, transportation, checkout, handling, installation, packing, storage, and disposal for all battery operations. Special requirements for LIB handling, transportation, storage, and disposal are included as part of certifying LIBs for Range usage.

Safety requirements also include the need to provide safety-related analysis, such as Failure Mode, Effects, and Criticality Analysis (FMECA) and Hazard Analysis Report (HAR). The FMECA consists in reviewing every component, part, and sub-assembly of a LIB in order to identify potential failure modes, their causes, effects, and criticality. The FMECA can be used to establish design recommendations in order to reduce the consequences of critical failures. The purpose of the HAR is to identify hazards associated with the design and operation of the LIB that are capable of causing injury to personnel, damage to the environment, or damage to critical hardware. Risk is then assessed and hazard controls needed to eliminate or reduce the associated risk to acceptable levels are established. Additional details on spacecraft LIB safety requirements and safe by design approaches are described in Chapter 7.

3.6.5.2 Safety Features and Devices

As described in Chapter 2, Li-ion cells should be designed so that they can be safely used under operating conditions consistent with the relevant application environment and intended battery configuration. Similarly, LIBs should have design features that allow safe use under the expected ground processing and on-orbit spacecraft EPS operating conditions. Selection of LIB safety features are commonly based on design compliance to the LIB specification performance, test, and quality control requirements. LIB safety features, including fault protection electronics, should also protect the battery cell and EPS from failure modes identified in the battery FMECA. Consideration of failure modes caused by credible or foreseeable battery misuse are typically included in the battery FMECA.

Table 3.3 lists various safety design features and devices commonly used in space-qualified LIBs for manned and unmanned spacecraft. Safety design features and devices protect the electrical, mechanical, or thermal function and operation of the battery and EPS. Electrical insulation, thermal insulation, and the use of non-flammable materials of construction in LIB design have been demonstrated to mitigate the risk of a catastrophic safety incident. Many space-qualified LIBs use BMS to monitor and report battery SOH parameters. Battery SOH parameters include voltage, load current, temperature, and relay (if any) status. BMS systems are commonly used to achieve cell balancing as required to maintain battery life and safety margins. Chapter 4 further discusses the role of LIB BMS in various spacecraft applications.

Table 3.3 Li-ion cell, module, battery, and EPS safety devices.

Safety feature or device type	Cell	Module	Battery	EPS
Software				
BMS	✓	✓	✓	✓
Electrical				
Fuse	✓	✓	✓	✓
Fusible link	✓	–	–	–
Blocking diode	✓	✓	–	–
Contactor	–	–	✓	–
Insulation	✓	✓	✓	–
Grounding	–	✓	✓	✓
Mechanical				
CID	✓	–	–	–
Pressure vents	✓	–	✓	–
Bypass switch	–	✓	–	–
Cell spacing	✓	✓	✓	–
MMOD shielding	–	–	✓	–
Thermal				
PTC	✓	✓	✓	–
Temperature cut-off switch	–	–	✓	–
Thermal insulation	✓	✓	✓	–
Flame trap	–	✓	–	–

3.7 Battery Testing

Battery testing verifies that the battery hardware end item meets specification requirements by measuring performance during and/or after exposure to laboratory-controlled environments. Battery test events are commonly categorized as development, acceptance, proto-qualification, or qualification testing for the purposes of demonstrating that the flight battery design meets specification requirements with positive margin. The battery test article design may vary from a prototype representative of a new custom design to a recurring follow-on unit manufactured in accordance with a design that has been previously qualified. Testing and analysis are the most common verification compliance methods used to demonstrate that battery requirement compliance meets relevant spacecraft mission-specific requirements. In addition to requirements verification, battery test data are used as inputs to the spacecraft EPS power budget, energy balance, voltage drop, mission life, mass properties, and reliability analyses. Test objectives commonly include defining mission-relevant environments and configurations representative of the intended on-orbit satellite application.

3.7.1 Test Requirements and Planning

Battery qualification, proto-qualification, and acceptance test event requirements are documented in the customer-approved battery specification. Battery specification test requirements are allocated to the responsible test organization to support development of test plans and procedures. The battery SOW specifies the battery type (such as development, proto-qualification, or qualification) and quantity of test articles required to support the battery test program. Battery specification test requirements include, but are not limited to:

- A description and purpose of all electrical, mechanical, and thermal tests
- Test sequence requirements consistent with expected launch and on-orbit flight operational events
- Test levels, durations, tolerances, and pass/fail requirements for all electrical, mechanical, and thermal tests used to accept and qualify the battery test article
- Safety test requirements to verify safe ground and flight operations

Other requirements for facilities, equipment, and personnel are also included in the test plan and procedure documentation. Finally, the battery SOW defines the entrance and exit criteria used to define the conditions that must be satisfied for the Test Readiness Review (TRR). Typical TRR requirements for flight LIB test programs include checks to ensure that test procedures have been assessed for compliance with test plans and descriptions; verification that new or modified test support equipment and facilities are available and satisfy requirements; test equipment calibration status is updated; and test emergency procedures are baselined.

3.7.2 Test Articles and Events

Battery flight hardware is intended to be used operationally in space. Subsets of battery flight hardware include development, prototype, recurring, and spare hardware. During the development phase, engineering model (EM) batteries are commonly used for early mitigation of technical risk associated with new or custom LIB designs. LIB EM designs are commonly manufactured with non-flight parts and are thus not typically required to comply with environmental test requirements. Program risk associated with a new custom battery design or changes to satellite bus platform layout interfaces can be mitigated by utilizing battery EMs to establish form, fit, and function characteristics. In some cases, battery EMs are also used as test batteries during satellite AIT activities. Lessons learned from analysis, manufacturing, and test of battery EMs may also be incorporated into the preliminary stages of a new battery design.

Prototype battery hardware is hardware that can serve as a pathfinder of a new or custom design. New or modified battery designs may be required to support new satellite bus platform interfaces or increases in satellite payload power requirements. Recent space industry experience with LIB raw cell material obsolescence, cell manufacturer product line improvements, and other supply chain constraints have caused changes to recurring space LIB product line designs. New battery designs are subject to a design qualification test program that demonstrates a design margin to account for expected variations in parts, material properties, dimensions, manufacturing processes, test cycles, and operating conditions. Due to the effects of wear-out caused by stressful qualification test conditions on battery design life, qualification battery hardware is not intended for flight. Typically, only one qualification battery is allocated per program. However, in some cases multiple batteries may be manufactured and qualification tested to mitigate program-unique technical, cost, or schedule risk.

Proto-qualification testing of flight LIBs is typically conducted when no test-dedicated program qualification LIB assets are available. Prior to program implementation, the technical risk associated with proto-qualification of first-flight LIBs is assessed to ensure that changes to mission performance requirement margins are quantified. Proto-qualification is a modified qualification strategy where design verification testing is performed at levels and/or durations reduced from qualification, with an analysis performed to the qualification requirements. Qualification test temperatures, environmental test levels and durations, and pass/fail requirements are tailored for proto-qualification testing. Proto-qualification battery hardware is commonly considered to be flight-worthy.

Recurring flight LIB hardware manufactured in accordance with a battery design previously qualified either as a new design or as proto-qualified hardware is subject to a flight acceptance test program. Acceptance testing is intended to screen flight LIBs for latent defects in parts, materials, and manufacturing workmanship. Spare LIB flight battery hardware is recurring hardware held in reserve in case of ground failure of battery flight hardware intended for flight operations.

3.7.3 Qualification Test Descriptions

Figure 3.6 lists the standard space industry qualification and proto-qualification LIB tests in a test-like-you-fly (TLYF) sequence that are consistent with expected launch and on-orbit flight operational events [6]. Specification BOL capacity, resistance, charge retention, and mission-specific performance testing are conducted before and after the most stressful electrical and mechanical tests to verify LIB design specification requirements. Space program ground and on-orbit experience have shown that LIB random vibration, thermal cycling, thermal vacuum, life cycle, and safety qualification testing are significant in demonstrating design performance margins.

3.7.3.1 Capacity

The objective of battery capacity testing is to measure the capacity, energy, power output, and voltage of the battery under specific operating conditions. Battery capacity test operating conditions include temperature, charge–discharge rate, EOCV, end-of-charge current, and end-of-discharge-voltage (EODV). Battery capacity is commonly measured at fixed charge–discharge rates, EOCV, and EODV values as a function of qualification or acceptance of temperature ranges. Figure 3.7 shows the results of a space-qualified LIB capacity test conducted between the manufacturer's recommended EOCV and EODV limits at varying temperatures. These data were used to support the EPS power budget and mission life expectancy analysis. Mission-specific performance testing is a TLYF approach to battery capacity testing. Using this TLYF approach, battery capacity is measured under predicted on-orbit charge–discharge rates and temperatures. Satellite payload peak pulse power requirements (if any) may also be incorporated into mission-specific battery performance testing.

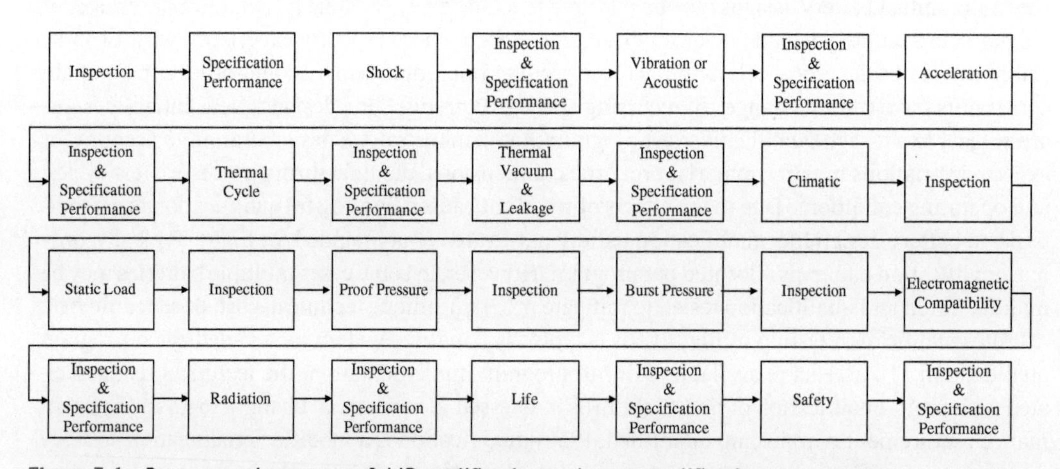

Figure 3.6 Representative spacecraft LIB qualification and proto-qualification test sequence.

3.7.3.2 Resistance

The objective of battery resistance testing is to measure the internal resistance of the battery under specific operating conditions. Battery resistance is commonly measured at constant temperature over a range of SOC consistent with expected on-orbit operating conditions. By interrupting the current and/or applying current pulses during the discharge, it is possible to calculate the LIB internal resistance, which is equal to the change in voltage divided by the change in current, as shown in Figure 3.8. The measured battery resistance includes the ohmic contributions from the

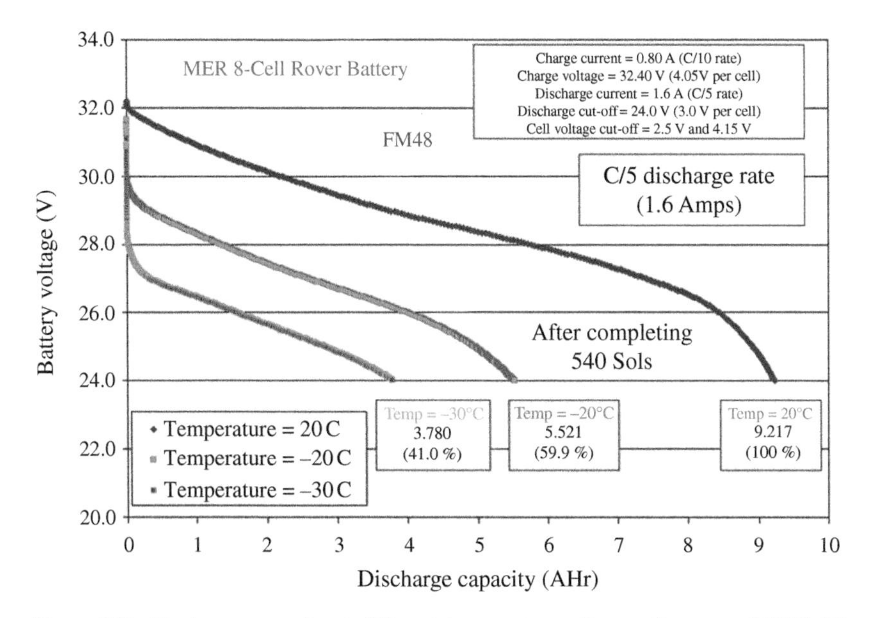

Figure 3.7 Discharge capacity at different temperatures. *Source:* Courtesy of NASA-JPL.

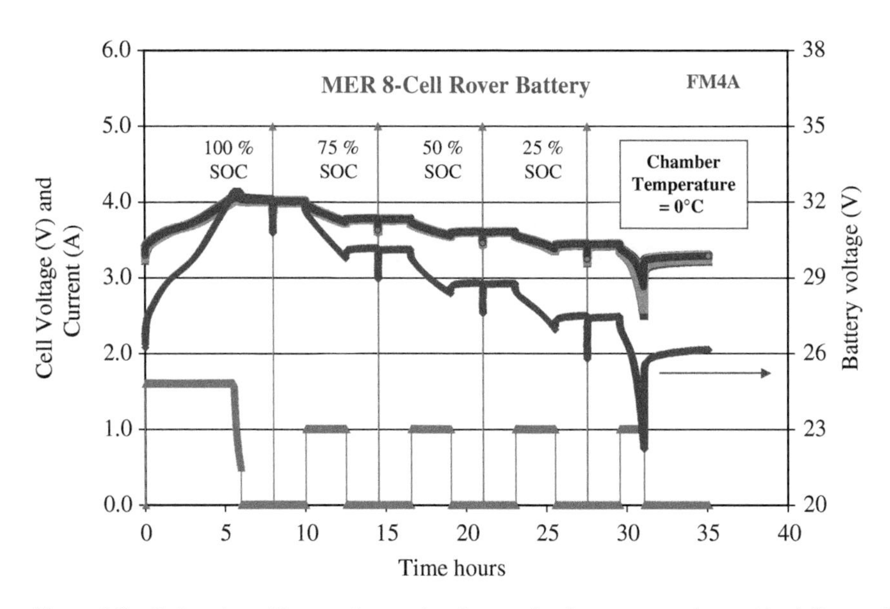

Figure 3.8 Estimation of battery internal resistance by the current pulse method. *Source:* Courtesy of NASA-JPL.

Li-ion cells, wire harness, terminal connections, switches, relays, internal heaters, integrated BMS, and other electronic and mechanical component sources. Battery source resistance data are used to support satellite harness voltage drop analysis, battery harness sizing, and mission-life expectancy analysis. Battery resistance data are also used as an input to the spacecraft EPS transient and dynamic stability analysis.

3.7.3.3 Charge Retention

The objective of battery charge retention testing is to measure the battery self-discharge rate in order to screen the battery assembly for internal shorts. As discussed in Chapter 2, the battery cells are screened for soft shorts as part of LAT. However, poor battery manufacturing workmanship, inadvertent introduction of FOD, or other latent defects introduced during the battery manufacturing process may cause battery-level internal shorts. Periodic battery charge retention testing is commonly used to screen for internal shorts induced after mechanical qualification or acceptance testing. Charge retention test durations may vary between 7 and 30 days, depending on test temperature requirements and the presence of integrated battery cell-level charge balancing circuitry.

3.7.3.4 Vibration

Launch vibration is the most severe mechanical environment imposed on spacecraft LIBs during their entire life cycle. As such, the objective of vibration testing is to verify the battery mechanical design margins when subjected to the predicted LV and ground transportation vibration environments. An example of a vibration test configuration with a 8s52p LIB module is illustrated in Figure 3.9. Battery vibration environments are derived from the mission-specific LV dynamics. Flight batteries can be required to be tested to the same levels and durations in all three axes, so that the battery design is not sensitive to launch orientation. When the launch orientation is frozen early enough in the satellite platform design phase, it is possible to define axis-specific levels from the LV spacecraft coupled load analysis. To adhere to a TLYF philosophy, flight batteries are vibration tested at the specified SOC and discharge rate expected during launch. Before and after each vibration run, it is common practice to perform a low-level resonance search and a charge retention test. Comparing the LIB response of these post-test diagnostics can then be used to verify that the LIB was not damaged during the vibration run.

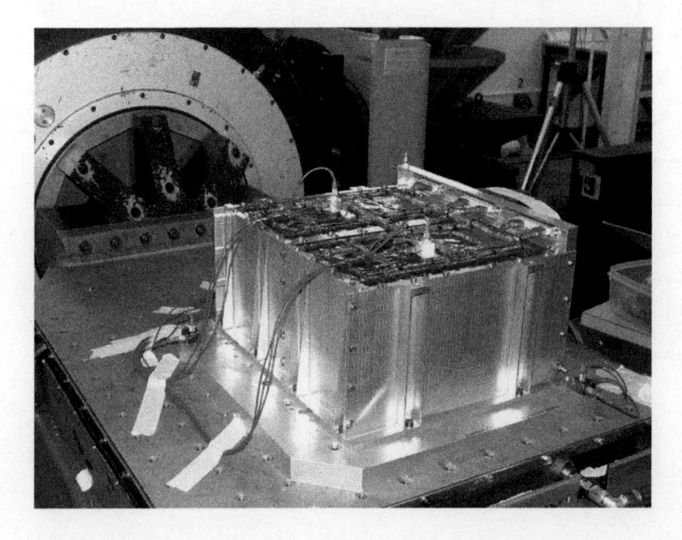

Figure 3.9 In-plane vibration testing of an 8s52p LIB manufactured by EnerSys/ABSL. Credit: EnerSys/ABSL.

3.7.3.5 Shock

The objective of battery shock testing is to determine battery sensitivity to LV and satellite-level shock events. Battery shock-sensitive components include connections, fasteners, relays, or mechanical switches. Electrical isolation between various internal battery components may also be susceptible to shock events. The purpose of battery shock testing is to further verify that there is no physical or performance degradation effects as a result of exposure to the specified shock environment.

Beginning with a spacecraft launch to the end of mission operations, satellite batteries are exposed to various types of shock events. Mission events that may induce shock events to the flight battery include [48]:

- LV induced shocks such as fairing or stage separation. LV shock events propagate in the launcher structure, reach the LV spacecraft interface and then propagate into the spacecraft structure. These shock environments are typically documented in the LV user manual.
- Spacecraft release shock events. Release shock events are directly generated at the LV spacecraft interface by the separation system (clamp band release or other system). Release shock event environments are also documented in the relevant LV user manual.
- Appendage release pyroshock or subsystem actuating pyroshock events. All appendages are stowed and clamped for launch. Once on-orbit, satellite appendages are deployed according to the mission-specific timeline via mechanical pyroshock devices. Typical appendages include solar arrays, antenna reflectors, thermal radiators, and science instruments. These shock events occur generally during the early orbit phase. Pyroshock events may also occur at much later times in the mission timeline, such as EDL for planetary missions.
- Flight batteries may also be subjected to shock events encountered during routine bench handling and transportation activities.

To adhere to a TLYF philosophy, if the LIB is operational during the shock event, it should be charged to a SOC consistent with the shock event scenario, discharged under a mission-relevant representative load and with voltage monitored continuously. This verifies the battery can provide continuous, uninterrupted power during and after the shock event. Several methods are available for shock testing, such as explosive detonation of a pyrotechnic device, or metal–metal mechanical impact. The test method should be selected considering its adequacy with respect to the test specification and physical properties of the battery test article. Figure 3.10 shows an example of a spacecraft LIB undergoing pyro-shock qualification testing.

3.7.3.6 Thermal Cycle

Thermal cycling is intended to impose environmental stress screening on the LIB design, Li-ion cells, parts, processes, and interfaces to detect flaws in both design and workmanship. To adhere to TLYF principles, test batteries are mounted in a test chamber configuration that simulates the flight thermal interface and environment. Qualification level thermal cycle testing is performed to the maximum and minimum qualification temperature limits for the specified number of qualification cycles to demonstrate robustness of the LIB design over the operational survival temperature range, and the ability to function and meet performance requirements during subsequent LIB acceptance testing. Thermal stress qualification cycling also accelerates the LIB aging process in order to demonstrate design margin. Acceptance level thermal cycling testing is performed to the maximum and minimum acceptance temperature limits for a reduced number of cycles to demonstrate adequate workmanship of each LIB. Thermal cycling may be conducted at ambient pressure and/or during thermal vacuum

Figure 3.10 Pyro shock testing of a spacecraft LIB. Credit: Saft/Thales Alenia Space.

(TVAC) testing. However, thermal cycle screening at ambient pressure is more effective than in a vacuum due to higher achievable temperature ramp rates, and thus is the preferred screening method.

Industry experience has shown that once Li-ion cell and electronic circuit-board subassemblies are integrated into the Representative spacecraft LIB, the allowable temperature range the LIB can be exposed to become severely limited. To mitigate this risk, electronic circuit-board subassemblies (such as integrated cell charge balancing electronics) can be first thermally cycled at a subassembly level prior to LIB integration. Although this approach is not consistent with TLYF principles, exposure to required temperature extremes meets environmental thermal stress screening requirements and mitigates risk associated with a temperature-limited battery level.

3.7.3.7 Thermal Vacuum

The objective of LIB TVAC testing is to verify battery specification performance and survival over combined thermal and vacuum conditions to demonstrate performance in the relevant space vacuum environment. A representative TVAC chamber and how the battery is configured within it is illustrated in Figure 3.11. The LIB design margin is demonstrated by qualification TVAC testing of the battery unit performance in the qualification thermal environment. In a similar manner to LIB thermal cycling testing, acceptance-level LIB TVAC testing detects material and workmanship defects that may impact the flightworthiness of the battery unit. Thermal cycling of the LIB test article is commonly performed during TVAC testing to characterize and verify battery specification requirements. In TVAC qualification testing, worst-case hot and cold test temperature limits are employed to establish LIB design margins. During qualification testing, consideration is given to placing temperature sensors at strategic locations on the LIB test article to confirm the detailed LIB thermal analysis results. This includes monitoring test temperature sensors for the battery cell minimum and maximum temperatures and battery-to-spacecraft interface mounting temperatures. These data are used to demonstrate compliance to the battery specification thermal gradient

Figure 3.11 Thermal vacuum testing of a spacecraft LIB. Credit: EnerSys/ABSL.

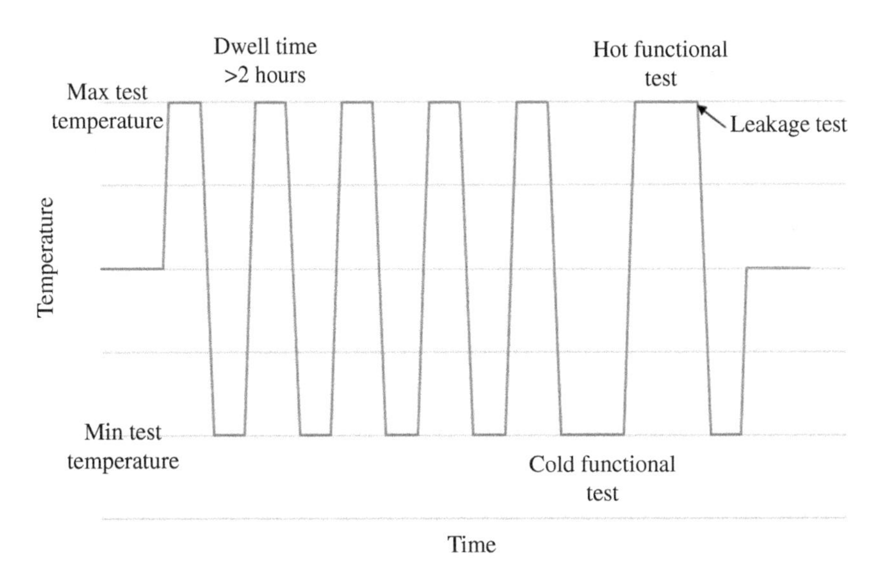

Figure 3.12 Representative spacecraft LIB thermal vacuum temperature profile.

and thermal control subsystem requirements. Industry experience has shown that the TVAC testing can be used as a convenient environment for detecting LIB leakage. A representative space LIB TVAC temperature profile is shown in Figure 3.12.

3.7.3.8 Electromagnetic Compatibility

The purpose of EMC testing is to demonstrate that the EMI characteristics of the battery do not result in degraded performance of the battery. EMC testing also demonstrates that the battery does

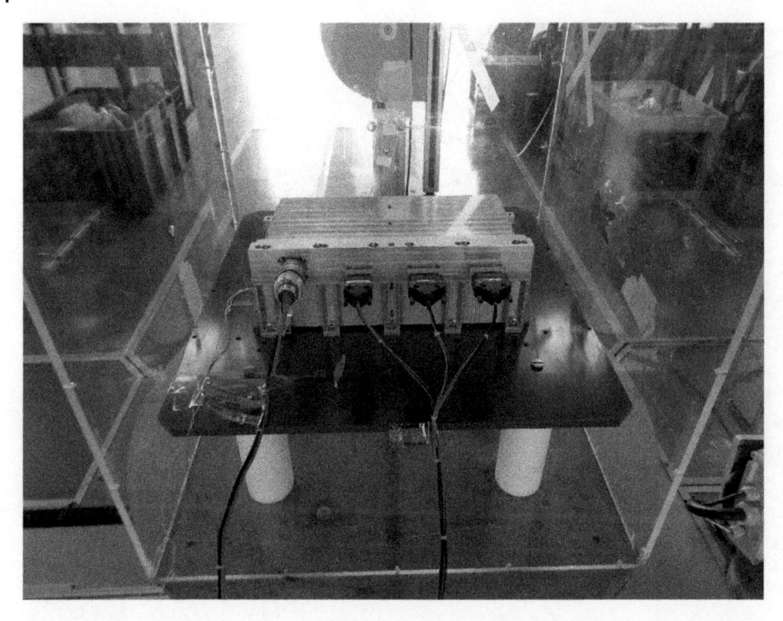

Figure 3.13 Magnetic moment testing of a spacecraft LIB. Credit: EnerSys/ABSL.

not emit, radiate, or conduct interference that may result in malfunction of other satellite units [49]. EMI/EMC requirements are verified by test and analysis according to the satellite program EMC control plan. In general, EMI/EMC testing is required when the LIB assembly contains integrated active and/or passive electronic components such as integrated cell-to-cell balancing, voltage, and/or temperature monitoring, protection, safety, or other electronic circuit-board electronics. Consistent with TLYF principles, EMC testing is commonly performed with flight-like battery power and signal harness cables connected to simulated flight loads that mimic actual load-switching conditions during test.

A specific subset of EMC testing relates to the measurement of dynamic (current-induced) and static (material) magnetic moments. Since this can have an impact on magnetically sensitive instrumentation or on satellite attitude, magnetic moment vectors are often measured as part of the unit flight testing. A representative magnetic moment characterization test set-up is shown in Figure 3.13. The dynamic test is performed with a load on the LIB to replicate the current flow expected in operational use. The unit may be depermed to remove the residual magnetic field induced in some components prior to the magnetic moment measurement.

3.7.3.9 Life Cycle

Verification of LIB specification life expectancy requirements is accomplished by real-time life cycle testing (LCT) and analysis. To demonstrate compliance to mission-life requirements, space industry LIB manufacturers commonly conduct LCT on combinations of Li-ion cells, modules, and/or batteries in a relevant mission-specific configuration and environment. Due to the shorter mission-life expectancy of LV spacecraft batteries, LCT is conducted under real-time conditions. However, due to the longer mission lifetimes of Earth-orbiting satellites, a combination of real-time and accelerated LCT conditions is commonly employed. A common space industry risk-reduction best practice is to combine real-time with accelerated LCT test results along with a model analysis. Details of LIB LCT test requirements, planning, execution, and modeling analysis are discussed in Chapter 8.

3.7.3.10 Safety

During various times in the LIB life cycle, spacecraft batteries may be exposed to unexpected abuse conditions resulting in an energetic safety incident. Safety incidents in market sectors using high specific energy LIBs have demonstrated that exposure of LIBs to abuse conditions may result in catastrophic consequences to personnel, hardware, and facilities. In order to maintain system safety during the entire LIB life cycle, LIB designs are now required to be safety tested for both crewed and unmanned spacecraft applications [6, 44].

The spacecraft LIB specification contains a list of test requirements, protocols, compliance methods, and verification events associated with the required safety testing to meet ground processing and launch site requirements. Due to the possibility of destructive outcomes for most space LIB safety test protocols, test articles are chosen from battery assets not intended for flight. These non-flight battery designs may include, but are not limited to, qualification units or flight-like EMs. Alternatively, approved tailoring of relevant safety requirements may enable the use of cell-level or module-level safety testing in lieu of battery-level safety testing. In all cases, safety testing is conducted during the qualification test event for the spacecraft program. Additional details of LIB and Li-ion cell safety testing are discussed in Chapter 7.

3.7.4 Acceptance Test Descriptions

Subsequent to completion of LIB manufacturing and prior to spacecraft integration, all flight LIBs undergo acceptance testing. The purpose of acceptance testing is to demonstrate that flight LIBs are free of workmanship defects and meet battery specification performance requirements. As such, acceptance testing is not intended to adversely impact mission design life by imparting electrical, mechanical, or thermal stress factors on to the flight LIB test article. Figure 3.14 lists the standard space industry flight LIB acceptance tests in sequence consistent with expected launch and on-orbit flight operational events [6]. The LIB acceptance test results provide a foundation for contractual flight battery delivery to battery customer for the next level of spacecraft integration. Typically, accepted flight LIB units are delivered to the spacecraft manufacturer factory for integration into the EPS and bus structure.

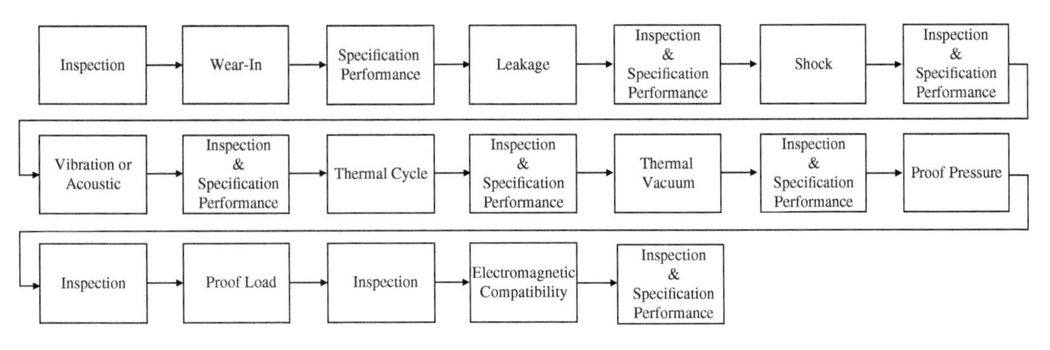

Figure 3.14 Space LIB acceptance test sequence.

3.8 Supply Chain

The space Li-ion cell and LIB markets are interconnected to supply chain dynamics impacting adjacent commercial consumer electronics and electrified transportation LIB marketplaces. These global markets generally rely on a unique set of subtier suppliers for raw cell materials, electronic parts, and other hardware components across a diverse geographical landscape. For example, global subtier suppliers for critical Li-ion cell electrode raw elemental materials are generally shared across the space and commercial market sectors. Other Li-ion cell and LIB-level parts and materials also have common subtier suppliers who create various supply-chain risks and opportunities. However, there are significant reliability and quality differences between space-rated LIB parts and materials and those used in terrestrial LIB applications.

3.8.1 Battery Parts and Materials

Space battery suppliers and manufacturers BOMs list the subassemblies, components, parts, materials, and quantities for the end-item LIB. The BOM also includes the various subtier supplier for each listed item, part number, model number, and quantities to produce a single finished LIB unit. A typical space LIB BOM will list the Li-ion cell types, high-reliability EEE parts, and raw materials. LIB raw materials include all the electrical, mechanical, and thermal stock materials needed for LIB production. Past LIB manufacturing experience has shown that space-qualified electrical power, signal, and test connectors as well as Li-ion cells are typical long-lead hardware items impacting LIB production scheduling. Integrated BMS electronic parts and components may also become long-lead procurement items due to requirements to survive exposure to on-orbit radiation environments. Finally, governmental restrictions and requirements to adhere to US International Traffic in Arms Regulations (ITAR) governing satellite components may also impact satellite LIB BOM procurement activities.

3.8.2 Space LIB Suppliers

LIB cells, EEE parts, safety devices, and other components may only be available from limited selection of space qualified subtier suppliers. These supply chain limitations create procurement risks associated with materials obsolescence, elevated unit cost, and supply. In order to meet demand while mitigating these market risks, procurement of space-qualified LIBs has been met with two general supply chain solutions. First, satellite manufacturers have procured space-qualified LIBs directly from original equipment manufacturers (OEMs) such as EnerSys, EaglePicher Technologies, GS Yuasa, and Saft [50–53]. A second approach used to meet LIB demand is for satellite manufacturers to design, build, and test LIBs in-house. Using this procurement strategy, the satellite manufacturer procures space-qualified Li-ion cells or cell module subassemblies directly from a wide variety of Li-ion cell OEMs for integration into in-house designed LIBs. This LIB procurement approach has risk and opportunities that are unique to each space program application.

3.9 Summary

Primarily due to a significant increase in specific energy and higher operating voltages, space LIBs have proven to be a disruptive energy storage technology for spacecraft applications. Modular LIB-based EPS architectures continue to be enabling for today's complex Earth-orbiting and planetary mission space systems requiring increased battery-level specific energy density and a reliable on-orbit service life. To meet these evolving on-orbit mission needs, systems engineering best practices are applied to develop compliant LIB design solutions to EPS ground and on-orbit satellite requirements. This includes allocating applicable spacecraft performance and environmental requirements to configuration-managed EPS and LIB specifications. Future LIB designs will prioritize safe-by-design principles for crewed and unmanned spacecraft missions to ensure safe ground processing and compliance to on-orbit mission assurance requirements.

Satellite manufacturers and battery suppliers continue to benefit from the wide variety of custom and COTS Li-ion cell designs readily available in the growing commercial Li-ion marketplace. Moreover, the diverse set of Li-ion cell design options has created new opportunities to tailor LIBs to be increasingly modular with changing EPS architectures and next-generation satellite bus design layouts. LIB requirements for spacecraft are expected to prioritize increases in cell-level specific energy to reduce total LIB mass and cost. It is expected that the space satellite market will continue to use space qualified LIBs based on both custom large and small-format COTS Li-ion cell design solutions.

Acknowledgments

Some of the work described here was carried out at the Jet Propulsion Laboratory, California Institute of Technology, under contract with the National Aeronautics and Space Administration (NASA) (80NM0018D0004).

References

1 Borthomieu, Y. (2014). Satellite batteries. In: *Lithium-Ion Batteries: Advances and Applications* (ed. G. Pistoia), 311–314. Amsterdam: Elsevier B.V.
2 Darst, J., Thomas, J.C., Finegan, D.P., and Darcy, E. (2018). Guidelines for safe, high performing Li-ion battery designs for manned vehicles. *Power Sources Conference*. Denver, CO.
3 Broussely, M., Biensan, P., Bonhomme, F. et al. (2005). Main aging mechanisms in Li ion batteries. *J. Power Sources* 146: 90–96.
4 Vetter, J., Novák, P., Wagner, M.R. et al. (2005). Aging mechanisms in lithium-ion batteries. *J. Power Sources* 147: 269–281.
5 Barré, A., Deguilhem, B., Grolleau, S. et al. (2013). A review on lithium-ion battery aging mechanisms and estimations for automotive applications. *J. Power Sources* 241: 680–689.
6 United States Space Force Space Systems Command Standard (2022). *Lithium-ion Battery Standard for Spacecraft Applications*. SSC Standard SSC-S-017.
7 Kodheli, O., Lagunas, E., Maturo, N. et al. (2021). Satellite Communications in the new space era: A survey and future challenges. *IEEE Commun. Surv. Tutor.* 23 (1): 70–109. https://doi.org/10.1109/COMST.2020.3028247.
8 Schwanbeck, E. and Dalton, P. (2019). International Space Station lithium-ion batteries for primary electric power system. In: *2019 12th European Space Power Conference (ESPC)*, 1.
9 Dietrich, F.J., Metzen, P., and Monte, P. (1998). The Globalstar cellular satellite system. *IEEE Trans. Antennas Propag.* 46 (6): 935–942.
10 Boiardt, H. and Rodriguez, C. (2009). The use of Iridium's satellite network for nanosatellite communications in low earth orbit. In: *2009 IEEE Aerospace Conference*, 1–5.
11 Kokez, H.A.-D.F. (2020). On terrestrial and Satellite communications for telecommunication future. In: *2020 2nd Annual International Conference on Information and Sciences (AiCIS)*, 58–67.

12 Thépaut, J., Dee, D., Engelen, R., and Pintym, B. (2018). The Copernicus Programme and its climate change service. In: *IGARSS 2018–2018 IEEE International Geoscience and Remote Sensing Symposium*, 1591–1593.

13 Kirby, K. and Stratton, J. (2013). Van Allen probes: successful launch campaign and early operations exploring Earth's radiation belts. In: *2013 IEEE Aerospace Conference,* 1–10.

14 Briskman, R.D. and Akturan, R. (2021). S-band commercial satellite usage in North America. In: *2021 6th Advanced Satellite Multimedia Systems Conference (ASMS) and 12th Signal Processing for Space Communications Workshop (SPSC)*, 166–167.

15 LaCava, S.N. (2013). TacSat-4 & ORS bus standards program 30Ah Li-ion battery. In: *The NASA Aerospace Battery Workshop*. Huntsville, AL.

16 Folta, D. and Beckman, M. (2002). Libration orbit mission design: applications of numerical and dynamical methods. In: *Libration Point Orbits and Applications*, 85–113. Girona, Spain: World Scientific.

17 International Electrotechnical Commission (2004). *International Electrotechnical Vocabulary (IEV) - Part 482: Primary and Secondary Cells and Batteries*. IEC 60050–482:2004

18 Prévôt, D., Castric, A.-F., and Pasquier, E. (2008). Batteries and bypass switch: Saft heritage and standard search. In: *8th European Space Power Conference (ESPC)*, 1–10.

19 Bouhours, G., Klein, E., Rebuffel, C. et al. (2011). Lithium-ion cell passivation in flight conditions. In: *9th European Space Power Conference (ESPC)*, 1–4.

20 Borthomieu, Y. and Pasquier, E. (2005). Li-ion battery by-pass removal qualification. In: *7th European Space Power Conference (ESPC)*, 1–6.

21 Powers, A., Mnatsakanian, A., Ang, V. et al. (2016). Lithium-ion battery aerospace design: Cell bypass considerations. In: *Space Power Workshop*, 1–21.

22 Laurance, M.R. and Brownlee, D.E. (1986). The flux of meteoroids and orbital space debris striking satellites in low earth orbit. *Nature.* 323 (6084): 136–138.

23 McKay, D.S., Barrett, R.A., and Bernhard, R.P. (1987). Impact damage to solar maximum satellite caused by micro-meteorites and microparticle orbital debris. *Meteoritics.* 22 (4): 455–456.

24 Hyde, J.L., Christiansen, E.L., and Kerr, J.H. (2001). Meteoroid and orbital debris risk mitigation in a low Earth orbit satellite constellation. *Int. J. Impact Eng.* 26: 345–356.

25 Ratnakumar, B.V., Smart, M.C., Whitcanack, L.D. et al. (2004). Behavior of Li-ion cells in high-intensity radiation environments. *J. Electrochem. Soc.* 151 (4): A652–A659.

26 Ratnakumar, B.V., Smart, M.C., Whitcanack, L.D. et al. (2005). Behavior of Li-ion cells in high-intensity radiation environments: II. Sony/AEA/ComDEV cells. *J. Electrochem. Soc.* 152 (2): A357–A363.

27 Qiu, J., He, D., Sun, M. et al. (2015). Effects of neutron and gamma radiation on lithium-ion batteries. *Nuclear Instruments and Methods in Physics Research Section B: Beam Interactions with Materials and Atoms.* 345: 27–32.

28 Palmerini, G.B. and Pizzirani, F. (2002). Design of the radiation shielding for a microsatellite. *Acta Astronaut.* 50 (3): 159–166.

29 Palmerini, G.B. and Pizzirani, F. (2003). Structural design in a space radiation shielding perspective. In *54th International Astronautical Congress of the International Astronautical Federation, the International Academy of Astronautics, and the International Institute of Space Law*, Bremen, Germany.

30 Kern, D.L., Gordon, S.A., and Scharton, T.D. (2005). NASA Handbook for spacecraft structural dynamics testing. *European Conference on Spacecraft Structures, Materials and Mechanical Testing.* Noordwijk, The Netherlands.

31 Berg, P., Spielbauer, M., Tillinger, M. et al. (2020). Durability of lithium-ion 18650 cells under random vibration load with respect to the inner cell design. *J. Energy Storage* 31: 101499.

32 Bach, T.C., Schuster, S.F., Fleder, E. et al. (2016). Nonlinear aging of cylindrical lithium-ion cells linked to heterogeneous compression. *J. Energy Storage* 5: 212–223.

33 Valentin, O., Thivel, P.-X., Kareemulla, T. et al. (2017). Modeling of thermo-mechanical stresses in Li-ion battery. *J. Energy Storage* 13: 184–192.

34 Xu, R. and Zhao, K. (2016). Electrochemomechanics of electrodes in Li-ion batteries: A review. *J. Electrochem. En. Conv. Stor.* 13 (3): 030803.

35 Hengeveld, D.W., Mathison, M.M., Braun, J.E. et al. (2010). Review of modern spacecraft thermal control technologies. *HVAC&R Res.* 16 (2): 189–220.

36 Ratnakumar, B.V., Smart, M.C., Ewell, R.C. et al. (2004). Lithium-ion rechargeable batteries on Mars Rovers. In: *2nd International Energy Conversion Engineering Conference (IECEC)*. Providence, Rhode Island.

37 Smart, M.C., Ratnakumar, B.V., Ewell, R.C. et al. (2018). The use of lithium-ion batteries for JPL's Mars missions. *Electrochim. Acta* 268: 27–40. https://doi.org/10.1016/j.electacta.2018.02.020.

38 Materials and Processes Technical Information System (MAPTIS). https://maptis.nasa.gov (accessed 25 May 2022).

39 NASA Parts Advisory. NA-044A (1998).

40 Brusse, A., Leidecker, H., and Panashchenko, L. (2008). Metal Whiskers: Failure modes and mitigation strategies. *2nd International Symposium on Tin Whiskers*. Tokyo, Japan.

41 Tribble, A.C., Boyadjian, B., Davis, J. et al. (1996). Contamination control engineering design guidelines for the aerospace community. In: *Proceedings of the SPIE 2864, Optical System Contamination V, and Stray Light and System Optimization*.

42 Martin, D.J. and Maag, C.R. (1993). The influence of commonly used materials and compounds on spacecraft contamination. *Acta Astronaut.* 30: 51–65.

43 White, L., Anderson, M., Blakkolb, B. et al. (2017). Organic and inorganic contamination control approaches for return sample investigation on Mars 2020. In: *2017 IEEE Aerospace Conference*, 1–10. https://doi.org/10.1109/AERO.2017.7943709.

44 NASA JSC-20793 (2017). Rev. D. *Crewed Space Vehicle Battery Safety Requirements*.

45 National Aeronautics and Space Administration (1995). *Safety Requirements Document*. SSP 50021, Section 3.3.6.11.

46 Jeevarajan, J.A. and Winchester, C. (2012). Battery safety qualifications for human ratings. In: *Interface*, 51–55. The Electrochemical Society.

47 Air Force Space Command Manual 91–710 (2019). *Range Safety User Requirements Manual*, vol. 3.

48 ECSS-E-HB-32-25A (2015). *Space Engineering – Mechanical Shock Design and Verification Handbook*. ESA-ESTEC Requirements & Standards Division.

49 MIL-STD-461G (2015). *Requirements for the Control of Electromagnetic Interference Characteristics of Subsystems and Equipment*.

50 EnerSys. http://www.enersys.com (accessed 30 May 2022).

51 EaglePicher Technologies. http://www.eaglepicher.com (accessed 30 May 2022).

52 GS Yuasa. http://www.gs-yuasa.com (accessed 30 May 2022).

53 Saft. http://www.saftbatteries.com (accessed 30 May 2022).

4

Spacecraft Electrical Power Systems

Thomas P. Barrera

4.1 Introduction

Today's Earth-orbiting and planetary spacecraft types represent a diverse combination of unmanned and human-rated (crewed) space missions. The diversity of spacecraft missions is largely derived from the growing variety of applications including communications, meteorology, navigation, Earth observation, scientific, planetary exploration, and transport of astronaut crew and cargo supplies. In addition, expendable and reusable launch vehicles are used exclusively to transport spacecraft for operation in Earth-orbit or interplanetary mission flight regimes. Each spacecraft type can be represented by a unique set of system-level requirements specific to mission objectives. Spacecraft system designs are commonly partitioned into unique functional subsystem architectures [1]. Figure 4.1 shows the common spacecraft subsystem elements most frequently employed in Earth-orbiting satellite or planetary mission spacecraft. The payload subsystem is represented by specific categories of payload operational and scientific functions dedicated to the given spacecraft mission objectives. The bus (or platform) subsystems consist of the various units and components necessary to support overall satellite functionality. Divided along specific systems engineering disciplines, each subsystem also forms a hardware and/or software interface internal to the spacecraft. In addition, operational crewed spacecraft such as the NASA International Space Station (ISS), SpaceX Crew Dragon, and Energia Soyuz also employ sophisticated environmental control and life-support subsystems. For example, the NASA Orion spacecraft service module electrical power system (EPS) consists of four solar array wings for power generation and four lithium-ion batteries (LIBs) for energy storage [2]. The EPS distributes power to all other subsystems and components by means of four independent 120 VDC unregulated power busses (Figure 4.2).

All spacecraft architectures utilize EPSs to provide continuous electrical energy to user loads under all mission phases and expected modes of operation for the duration of the spacecraft lifetime. To meet spacecraft power requirements, the EPS consists of all the hardware and software required to generate, store, manage, and distribute electrical energy to all bus and payload user loads. Electrical power generation (solar arrays), energy storage (batteries), power management and control (electronics), and distribution (wire harness and switchgear) components constitute the main elements of a satellite EPS (Figure 4.3). Today, rechargeable LIB energy storage technologies are used exclusively to provide primary power capability for all types of spacecraft vehicles. As

Spacecraft Lithium-Ion Battery Power Systems, First Edition. Edited by Thomas P. Barrera.

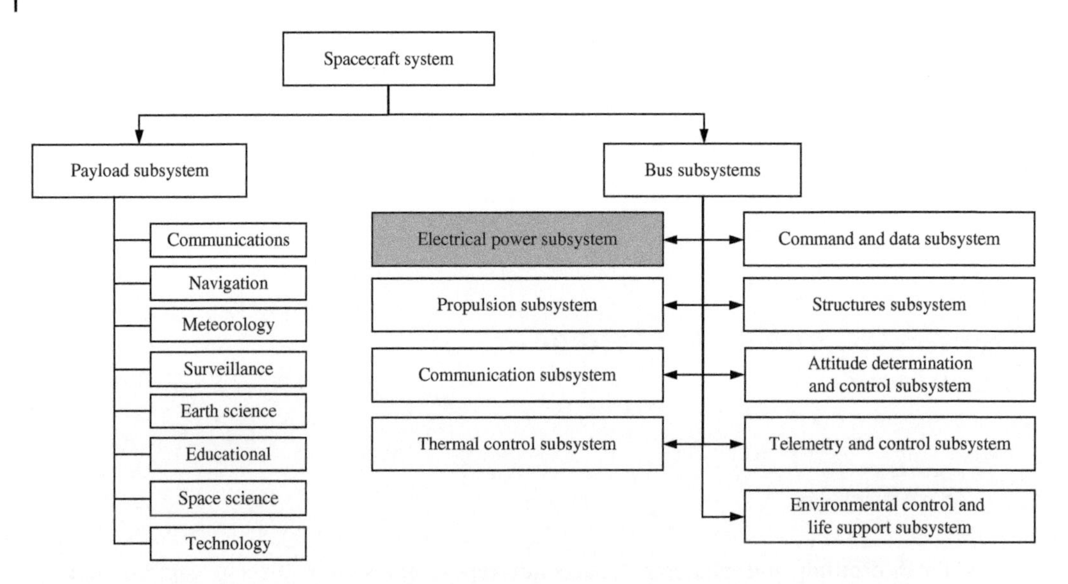

Figure 4.1 Unmanned and crewed spacecraft subsystem elements.

Figure 4.2 The NASA Orion Artemis I spacecraft with ESA's service module shown during ground processing. The service module sits directly below Orion's crew capsule and provides power, propulsion, thermal control, water, and air for the astronaut crew life support subsystem. Credit: Radislav Sinyak/NASA.

discussed in Chapter 6, in some planetary missions, secondary payload EPS architectures employ multiple LIBs to maintain safety during ground processing and during spacecraft cruise periods.

This chapter describes the general operational and functional characteristics of LIB-based EPS for spacecraft systems. The importance of understanding EPS-level requirements in terms of meeting spacecraft orbital power demands throughout the mission life using LIBs is emphasized. Special EPS topics such as a battery management system (BMS), dead bus recovery, assembly integration and test (AI&T), and specialty analyses are also discussed.

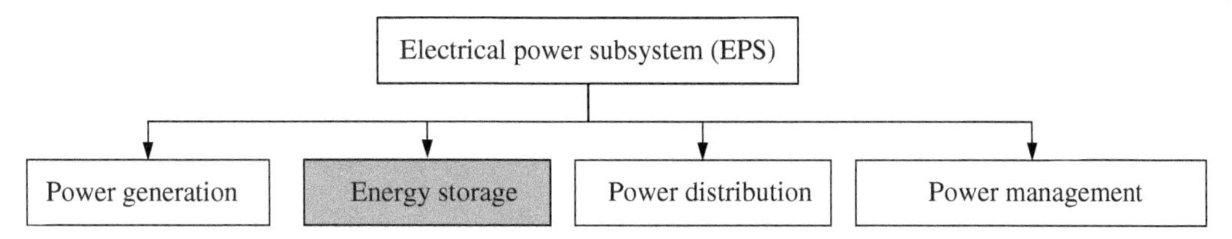

Figure 4.3 Fundamental elements of a spacecraft electrical power subsystem.

4.2 EPS Functional Description

A simplified functional block diagram of a spacecraft electrical power subsystem. is shown in Figure 4.4. Electric power for conventional satellites and many planetary probes can be generated by using either solar photovoltaic or solar thermal systems depending on a spacecraft load power duty cycle, mission orbit, and spacecraft lifetime requirements. As further described in Chapter 6, the Mars Science Laboratory (MSL) and Mars 2020 Perseverance Rover planetary missions utilized radioisotope thermoelectric generators (RTGs) and LIBs to meet spacecraft EPS requirements.

4.2.1 Power Generation

The most commonly used power generation device in satellite applications is the solar photovoltaic direct current (DC) power generation system, which employs arrays of photovoltaic solar cell building blocks electrically connected in series–parallel combinations to convert sunlight (solar energy) into DC electricity. Earth-orbiting satellites ranging in power from watts (CubeSats) to hundreds of kilowatts (NASA-ISS) commonly employ photovoltaic solar array technologies as a DC current source for meeting satellite user load demands. Solar array architectures range from traditional rigid panel substrates to more compact, flexible, and lower mass technologies [3]. Solar arrays are designed to meet bus and payload power demands during each mission phase, including battery recharge power, during the entire satellite operational life in sunlight. Solar arrays are sized to meet satellite power requirements under worst-case (June solstice) end-of-life (EOL) orbital operating conditions. Other sizing parameters include mass limitations, environmental

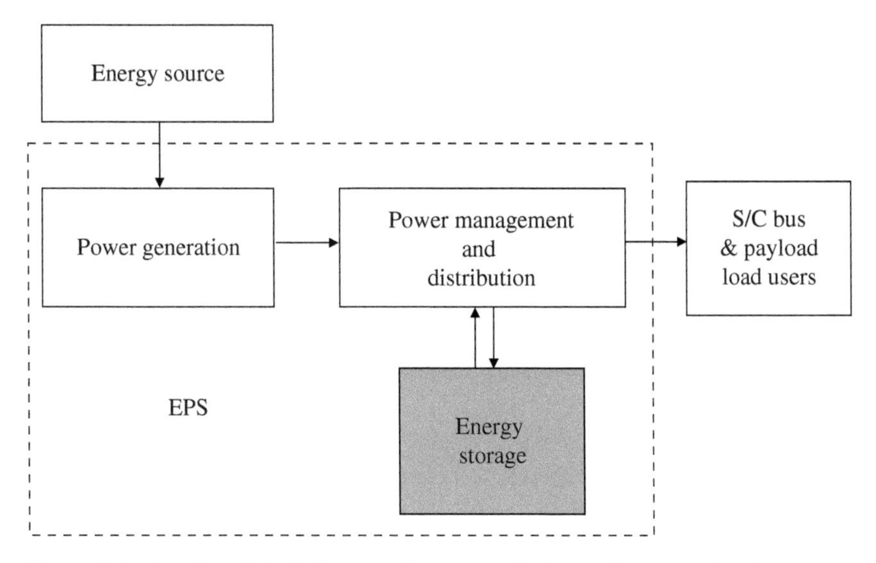

Figure 4.4 Functional block diagram of a spacecraft electrical power subsystem.

degradation factors (such as radiation, atomic oxygen, micrometeoroid damage, and thermal cycling), geometric area, EOL power margin, and number of failed cell string circuits to meet reliability requirements.

4.2.2 Energy Storage

As described in Chapter 3, rechargeable LIBs provide primary electrical energy storage capability for nearly every satellite, planetary, launch, crew transfer, and cargo vehicle used in a space application. Rechargeable LIBs and non-rechargeable lithium batteries are also used on the US astronaut extravehicular mobility unit (EMU) and portable astronaut crew electronics equipment aboard the NASA-ISS. Non-rechargeable silver-zinc (Ag-Zn) batteries used in expendable launch vehicle (LV) avionics and flight termination systems are now being displaced by compliant LIB technology options [4]. LIBs are generally characterized by the ability to store and deliver energy at a characteristic DC voltage to the spacecraft user loads. To maintain a positive energy balance during all mission phases, energy must be stored to meet user power demands during eclipse (no or partial sunlight) when the solar arrays are not capable of converting sunlight into electrical energy. During non-eclipse (full sunlight) orbital time periods, the spacecraft user loads may exceed the solar array power generation capability. During these periods of high load demand, the spacecraft batteries may be required to assist the solar arrays meet spacecraft user load demands. Spacecraft LIBs are sized to support mission-specific operations during geosynchronous orbit (GEO) transfer orbit, on-orbit operational (station-keeping) periods, peak power demands, and worst-case spacecraft contingencies. Operationally, the EPS LIBs are discharged when meeting user load demands during eclipse and peak load demands during the sunlit portion of the satellite orbit. Regardless of the satellite Earth-orbit type, the batteries are fully recharged before re-entering the next eclipse period (or other period such as a number of orbits) to maintain an orbit-to-orbit positive power energy balance. Certain science missions may take a different approach to fully recharge after a science encounter, series of maneuvers, and transmission of data. Each mission uniquely determines how to manage the energy balance to meet spacecraft performance requirements.

4.2.3 Power Management and Distribution

Electrical power management and distribution (PMAD) includes the conditioning, regulation, control, and distribution of electrical power for the entire spacecraft. More specifically, the spacecraft EPS PMAD function is to:

- Control and direct electrical power from the solar array units to (charge) and from (discharge) the batteries and to the power distribution subsystem,
- Distribute power from the main power bus to primary and secondary distribution points,
- Provide a DC voltage source (regulated or unregulated) to all satellite bus and payload module unit loads, and
- Provide fault protection functionality to ensure that no single failure results in loss of EPS capability to meet minimum mission-specific requirements for the intended mission life of the spacecraft.

Typically, there is one centralized PMAD unit per spacecraft bus and possibly separate PMAD-type components dedicated to the payload subsystem. PMAD architectures are often custom-designed depending on mission-specific power requirements. However, commercial off-the-shelf

(COTS) electronics components for short duration small satellite (<500 kg) applications have become a cost-effective EPS PMAD design alternative [5]. Examples of typical PMAD power electronics components include electronics for managing solar array power, battery charge–discharge power converters, switchgear for load distribution and ground fault protection, voltage–current regulators, magnetics, filter circuits, power relays, and switching devices and control electronics to ensure that the load demand is satisfied while maintaining a safe range of battery state-of-charge (SOC). Integrated PMAD software is used to monitor EPS telemetry, provide fault protection, and other mission critical functions. The PMAD may also contain flight software for battery charge and discharge management, autonomous load shedding, EPS telemetry management, and command verification to ensure an uninterrupted flow of power to and from energy storage and power conversion devices.

LIB-based EPS BMS may be integrated into the EPS PMAD functionality or directly into the LIB as an electronics subassembly. The BMS design may include control and monitoring of LIB on-orbit cycling characteristics such as end of charge (EOCV) and discharge voltages (EODV), depth-of-discharge (DOD), and charge/discharge rates. Ancillary EPS-related mechanical components such as solar array drives, positioners, and deployment mechanisms are commonly allocated to the satellite attitude determination and control subsystem.

4.2.4 Harness

All satellite bus and payload module units must be interconnected (unit-to-unit) to transfer electrical power from the EPS to unit loads. Wire harness and cabling is used to provide signal, power, electrical test access, and other interconnections between subsystems and units. In general, wire harnesses are bundles of individually insulated copper wire conductors mechanically held together by lacing cord, ties, straps, or clamps and then terminated to connectors or terminal lugs. All wire harness interfaces include space-rated connectors and may contain other electrical hardware components such as diodes and fuses. Co-axial cabling is used for transmission of radio frequencies commonly used in satellite payload high-rate communications and networking applications. Use of co-axial cable assists in avoiding cross-talk interference when transmission occurs at high-rate transmission speeds in twisted-wire pair wire harnesses. NASA, Society of Automotive Engineers (SAE), and US Space Force (USSF) harness and cable wiring standards contain general design practices for interconnecting electrical and electronic equipment installed in spacecraft [6–8].

4.3 EPS Requirements

Consistent with systems engineering best practices, EPS requirements are developed by systematically decomposing and allocating top-level customer parent requirements into the satellite bus subsystem specifications. The EPS specification requirements establish and define the hardware and software design configurations that comply with higher hierarchical system level environmental, interface, and other program-unique requirements. Additional requirements related to specific EPS performance, functionality, interfaces, and operating environments are derived from the customer and program-unique parent requirements. The fundamental satellite customer-level mission-specific parent capability requirements impacting the EPS design include orbit type, mission duration, payload unit power demand, configuration, mass, reliability at EOL, and LV type. The baseline set of customer requirements are used to derive performance-based EPS requirements, which are documented in the EPS requirements specification.

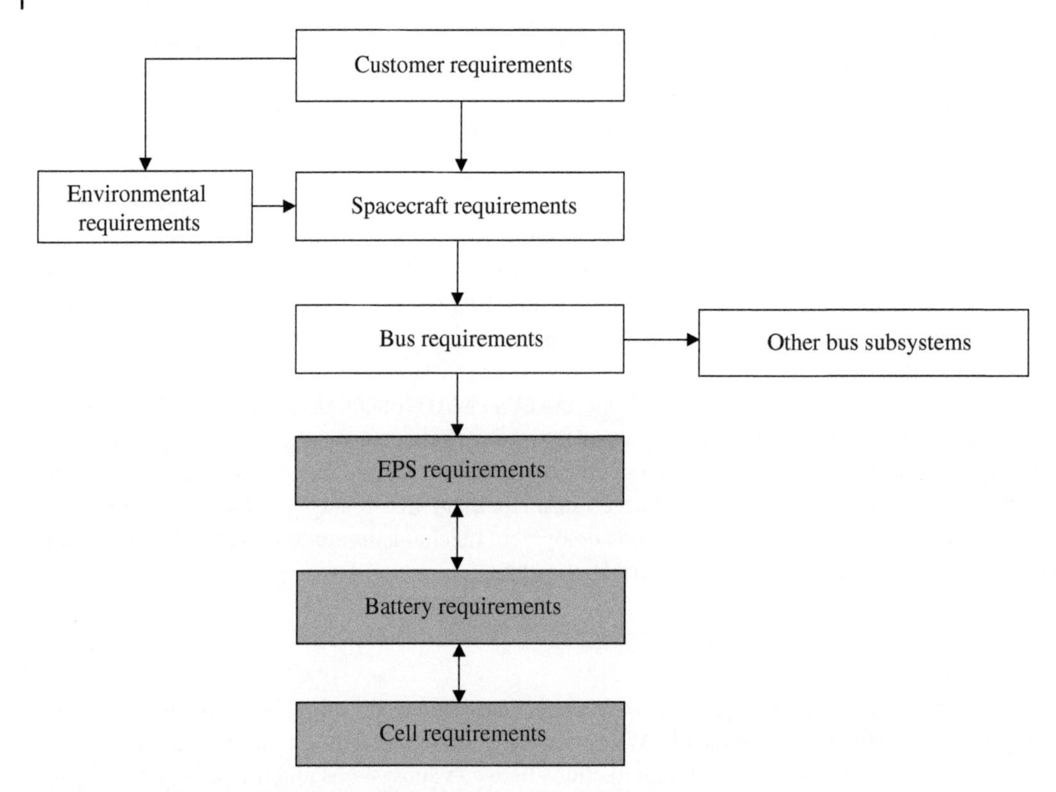

Figure 4.5 Spacecraft EPS requirements engineering flow diagram.

Bus and EPS electrical, mechanical, thermal, and software interface control documents (ICDs) are also derived from customer-defined specifications to ensure spacecraft-to-subsystem compatibility. ICD requirements also aid in establishing the spacecraft-to-LV design solutions. Figure 4.5 shows a requirements hierarchy flow diagram that depicts how spacecraft customer performance and capability design requirements are allocated to a spacecraft battery unit. Early in the preliminary design phase of an LIB power system there may be an iterative engineering requirement exchange between the EPS and LIB unit engineering teams. This iteration process allows for specific derived battery requirements to be documented in the EPS specification to optimize performance and compatibility at the bus and spacecraft level of integration. Requirements iteration between the LIB and cell subassembly design progression is less commonplace due to the COTS nature of most space-qualified Li-ion cell designs. As a result, battery cell designs are selected from available cell manufacturer product-line offerings to meet battery assembly level requirements.

Aerospace industry requirements resources for high-reliability Earth-orbiting and planetary mission spacecraft EPS include the American Institute of Aeronautics and Astronautics (AIAA) S-122 and European Cooperation for Space Standardization (ECSS) standard ECCS-E-ST-20C [9, 10]. EPS requirements specified in the AIAA and ECSS standards are commonly tailored by spacecraft manufacturers for incorporation into the relevant EPS requirements specification supporting the applicable spacecraft program.

4.3.1 Requirements Specification

High-reliability spacecraft EPSs are designed to meet power requirements under all non-operating and operating mission-specific phases of the spacecraft lifecycle. EPS non-operational and operational states and modes include ground processing testing, storage, pre-launch, launch, ascent, transfer orbit, deployments, normal operational states, contingency modes, decommissioning, de-orbit, and disposal phases. In order to develop the EPS design, an EPS requirements specification is used to establish the performance, design, development, test, and verification requirements for the spacecraft EPS. The EPS requirements set defines the baseline configuration in terms of the system-level requirements, which support bus and payload module unit design. For example, the EPS specification defines bus power quality characteristics such as bus voltage characteristics, impedance, ripple, transients, regulation, grounding, and power consumption requirements. EPS telemetry points and quantities for satellite monitoring and control throughout all ground and on-orbit mission phases are also specified. EPS bus power distribution, battery charge management, autonomy, mass, and reliability requirements are typically documented in the EPS specification. Key EPS sizing requirements include orbital mission profile, power (average and peak) capability, and mission lifetime [11]. Table 4.1 shows the effect of spacecraft system requirements on EPS specification and design requirements. EPS component sizing also includes a design margin to meet satellite level EOL power and reliability requirements.

4.3.2 Orbital Mission Profile

Spacecraft EPS architectures are designed to meet the Earth-orbit or planetary mission-specific operational and environmental requirements unique to the orbital profile. The operational orbit type is used to derive battery and solar array design input parameters such as inclination, altitude, solar intensity, sun angle, eclipse, and solstice durations. As discussed in Chapter 3, LIBs deployed on GEO-specific missions are sized based on many factors including orbit-specific eclipse and solstice period characteristics. For most Earth-orbiting spacecraft, the EPS LIBs are sized to provide all the required satellite system power during the longest expected Earth shadow eclipse period. For GEO and low Earth orbit (LEO) satellite systems, the effect of Moon (lunar) shadow eclipses (up to four per calendar year) on battery DOD must also be included in the sizing analysis [12]. As shown in Table 3.1, the orbit type has a significant impact on battery DOD and cycle life

Table 4.1 Effect of key spacecraft requirements on selected EPS specification requirements.

Spacecraft requirement	EPS specification requirement
Mission profile	Space thermal/radiation/orbital debris environments; battery size (calendar and cycle life, DOD, heater size, BMS); solar array size (degradation factors and sun intensity); PMAD architecture (power bus regulation, redundancy, autonomy).
Mission life	Component design life, mass, reliability; battery size (EOL power and capacity, calendar and cycle life, BMS); solar array size (EOL power margin and power generation).
Power capability	Battery size (capacity, cell design type, number of cells and batteries, cell topology); solar array size (current output, solar cell design type, number of circuit groups, area, EOL power margin and operating voltage, mass); PMAD architecture (bus voltage, regulation, fault protection, power consumption, harness design).

Source: Adapted from Ref. [11].

requirements. Different spacecraft mission orbital periods also impact battery charge rate requirements, which affects the EPS PMAD architecture. For example, when compared to satellite batteries in GEO orbit, LEO satellite batteries generally require higher charge rates due to the shorter LEO sunlight period. Orbit type also determines the radiation and thermal environments to which EPS components are exposed.

4.3.3 Power Capability

Satellite power capability is bounded by the worst-case predicted operational modes and phases of the payload module and bus unit power requirements. For most GEO and LEO spacecraft, payload module unit power requirements are much greater than the bus unit power demands. For example, the average payload power for a large commercial GEO communications satellite (COMSAT) can vary between 5 and 25 kW, while the total satellite bus housekeeping average power seldom exceeds 1–2 kW. The satellite payload peak power demand also has a significant impact on the EPS battery, solar array, and PMAD component sizing. On-orbit experience has demonstrated that spacecraft contingencies such as safe mode or other types of unexpected anomalies that cause a reduction or loss of spacecraft power are credible bus subsystem failure modes and effects. As such, in addition to EPS sizing based on predicted average and peak power demands, battery sizing analysis also includes a power margin to support on-orbit anomaly resolution.

4.3.4 Mission Lifetime

The on-orbit spacecraft mission lifetime impacts system requirements for redundancy, mass, and reliability. Operational mission lifetimes vary from a few hours for expendable launch vehicles to more than 20 years for planetary and GEO Earth-orbiting spacecraft. Single-string EPS designs are characteristic of short duration (<3 years) low-cost missions with selected added redundancy for critical EPS functions. Longer on-orbit mission lifetimes expose spacecraft EPS hardware to wear out mechanisms such as high space environmental radiation dosages, additional parts stress due to increased thermal cycling, random component failures, and increased battery charge–discharge cycling capacity loss. The added risk associated with long-term wear-out mechanisms creates more stringent EPS performance and reliability requirements. As a result, functional redundancy is commonly added to the EPS battery, solar array, and PMAD hardware and software design solutions.

4.4 EPS Architecture

Today's Earth-orbiting satellite power levels vary over a wide range of values depending on the mission-specific objectives. An EPS power demand can range from 3 to 5 W for a 1 U CubeSat to greater than 30 kW for some large commercial GEO satellite platforms. In addition to cost, selecting an EPS architecture depends on system trade studies focused on power source conversion efficiency, mass, and power electronics complexity. Since satellite mass and cost are directly proportional to power demand requirements, COMSAT manufacturers have transitioned to highly optimized modular satellite bus platform designs. As described in Chapter 3, bus platform EPS modularity is based on the ability to vary the LIB building blocks used to meet spacecraft power requirements. Table 4.2 lists various commercial satellite provider bus LIB-based EPS architectures for LEO and GEO mission applications.

Table 4.2 Representative commercial satellite LIB-based bus platform providers and types.

Bus provider	Bus family	Primary bus voltage (V)	Payload power range (kW)
Airbus Defense and Space	Eurostar 3000	100	7–25
Chinese Academy of Space Technology	DFH	100	Up to 11
Information Satellite Systems Reshetnev	EXPRESS 2000	100	Up to 15
Lockheed Martin	AS2100	70	Up to 20
Maxar	SSL 1300	100	5–25
Mitsubishi Electric Corp.	DS2000	100	Up to 15
Northrop Grumman Corp.	GEOStar	36	Up to 8
OHB System AG	Small GEO	50	Up to 10
Thales Alenia Space	Spacebus 4000	100	5–20
The Boeing Co.	702	100	8–25

4.4.1 Bus Voltage

EPS power regulation and control topology is dependent on an energy source feeding the main satellite bus electronics. Power regulation of the operating bus voltage includes controlling solar array current and battery-charging requirements to meet user load demand. First-generation Earth-orbiting spacecraft EPS primary bus voltages were commonly specified at a nominal 28-V DC due to heritage design practices inherited from US military aircraft EPS equipment requirements [13, 14]. However, growing market demands for lower mass spacecraft have driven COMSAT manufacturers to higher EPS primary bus voltage levels. Common COMSAT satellite bus voltages are 28, 50, 70, and 100 V. The highest Earth-orbiting spacecraft primary bus voltage architecture belongs to the NASA-ISS EPS PMAD system. The ISS PMAD delivers power at 160 VDC throughout the ISS through a series of computer-controlled switches. To meet operational requirements, DC-DC converter units step down and condition the voltage from 160 to 120 VDC to form a secondary bus power system to service the ISS load demands. The converters also serve to isolate the secondary bus system from the primary bus system, which aids in maintaining uniform power quality throughout the ISS.

In general, high EPS bus voltages (>200 V) are not employed in conventional space applications due to the risk of arcing or corona discharge and breakdown effects in space vacuum environments [15]. Other considerations to the use of high-voltage systems include orbital space environment, safety, and availability of space-qualified high-voltage electrical components. Nonetheless, higher bus voltage designs enable lower load current demands to bus and payload equipment operating at a given power demand. Decreases in spacecraft user load current permits the use of lower mass harness wiring, which also decreases power loss. In addition to meeting spacecraft mass constraints, EPS power bus voltage is also selected based on the mission power profile requirements, solar array capability, battery energy storage characteristics, bus ripple, and electromagnetic compatibility (EMC) requirements.

Dual-voltage (secondary) power bus architectures are also commonly used when specific load groups (critical or essential) need to be isolated from other load groups (non-critical). Heritage or unique satellite bus equipment requiring lower 28 or 50 V services combined with 100 V payload units will also require secondary power bus voltage architectures. COMSAT provider examples utilizing dual-voltage power bus architectures include The Boeing Co. 702SP 100 V/30 V and Airbus Eurostar E3000 100 V/50 V LIB-based EPS systems [16, 17].

4.4.2 Direct Energy Transfer

Satellite LIB-based EPS architectures are based on how the solar array, LIBs, and PMAD components are interconnected. As previously described, the solar array current is controlled and conditioned by the PMAD to charge and discharge the LIB during orbital solstice and eclipse periods. More specifically, the EPS architecture dictates how the LIB and solar array are configured into the PMAD portion of the EPS design. To meet the wide range of satellite operational mission requirements, Earth-orbiting and planetary mission spacecraft EPS PMAD systems are classified into two general types of primary bus power control architectures.

The first general type of primary bus power control is known as a direct-energy-transfer (DET) architecture. DET power systems are characterized by the direct transfer of solar array power to the EPS bus loads without the use of any series connected regulator or DC–DC converters. Since DET EPS architectures do not extract maximum power from the solar array, unused solar array power is dissipated through a shunt regulator that operates in parallel to the solar array. As such, when the battery does not need charging or when satellite loads do not require additional power, unused solar array power is dissipated through the shunt regulator, or segments of the solar array are shunted directly to ground to reduce power dissipation. The shunt regulator also maintains the power bus voltage at the same level as the solar array.

4.4.2.1 Unregulated Bus

An unregulated spacecraft EPS power bus is characterized by a "battery-on-bus" architecture whereby the bus voltage range is determined by the battery voltage. Notwithstanding the harness voltage drop between the battery electrical interface and spacecraft power bus electronics, the bus voltage approximates the battery voltage during charge and discharge operating modes. Hence, in a single-battery unregulated bus architecture, the bus voltage will increase or decrease with battery SOC during charge (sunlight) and discharge (eclipse) orbital periods. The voltage range for an unregulated EPS bus architecture is therefore dependent on the number of battery cells (or series strings) electrically connected in series within the battery. For battery SOC regulation, solar array power is typically limited by series switches (such as MOSFETs) within the PMAD, which switch sections of the array on or off depending on battery charge demands. A battery-on-bus EPS architecture is characterized by simpler battery charge–discharge control electronics and low bus impedance performance, but in general requires a larger solar array to provide the required load current at minimum battery voltage.

4.4.2.2 Partially-Regulated Bus

In a partially-regulated (sun-regulated) DET architecture, the bus voltage is regulated by the solar array regulator during sunlight operations and is unregulated during an eclipse. Thus, a sun-regulated bus regulates the battery charge rate during sunlight operations and then shunts excess power at the end of the battery charge. As a result, the battery is disconnected from the power bus during sunlight while the solar array regulator controls the bus voltage. In this architecture, blocking diodes are used to block any uncontrolled solar array charge current from charging the battery. Since a sun-regulated bus architecture does not have a battery discharge regulator, the battery discharges directly to the spacecraft power bus during eclipse operations. As the battery voltage and SOC decreases during discharge in eclipse, the bus voltage can vary over a wide voltage range during the service life of the battery. In this architecture, the bus voltage range may also be impacted by a cell or bank-level failure in either a s-p or p-s battery architecture. This architecture also requires a larger solar array to provide the required load current at minimum battery voltage.

4.4.2.3 Fully-Regulated Bus

DET systems may be regulated or unregulated depending on the EPS bus requirements. In a fully-regulated DET architecture the bus voltage is controlled during the entire orbital period. During sunlight operations, bus voltage regulation is achieved via the solar array regulator, which maintains the bus voltage at a fixed value while the battery charge controller charges the battery. If the solar array power exceeds the user load demand and the battery's maximum charge rate, the shunt regulator will thermally dissipate the excess power or series switches will disconnect sections of the solar array. During the eclipse, the bus voltage is regulated by the battery discharge control electronics. Battery charge–discharge regulation is necessary to maintain a regulated bus voltage since LIB cell voltage varies between 4.1 V at 100% SOC and 3.0–2.7 V at 0% SOC conditions. A major advantage of a fully regulated bus architecture is the capability to maintain a narrow bus voltage range over all satellite operating conditions. This permits user loads to operate directly from the bus while not being subject to significant bus voltage variations. Fully-regulated DET bus architectures are most common in high-power (>3 kW) GEO and LEO satellite applications [18].

4.4.3 Peak-Power Tracker

A common design feature of spacecraft EPS is the voltage regulation approach for the solar array and batteries. As previously described, DET architectures are characterized by a battery-dominated power bus, which causes the solar array voltage to be clamped to the bus/battery voltage and thus operate with a constant current. As a result, the solar array will never operate at its peak power point, but will always provide a constant current slightly below its maximum power point. As the battery voltage decreases during discharge, the solar array voltage will track the battery-dominated power bus voltage, which causes the solar array to operate even further below its peak power design point when it is most needed.

Peak-power tracker (PPT) system architectures take advantage of the maximum solar array power capability by employing circuitry that maintains the solar array voltage near the maximum power point. Since PPT architectures extract maximum power from the solar array, there is no need for shunt regulation design features, which greatly reduce power dissipation characteristics when compared to DET-type EPS architectures. When the array power exceeds the load and battery charge demand, the PPT algorithm causes the PPT power train to reduce its output current to the bus.

The advantages of a PPT bus architectures include maximizing solar array output during the entire sun period to power user loads and recharge the batteries. This design feature is most useful after an eclipse when the batteries are at their lowest SOC and the cold solar arrays can generate the most power. Low power (<500 W) small satellite (mass less than 500 kg) systems in LEO environments or deep-space planetary missions benefit greatly from PPT bus architectures due to the wide range of operating temperatures and variations in solar illumination intensity [19].

PPT bus architectures may be fully regulated by employing buck-boost (DC-DC conversion) or employ unregulated bus voltage regulation. A PPT EPS architecture is characterized by the adjustment of the solar array operating voltage to a value on the "knee" of the array's IV curve that corresponds to its maximum power point to enable the maximum transfer of power to satellite user loads. In this mode, the solar array peak power point is continuously tracked as the array temperature and solar radiation characteristic change during operation. Both DET and PPT bus architectures commonly utilize buck-boost regulation to maintain a narrow regulation of the bus voltage.

4.4.4 Direct Energy Transfer and Peak-Power Tracker Trades

Trade studies used to select a spacecraft EPS architecture are typically based on requirement criteria established by the satellite program stakeholders. EPS system trade study criteria may vary widely due to the diversity of spacecraft mission requirements. In addition to mission-specific requirements, it is common to prioritize EPS mass and cost as a primary requirement criterion for selecting a spacecraft EPS architecture. Other selection criteria include thermal control subsystem limitations, EMC requirements, and relevant on-orbit heritage. Technical risk factors associated with spacecraft-unique LIB power systems requiring new battery charge–discharge electronics, flight software capability, or use of COTS EPS components is also evaluated and included in EPS trade study recommendations. Table 4.3 summarizes the advantages and disadvantages associated with DET and PPT EPS architecture types. In general, PPT EPS architectures are optimal for spacecraft missions operating over wide orbital eclipse-sunlight conditions. Fully-regulated and

Table 4.3 Advantages and disadvantages of various spacecraft EPS architectures.

EPS architecture	Advantages	Disadvantages
Unregulated DET (battery-on-the-bus)	Solar array connected directly to the power bus; preferred for EMI-quiet bus; simplicity of power electronics results in fewer EPS components; more resilient to shadow effects on solar array; low power bus source impedance; high pulse load capability; mass savings due to no battery charge–discharge electronics. Common for single-battery power systems.	Increased thermal dissipation due to shunt regulation of excess solar array power; susceptibility to solar array-battery latch-up risk requires solar array oversizing and EPS mass; bus user load equipment must operate over wide power bus voltage range, which increases load component mass and complexity. In multiple-battery power system may require diode isolation from bus.
Partially regulated DET (sun-regulated)	Simple battery discharge electronics results in fewer EPS components; high power transfer efficiency from solar array and battery to bus user loads beneficial in GEO orbits where solstice periods are long and eclipse periods are short.	Similar disadvantages to unregulated DET; harness mass penalty due to increased current requirements at eclipse exit.
Fully regulated DET	Constant operating bus voltage range over all operating conditions; bus user loads isolated from source voltage variations; reduced solar array mass due to resilience to battery latch-up risk; battery mass savings due to reduced need for additional battery cells or modules; harness mass savings due to constant bus voltage operation.	Increased thermal dissipation (similar to unregulated DET); mass penalty due to need for battery charge–discharge electronics; increased solar array size (additional series cells) due to poor sun angle and illumination capability.
PPT	Maximizes use of solar array capability resulting in smaller solar array size and mass; solar array voltage is not clamped to battery voltage; tolerates sun angle and solar intensity variations without oversizing solar array; no risk to solar array-battery lockup; no battery charge–discharge electronics (unregulated); leaves un-needed power on the solar array resulting in low thermal dissipation characteristics.	May have greater EMI noise due to buck converter in series with solar array; more complex power electronics; less power efficiency than DET architectures at EOL, solar array isolated from power bus (spacecraft anomaly recovery more complex); solar array shadowing may cause PPT algorithm anomalies; additional power loss factor between solar array and power bus.

sunlight-regulated DET EPS architectures are well-suited for high-power Earth-orbiting satellites. Since PPT systems are significantly more complex than DET systems, the DET is typically preferred unless there are constraints on the array size that do not allow for sufficient power [20].

4.5 Battery Management Systems

As described in Chapters 2 and 3, LIBs are intolerant to overcharge and overdischarge conditions. In order to avoid overcharge and overdischarge conditions, the spacecraft EPS BMS monitors individual cell, bank, and battery-level voltages to verify that EOCV and EODV control limits are not exceeded. In addition to monitoring LIB voltage characteristics, the BMS monitors the temperature of one (or more) cells (or banks) to maintain battery thermal control. Charge–discharge management also includes a load current management function to ensure that battery SOC is maintained within safe operating limits. Commonly available to ground control personnel via telemetry, the combination of battery voltage, temperature, and load current measurements constitutes an effective EPS BMS implementation to assess overall battery state-of-health (SOH) conditions.

To ensure that EPS LIBs can meet mission-specific requirements, on-orbit satellite experience has shown that

- During battery charge, the BMS must limit battery SOC and avoid an overcharge condition and
- During battery discharge, the BMS must limit battery DOD and avoid overdischarge conditions.

For GEO satellite missions, maintaining battery voltage at a low SOC during non-operational (such as solstice) periods also aids in minimizing LIB capacity loss. Maintaining battery SOH enables meeting mission-specific battery service life requirements in a reliable and safe manner.

4.5.1 Autonomy

The spacecraft EPS provides the capability to autonomously and continuously maintain LIB SOH. Autonomous EPS functionality is commonly integrated together with spacecraft capabilities to receive ground commands for updating battery maintenance parameters throughout the mission service life. Spacecraft autonomous functionality is required since ground control may be unavailable to provide commanding, the spacecraft is out of view, or other ground-to-space communication system constraints. Most importantly, spacecraft employ autonomous functionality for subsystem fault detection and safing to protect spacecraft SOH. Incorporating fault detection and isolation design features protects the spacecraft from critical subsystem failures that threaten spacecraft SOH and safety. For example, the impact of spacecraft loss of attitude control resulting in an inadequate EPS energy balance can be mitigated by incorporating fault isolation and detection protection. EPS autonomy is typically implemented as part of the spacecraft flight software subsystem or may be designed into hardware components. Design requirements for autonomous fault response is a primary and critical function of a spacecraft EPS. Performance requirements for EPS and LIB autonomy are documented in the relevant product specifications.

4.5.2 Battery Charge Management

A primary function of the EPS BMS is to control battery charge–discharge characteristics during all mission operating modes and states. To meet these requirements, EPS architectures typically include autonomous functionality to manage LIB SOC, charge–discharge rates, balance battery cell voltages, or activate battery cell (or bank) bypass devices without ground commands. To accomplish autonomous battery charge management, cell (or bank) voltage, temperature, and load current data are included in the EPS telemetry budget. Although battery charging is typically performed autonomously, some EPS architectures incorporate variable charge rate capability via uplink commanding by ground control personnel. A variable charge rate capability is advantageous for supporting fault recovery scenarios, mission extension protocols, and spacecraft decommissioning. Battery charge rates may be dependent on LIB cell chemistry type, capacity, age, SOC, and other operational on-orbit conditions. LIB charge rates are selected based on the orbital time available to recharge the battery to 90–100% SOC. This approach minimizes solar array power and demands on the power electronics required to recharge the battery.

Historically, the temperature-compensated cell internal operating pressure was used as an indication of heritage Ni-H_2 battery cell SOC. However, Li-ion cells do not experience significant temperature increases or pressure growth during charge and discharge cycling periods. As a result, Li-ion cell voltage is commonly used to estimate LIB SOC. However, certain Li-ion space cell chemistries exhibit flat discharge (cell voltage versus discharge time) characteristics, which presents a challenge when estimating on-orbit LIB SOC. Estimating on-orbit LIB SOC (and capacity) is critical to assessing pass/fail thresholds for ground testing, battery SOH, power margins, fault protection thresholds, and many other battery and EPS characteristics.

The effect of Li-ion cell temperature variations on voltage is also a consideration when using cell voltage as a battery SOC indicator. Cell temperature variations within a battery may be induced by internal heat generation, active thermal management systems, or environmental temperature effects external to the battery. As an alternative, amp-hour integration is used to determine LIB SOC. For example, LIB cell (or bank) voltage is measured for battery charge control. LIB cell voltage, which is an indicator of SOC, provides information to battery charge electronics and/or spacecraft computers to control battery charge current and charge termination. To minimize the dynamic effects of battery cell internal resistance on battery charge management, taper charge control protocols are commonly employed.

4.5.3 Battery Cell Voltage Balancing

Heritage space Ni-Cd and Ni-H_2 battery charge management strategies included maintaining uniform battery cell voltage and temperature to meet cycle life requirements. Due to the tolerance of Ni-Cd and Ni-H_2 cell chemistry to limited overcharge conditions, balancing cell voltages and SOC was commonly achieved during each charge cycle. However, unlike space Ni-Cd and Ni-H_2 battery cell technologies, Li-ion cell chemistry is intolerant to overcharge conditions [21]. In order to operate within LIB safe SOC limits and meet capacity utilization requirements, individual LIB cells are operated at uniform voltage, SOC, and temperature. Although cell screening and matching processes are designed to mitigate the effect of cell-to-cell variations on LIB performance, cell-to-cell variations in capacity, resistance, and self-discharge characteristics that increase with cell aging are sources of cell-to-cell voltage divergence [22]. Internal LIB temperature gradients and parasitic cell loads may also contribute to cell-to-cell divergence characteristics. In addition, LIB cell voltages may become unbalanced during storage or non-operational periods, which precede battery integration on to the satellite bus structure.

Cell-to-cell balancing is the process of controlling LIB cell voltages within a specified voltage range to ensure cell voltages remain uniform under on-orbit battery operating conditions.

Individual cell voltages within an LIB with p-s cell topology are generally balanced since the cell voltages of parallel-connected cell strings are inherently equalized [23]. However, as described in Chapter 3, LIBs with s-p cell topologies require individual cell voltage monitoring and balancing. Cell voltage balancing serves to equalize the voltage of LIB cell series connected in a battery. Failure to equalize cell voltages over long duration missions may cause cell divergence, leading to degraded LIB performance, reduced cycle life, and/or increased risk to safety hazards. Cell balancing also assists in maximizing LIB capacity utilization by allowing LIBs to be cycled within a wide voltage and SOC range while minimizing the risk to overcharge or overdischarge conditions [24].

A primary function of the EPS BMS integrated with an LIB with s-p cell topology is to balance and equalize the voltages of series connected cells and/or parallel cell strings. The need to include LIB cell-to-cell voltage balancing electronics as part of the EPS BMS functionality depends on mission-specific LIB requirements such as orbit type, lifetime, cycle life, DOD, and operating temperature. Passive (dissipative) and active (non-dissipative) energy transfer cell balancing protocols are the most commonly used LIB BMS protocols employed in satellite EPS architectures [25]. Selection of an LIB cell balancing BMS protocol is dependent on mission-specific orbit type, cycle life, DOD, mass, and cost requirements. Table 4.4 lists the general characteristics of passive and active cell-to-cell balancing techniques used in Earth-orbiting and planetary mission spacecraft LIBs.

4.5.3.1 Passive Cell Balancing

Passive cell-to-cell balancing techniques shunt and dissipate excess charge current around individual cells during the battery charging process in order to achieve cell-to-cell charge balancing. In a passive cell balancing system, series-connected cells (or banks) are charged at the same constant-current rate. As individual cell (or banks) voltages increase toward the battery EOCV, a switch and a high-precision shunt resistor are used to bypass excess shunt charge current around individual cells. The cell balancing function is implemented toward the end of a charge cycle since the shunt threshold is set to a high cell voltage threshold. When the first cell reaches a certain charge voltage threshold, the associated switch will close to shunt excess charge current around the cell.

Table 4.4 Operation and characteristics of common satellite BMS cell balancing techniques.

Cell balancing technique	Operation	Characteristics	Applications
Passive (dissipative)	Individual cell voltages balanced during battery charge only; shunt resistors used to reduce the charging current for higher SOC cells; resistive cell voltage equalization.	Shunt resistors dissipate energy into waste heat; cell voltages measured and maintained within specified limits; battery capacity limited by lowest capacity cell.	All Earth-orbiting satellite orbits; planetary missions.
Active (non-dissipative)	Continuous cell SOC balancing during charge and discharge; active balancing transfers charge from cells with higher SOC to cells with lower SOC; switched equalization.	Cell SOC of individual cells managed by transferring energy from cell to cell via non-dissipative electronics; cell voltages measured and maintained within specified limits; battery capacity not limited by lowest capacity cell.	All Earth-orbiting satellite orbits; planetary missions.

The excess charge is dissipated in the high-precision resistor in series with the switch. The power in the resistor is dissipated as waste heat energy during every battery charge cycle. Individual cell current shunting continues until the last cell in the series string is equal to the cell charge voltage threshold limit. All shunt switches will open when the battery begins to discharge. The magnitude of the shunt current required is dependent on the self-discharge rates, battery capacity, and time needed for balancing. Passive cell balancing protocols are commonly used in satellite LIB BMS, where there is sufficient time in each orbital period to balance cells and completely re-charge the battery to 100% SOC.

4.5.3.2 Active Cell Balancing

Active cell balancing is a non-dissipative cell balancing technique that transfers excess energy from individual cells with a higher SOC to cells with a lower SOC until all the cells are balanced. In this manner, excess energy is not shunted as waste heat, but is transferred into other individual cells. This non-dissipative method reduces waste heat generation by causing current to flow from cells with higher SOCs to cells with lower SOCs rather than shunting current from high-voltage cells. This method uses capacitors, inductors, or active power electronic DC-DC converters to transfer energy between cells [26]. Active cell balancing is further characterized by continuously balancing cells during charge–discharge cycling.

4.5.4 EPS Telemetry

Spacecraft EPS telemetry includes, but is not limited to, battery and battery cell (or bank) voltages and temperature, battery charge and discharge current, solar array temperature, solar array output current, bus voltage, power electronics operating voltages and temperatures, and fault protection status. As discussed in Chapter 3, battery requirements for voltage and temperature monitoring are generally documented in the battery specification electrical ICD. The rationale is based on battery SOH assessments by trending battery voltage, temperature, and load current data throughout the operating and non-operating lifetime of the spacecraft EPS. Requirements for independent cell voltage and temperature sensor redundancy increases battery-level reliability and fault tolerance to anomalies. LIB specifications also document requirements for BMS cell-to-cell balancing to meet mission performance and lifetime requirements. Depending on the complexity of the system, the EPS may provide telemetry capability to receive uplinks from ground stations to issue commands for adjusting battery charge voltage limits, initiate charge balancing, or activate cell bypass relay circuitry.

4.6 Dead Bus Events

A dead bus event is an anomalous condition where a spacecraft's primary bus decreases to a voltage below the minimum operational level (including zero volts) following the loss of primary power (usually loss of solar array power due to insufficient sun illumination) and depletion of available battery capacity [27]. A battery overdischarge condition may cause the primary satellite bus voltage to decrease below the minimum operating satellite bus voltage threshold, where the EPS power controller may no longer be able to regulate bus voltage within specified tolerances. This anomalous bus voltage condition results in a dead bus condition, but if short enough in duration, may also be referred to as a bus undervoltage event. Satellite units are typically designed and validated to survive bus undervoltage events down to a required minimum voltage threshold over a required time duration or persistence.

The risk of a dead bus condition may vary depending on the satellite EPS fault protection architecture, operational phase, ground intervention capability, and other unique mission-design characteristics. The most common causes for a dead-bus condition include safe mode contingencies, component hardware single-point failures, flight software anomalies, and ground operator errors.

4.6.1 Orbital Considerations

Each spacecraft mission has unique orbital characteristics that determine the EPS and fault management system design features needed to survive and then recover from a dead bus event. For example, missions that have transfer orbit phases (such as GEO and MEO missions) may not have fully deployed the solar panels and therefore have limited power available to recharge the battery power system. Hence, a dead bus event during the time period between LV separation and full solar array deployment presents increased risk to dead bus survival and recovery. After a transfer orbit is completed and on-orbit operations have commenced, GEO mission satellites experiencing a dead bus event have a greater chance of recovery due to the long solstice (insolation) orbital periods of operation. Alternatively, LEO satellite missions utilizing LIBs whose BMS is designed to protect the battery from overdischarge may require full autonomous recovery capability, including the ability to recharge, in order to survive a dead bus event.

4.6.2 Survival Fundamentals

In principle, a safe mode anomaly followed by a dead bus event may occur during any mission phase from the beginning of LV separation through satellite disposal. Industry lessons learned from on-orbit safe mode and dead bus anomalies have been used to baseline fundamental spacecraft and subsystem requirements for surviving and recovering from a dead bus event. For high-reliability satellite missions utilizing LIB-based EPS, these requirements include but are not limited to:

- In order to survive and recover from a dead bus event, the EPS shall have the capability to autonomously restore the satellite power bus to within its operational voltage once there is sufficient solar array power available to provide essential bus power for mission critical loads necessary for satellite recovery.
- LIB cells that can suffer irreversible damage by overdischarge should be protected either by reducing the discharge current to safe levels or completely inhibiting discharge before reaching the overdischarge voltage limit. As discussed in Chapters 2 and 3, space Li-ion cell manufacturers commonly provide recommended safe EODV limits.

Mission critical loads include the command and data handling, communications, and EPS, which perform all necessary spacecraft recovery functions required to return bus power for normal operations. Chapter 9 further discusses on-orbit satellite safe mode and dead bus contingency operations.

4.7 EPS Analysis

Beginning with a spacecraft contract award through on-orbit end-of-mission (EOM) operations, EPS analyses supports each phase of a spacecraft lifecycle. During spacecraft development, various EPS design analyses are conducted to support technical trade studies in order to size the solar array, LIB, and PMAD components. EPS component sizing is focused on meeting EPS specification requirements with a special focus on mass and cost savings opportunities. Key outcomes for EPS-level design trade studies include EPS voltage levels, EPS architecture (DET or PPT), solar array cell type, and LIB capacity. As discussed in Chapter 3, LIB sizing analyses are interrelated to EPS architecture type, bus voltage levels, fault protection design, and many other related EPS requirements. Each set of EPS analyses is used to verify specific spacecraft and EPS-level requirements under various nominal and off-nominal mission operating conditions. Typical EPS analyses include component electronic parts stress analysis, voltage drop analysis (VDA), failure modes, effects, and criticality analysis (FMECA), as well as worst-case circuit analysis (WCCA). However, it is an industry best practice to also conduct the following unique spacecraft EPS-level analyses:

- Energy balance analysis (EBA) – Dynamic time-domain analyses of energy balance between energy generated (or available) and energy consumed during a defined orbital period for a given load profile, and
- Bus stability and transient analysis – Analysis of EPS power bus voltage stability to load switching, operational mode changes, and fault modes.

4.7.1 Energy Balance

Electrical power is a critical resource for all spacecraft bus and payload subsystem components. As such, a detailed orbital EBA is performed to demonstrate that the EPS design is capable of providing sufficient solar array power margin to all user loads and for recharging the batteries over the entire mission lifetime of the spacecraft. The EBA is based on EPS specification power requirements allocated to the bus and payload components as a function of relevant orbital periods under beginning-of-life (BOL) and EOL mission operating conditions. In general, the EPS is required to provide sufficient solar array power to maintain a positive energy balance at a specified battery minimum SOC under worst-case operating conditions.

Worst-case spacecraft operating conditions may include hot-cold thermal stressing loads, min-max power loads, sun angles, and longest-day eclipse season cases. Types of spacecraft or EPS off-nominal contingencies that create worst-case operating conditions include safe mode, battery cell failures, loss of a solar array circuit, or a PMAD fault condition. Since GEO satellites have an orbital period of 24 hours, the EBA is established over the course of one orbital day. The EBA for LEO satellites may be established over the course of an integral number of orbits in a 24-hour period. In some LEO applications, the goal may be to maintain a positive energy balance from orbit-to-orbit, but it may be acceptable to have several orbits in a 24-hour period where a positive energy balance is not achieved. In summary, for both GEO and LEO orbits, an energy balance is achieved when the battery SOC is returned to 100%. Figure 4.6 illustrates how the battery SOC changes during a single charge–discharge cycle for a given orbital period. An energy balance is achieved when the solar array provides sufficient power to meet user load demand and re-charges the battery during each orbital period throughout the satellite mission life. As such, a key EBA output is battery SOC as a function of satellite operational modes, phase, and mission life.

4.7.2 Power Budget

In order to maintain spacecraft SOH, a positive energy balance must be achieved during nominal and contingency mission operations. A spacecraft negative energy balance occurs if the battery SOC cannot be returned to 100% prior to the start of the next eclipse period or whatever interval is required. If allowed to continue for a given period of on-orbit time, a spacecraft negative energy

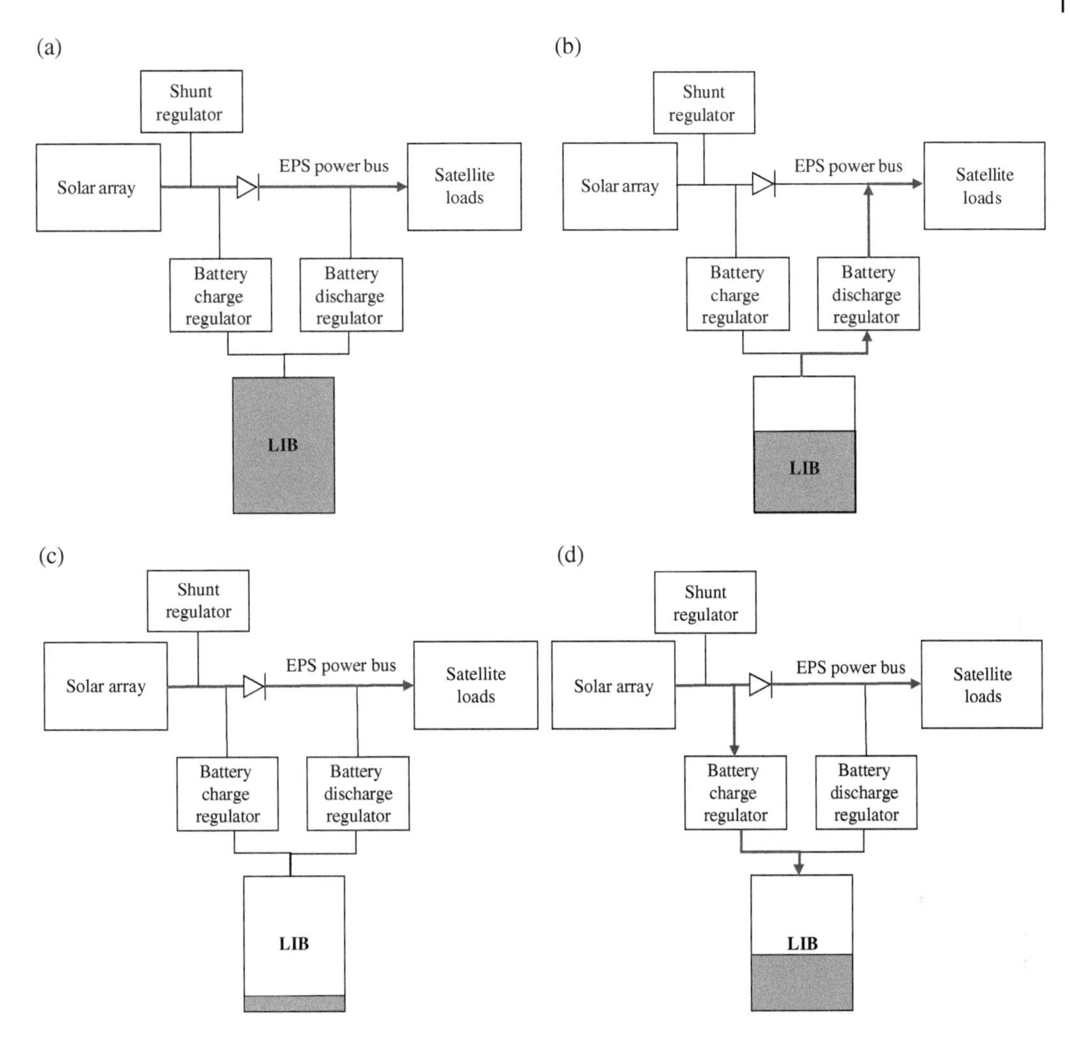

Figure 4.6 Simplified energy balance and power flow for a fully-regulated DET satellite EPS architecture. (a) End of the sunlight period and beginning of the eclipse period. The battery is fully charged and the solar array provides power to all satellite loads. (b) During the eclipse period. The battery is discharging and the SOC is decreasing and the battery provides power to all satellite loads. (c) End of the eclipse period and beginning of the sunlight period. The battery is discharged to a minimum SOC (maximum DOD); the solar array begins to provide power to all satellite loads. (d) During the sunlight period and before the eclipse period. The battery is charging and the SOC is increasing; the solar array provides power to all satellite loads.

balance will deplete the EPS battery capacity, which may result in a dead bus condition [28]. This implies that battery SOC (or DOD), sun intensity on the solar arrays, and EPS DC power bus user loads must be known in any given spacecraft operating state or mode. For all spacecraft, the EPS power budget is a detailed accounting of spacecraft power consumption and generation characteristics during design reference cases (DRCs) such as ascent, transfer orbit (GEO and MEO only), and on-station operating modes. The ascent EPS power budget predicts the battery SOC during the LV ascent as a function of the LV sun angle condition and satellite EPS load profile. The transfer orbit power budget analysis includes estimates for any satellite spin-rate, sun angle changes, and heater load variations to predict battery SOC. Satellite on-station power budgets are calculated for BOL and

EOL operations under worst-case Spring equinox and Fall equinox (eclipse) season (GEO only) cases. For the purposes of power budget analyses, EOL is defined as the contracted orbital lifetime of the satellite system. Due to the dynamic nature of the on-orbit satellite power demand, changes to the on-orbit concept of operations (CONOPS), battery aging effects, and solar array degradation, the EPS EBA is actively managed throughout the on-orbit life of the satellite.

4.7.2.1 Inputs

The majority of power budget analysis inputs are derived from satellite component qualification test data, specification limits, component ratings, and on-orbit data from similar components. BOL battery input data are obtained from the flight battery acceptance and qualification test results obtained from the flight battery manufacturer. EOL battery performance characteristics are derived from capacity loss measurements obtained from relevant ground life cycle test data as well as on-orbit experience, if available. Solar array BOL performance is derived from the solar array manufacturer data on the as-built solar panel characteristics and relevant mission-specific parameters. EOL solar array characteristic predictions are based on specification solar cell parameters degraded for mission-specific parameters such as mission duration, radiation environments, thermal loads, and longitude. In addition, the EPS power budget analysis inputs also include:

- Satellite DC power consumption – Estimate of the average steady-state DC power consumption of all satellite payload and bus user ascent, transfer orbit, and on-station loads. Includes power required from solar arrays to re-charge the batteries.
- Satellite heater power consumption – Computed time average power consumption of the bus and payload heaters.
- Power distribution losses – Commonly based on a comprehensive spacecraft-level VDA, power dissipation is estimated from losses due to power flow through the distribution harness subsystem, DC-to-DC conversion electronics, and switching losses.
- Battery charge–discharge characteristics – Estimate of battery DOD based on spacecraft power demand during eclipse periods, average battery discharge current, and average battery charge–discharge voltage. Includes an estimate of power dissipation in battery charge–discharge units and battery harness.
- Solar array power generation – Estimate of required solar array power to maintain a positive spacecraft power energy balance. Includes power loss assumptions for solar array circuit failures, solar array harness distribution loss, degradation losses due to radiation, and shadowing.

Power budget inputs are based on a conservative or worst-case approach to estimating the spacecraft-level power margins at EOL. Simulating a worst-case battery condition may include maximum DOD (minimum SOC) operation, loss of a battery cell (or cell string), and heater power load variations. Similarly, worst-case solar array operating conditions may include loss of a solar array circuit(s), elevated radiation dosages, or high thermal loads. Typically, a GEO COMSAT EPS is required to provide sufficient solar array power to maintain a positive energy balance with a battery maximum DOD of 70–80% under worst-case operating conditions. Conventional high-reliability LEO spacecraft EPS are commonly required to maintain a positive energy balance with a battery maximum DOD ranging from 20–40% under worst-case operating conditions.

4.7.2.2 Outputs

At the program level, power margins are continuously balanced against minimizing EPS mass, cost, and analysis uncertainties. As such, power budget analyses are intended to demonstrate that the EPS will provide reliable power to support all necessary bus and payload user loads under worst-case operating conditions. To verify these requirements, the power budget analysis is formulated to provide an estimate of the satellite load and source margins. The power budget should be capable of estimating the spacecraft source power (lowest possible solar array output power) and load power (highest possible load power) at EOL. The load margin is the difference between the load capability and required load for a fixed battery and solar array size. A positive load margin indicates that required user loads can be increased while maintaining a positive energy balance.

The source margin is the difference between the predicted solar array output and the required solar array output. Both the load and source margins are commonly used to size the battery, solar array, and other spacecraft power components. Power budget results are used to support on-orbit satellite CONOPS, mission planning, and contingency analyses. The outcome of the power budget analysis is a predicted minimum spacecraft-level power budget margin (%) at mission EOL. In general, prior to launch, GEO COMSATs are required to demonstrate a 5–10% energy margin at EOL for all EPS operating states and modes.

A worst-case power budget analysis output also includes an assessment of sufficient battery and solar array sizing to permit an adequate on-orbit time for fault recovery and a return to normal on-orbit operations. As further discussed in Chapter 9, a safe mode contingency is a DRC for any Earth-orbiting or planetary mission spacecraft where the EPS must also meet certain energy balance requirements. A spacecraft safe (or survival) mode is an on-orbit contingency condition resulting from an on-board processor fault detection response. A common cause of a satellite safe mode condition is an anomalous loss of sun on the solar arrays leading to loss of the battery recharge capability. Depending on available power margins, loss of sun on the solar arrays may result in a negative spacecraft energy balance. Under these contingency operating conditions, battery energy must be conserved to maintain power to critical satellite user loads. Surviving from a safe mode contingency commonly includes shedding non-essential (non-critical) user loads to reduce battery discharge rates in order to avoid battery overdischarge conditions. Load shedding may include discontinuing payload operations, reconfiguring satellite loads for minimal power usage, or other actions necessary to maintain a positive spacecraft energy balance.

4.8 EPS Testing

As described in Chapter 3, flight LIB acceptance, proto-qualification, and qualification testing is the final process step in flight LIB readiness for spacecraft EPS integration activities. Satellite EPS LIB ground processing includes all AI&T operations performed on the ground at the satellite manufacturing, system test, and launch site facilities. Ground processing of LIBs during satellite AI&T also requires a multi-disciplinary understanding of all phases of LIB-unique storage, handling, and transportation procedures. As further discussed in Chapter 9, adherence to approved ground handling and transportation protocols ensures that ground personnel, flight hardware, and processing facilities are properly protected and maintained. Finally, satellite-level AI&T activities are intended to demonstrate that the satellite system will meet mission-specific ground, launch, and mission operational service life requirements. This includes verification of EPS functions at the satellite bus level of integration while exposed to various operational and environmental test conditions.

4.8.1 Assembly, Integration, and Test

The purpose of manufacturing and AI&T activities is to assemble the satellite and test the integrated system to verify that the satellite will operate as intended while exposed to simulated mission-specific space environments. Satellite AI&T represents a critical phase of the satellite life-cycle while conducting end-to-end compatibility testing and demonstrating verification of flight components and subsystem functions. AI&T begins with delivery of flight-qualified hardware and software configured items to the satellite manufacturing facility. Delivered flight hardware is then assembled into their final spacecraft configuration while adhering to validated processes and procedures unique to the satellite manufacturer. Certified and compatible mechanical and electrical ground support equipment (EGSE) (such as power supplies, battery simulators, and solar array simulators) are also integral in supporting the AI&T workflow.

Satellite component and subsystem AI&T is also characterized by the execution of comprehensive test plans in order to verify compliance to subsystem and system-level requirements. During AI&T planning, established test-like-you-fly (TLYF) principles are adopted to ensure that ground test conditions simulate on-orbit conditions in a flight-like manner. TLYF testing includes exposure to expected flight-like environmental test conditions, as well as simulated on-orbit operational sequences [29]. For example, ground test sequences should demonstrate compliance to launch conditions before exposure to on-orbit conditions. The final phase of satellite AI&T operations includes conducting a countdown launch procedure, integrating the stowed satellite into the shipping container and then transporting the integrated shipping container to the launch site. As further discussed in Chapter 9, ground processing personnel must also comply with stringent LIB safe storage, handling, and transportation requirements during all phases of AI&T operations.

4.8.2 Bus Integration

Bus-level AI&T activities include assembling all subsystem components into the bus mechanical structure. This includes mating and validation of satellite subsystem electrical interfaces with flight harnesses. Electrical sizing and mechanical routing of harnesses is baselined prior to AI&T and verified during subsystem-level hardware integration activities. Typically, EPS components, such as the test battery and PMAD electronic EPS components, are integrated early in the bus AI&T workflow to facilitate functional EPS test objectives. Since solar array integration commonly occurs after battery and PMAD integration, solar array EGSE simulators are commonly used to simulate solar array current generation inputs. Flight battery integration into the satellite bus also begins ground-level battery processing operations requiring compliance to a unique set of transportation, handling, and storage requirements.

Batteries employed for systems AI&T are used to verify the electrical, mechanical, thermal, and software interfaces of the satellite EPS and other subsystems. The classification of these batteries may be either flight or non-flight (system test) depending on program requirements. In general, system test batteries meet the form, fit, and function of the mission-specific flight battery hardware. Use of system test batteries is a risk-reduction opportunity commonly used to minimize flight battery exposure to satellite-level AI&T ground activities prior to launch. In order to meet satellite-level AI&T requirements, system test batteries are certified to support flight spacecraft environmental and non-environmental ground testing. Depending on satellite AI&T workflow, flight-like test batteries may be used during the initial phase of AI&T and later replaced by flight batteries for final AI&T activities. As such, flight batteries replace system test batteries in a manner that permits flight battery functional verification prior to spacecraft delivery to the launch site.

Whether the LIB is integrated into the satellite bus at the manufacturing facility or the launch site, integration is performed at low battery SOCs to avoid possible safety hazards associated with fully charged LIBs. To mitigate inadvertent hot-mate or electrical shorting hazards, the LIB and/or

PMAD components are designed with integrated ground safety relays that function to electrically disconnect the batteries during ground handling or transportation activities. Furthermore, easily removable non-conductive covers are typically used to protect LIB power, signal, or test connections at the flight harness interfaces. To meet applicable occupational safety and health requirements, dedicated battery tooling is commonly used by AI&T personnel to safely handle, lift, and install flight LIBs into the satellite bus structure.

4.8.3 Functional Test

In general, bus-level AI&T activities signify the first time that bus subsystems, including all EPS flight components, interface with the satellite bus structure. This integration event presents the opportunity to conduct end-to-end verification testing of all satellite bus subsystems. As such, the integrated EPS is functionally tested to verify EPS-to-satellite bus electrical, thermal, and mechanical interface requirements. In addition to bus-level EPS requirements verification, a significant portion of EPS functional testing is conducted to characterize expected satellite operational modes. Specific EPS functions, such as energizing PMAD relays interfacing the LIB, LIB charge–discharge characteristics, and other operational states, are sequentially tested and data are acquired for subsequent performance analyses. Verification of flight software command and telemetry acquisition interfaces are also a key component of EPS functional test activities.

After EPS and other bus subsystem functional testing are completed, mission payload module units are integrated on to the satellite bus structure to verify EPS-to-payload functional requirements. At this next level of integration, the EPS solar arrays and other deployable subsystems are typically installed on to the satellite structure. Figure 4.7 shows deployment testing of an Earth-orbiting satellite solar array panel under ambient conditions during the AI&T workflow. The solar panel is suspended from above to simulate zero-gravity space environments. After bus-level AI&T completion and payload integration, satellite system-level integrated testing is performed using a TLYF test sequence approach. In this systems configuration, EPS functional tests are conducted before, during, and after fully integrated satellite system environmental test events in order to demonstrate

Figure 4.7 The first Meteosat Third Generation-Imaging (MTG-I) satellite undergoing solar array deployment testing in AI&T. Credit: Thales Alenia Space/Imag[IN].

compliance after exposure. Systems-level thermal vacuum testing in a flight-like configuration is a common environmental test event where comprehensive EPS functional testing is conducted.

4.9 Summary

The wide range of spacecraft power demand, reliability, and mission lifetime requirements creates a broad trade space for EPS design solutions. Nonetheless, all spacecraft EPS architectures employ the capability to generate power and store energy throughout the lifetime of the specified mission. Generated power and stored energy are managed, controlled, and distributed by the EPS PMAD components to meet spacecraft user load requirements. EPS mass and cost remain significant design requirements in order for spacecraft manufacturers to meet evolving market affordability targets. Satellite mass saving opportunities have been achieved by adopting modular EPS components scalable to varying spacecraft power requirements. Lower non-recurring cost targets have been achieved via modular fixed solar array panel designs and LIBs with common cells, banks, or modules. Power regulation units that combine multiple functions into a single unit may decrease parts count, minimize harness interfaces and mass, and reduce manufacturing and test costs over the lifecycle of the satellite EPS. Emerging requirements for higher-power and lower-mass Earth-orbiting satellites and planetary mission spacecraft require new EPS architectures employing next-generation solar and Li-ion cell technologies. New operational modes for LIBs for longer eclipse periods, deeper DODs, and higher pulse power will also challenge existing state-of-the-art technologies. Finally, as customer demands for higher reliability and more resilient spacecraft architectures evolve, EPS requirements for added autonomous functionality and dead bus recovery capability will be baselined.

References

1 Everett, D.F. (2011). Overview of spacecraft design. In: *Space Mission Engineering: The New SMAD*, Chapter 14 (ed. J. Wertz, D. Everett and J. Puschell), 397–438. Microcosm Press.

2 Furin, S.C. (2021). Orion power transfer: Impacts of a battery-on-bus Power System Architecture. *AIAA SCITECH 2022 Forum*, AIAA 2022-0313 (29 December 2021). https://doi.org/10.2514/6.2022-0313.

3 Chamberlain, M.K., Kiefer, S.H., LaPointe, M., and LaCorte, P. (2021). On-orbit flight testing of the roll-out solar array. *Acta Astronaut.* 179: 407–414. https://doi.org/10.1016/j.actaastro.2020.10.024.

4 Barrera, T.P. and Wasz, M. (2018). Spacecraft Li-ion battery power system state-of-practice: A critical review. In: *2018 International Energy Conversion Engineering Conference, AIAA Propulsion and Energy Forum* (AIAA 2018-4495). https://doi.org/10.2514/6.2018-4495.

5 Cappelletti, C. and Robson, D. (2021). CubeSat missions and applications. In: *CubeSat Handbook: From Mission Design to Operations*, Chapter 2 (ed. C. Cappelletti, S. Battistini and B.K. Malphrus), 53–65. Academic Press.

6 NASA Technical Standard (2016). *Workmanship Standard for Crimping, Interconnecting Cables, Harness, and Wiring*, NASA-STD 8739.4A (30 June 2016).

7 Society of Automotive Engineers (2019). *Wiring Aerospace Vehicles*. SAE AS50881G (6 August 2019).

8 United States Space Force (2009). *Technical Requirements for Wiring Harness, Space Vehicle*. SMC-S-020 (3 June 2009).

9 American Institute of Aeronautics and Astronautics Standard (2007). *Electrical Power Systems for Unmanned Spacecraft*, S-122-2007 (5 January 2007).

10 European Cooperation for Space Standardization Standard (2022). *Space Engineering: Electrical and Electronic*, ECCS-E-ST-20C (8 April 2022).

11 McDermott, J.K., Schneider, J.P., and Enger, S.W. (2011). *Space Mission Engineering: The New SMAD*, Chapter 21.2. In: *Power* (ed. J. Wertz, D. Everett and J. Puschell). Microcosm Press.

12 Patel, M.K. (2005). *Spacecraft Power Systems*. CRC Press.

13 Department of Defense Interface Standard. *Aircraft Electrical Power Characteristics*, MIL-STD 704F.

14 Department of the Air Force. *Electrical Power, Direct Current, Space Vehicle Design Requirements*, MIL-STD-1539.

15 White, R.A. (1978). *High Voltage Design Criteria*. MSFC-STD-531 (September 1978).

16 Arastu, A. (2013). Boeing 702SP evolution. *Space Power Workshop*. The Aerospace Corporation, Manhattan Beach, CA (23 April 2013).

17 Poussin, J. and Berger, G. (2013). Eurostar E3000 three-year flight experience and perspective. In: *25th AIAA International Communications Satellite Systems Conference*, AIAA 2007-3124. https://doi.org/10.2514/6.2007-3124.

18 Chetty, P.R.K. (1990). Electrical power system for low-earth-orbit spacecraft applications. *J. Propul.* 6 (1): 63–68.

19 Bouwmeester, J. and Guo, J. (2010). Survey of worldwide pico- and nanosatellite missions, distributions and subsystem technology. *Acta Astronaut.* 67: 854–862. https://doi.org/10.1016/j.actaastro.2010.06.004.

20 Freeman, W. (1992). Peak-power tracker versus direct energy transfer electrical power systems. In: *27th Intersociety Energy Conversion Engineering Conference*, SAE Technical Paper 929456. https://doi.org/10.4271/929456.

21 Dudley, G.J., Olsson, D., and Brochard, P. (2002). Maintaining cell state of charge balance in lithium ion batteries, ESA SP-502. In: *Proceedings of the Sixth European Conference*, 501, Porto, Portugal (6–10 May 2022).

22 Loche, D. and Barde, H. (2002). Lithium-ion battery cell balancing in LEO, ESA SP-502. In: *Proceedings of the Sixth European Conference*, 507, Porto, Portugal (6–10 May 2002).

23 Pearson, C., Thwaite, C., Curzon, D., and Rao, G. (2004). The long-term performance of small-cell batteries without cell balancing electronics. In: *NASA Aerospace Battery Workshop*, Huntsville, AL (16–18 November 2004).

24 Altemose, G., Hellermann, P., and Mazz, T. (2011). Active cell balancing system using an isolated share bus for Li-Ion battery management: Focusing on satellite applications. In: *2011 IEEE Long Island Systems, Applications and Technology Conference*, 1–7. http://dx.doi.org/10.1109/LISAT.2011.5784237.

25 Hemavathi, S. (2021). Overview of cell balancing methods for Li-ion battery technology. *Energy Storage* 3 (2). https://doi.org/10.1002/est2.203.

26 Canter, S., Choy, W., and Martinelli, R. (2003). Autonomous battery cell balancing system with integrated voltage monitoring. US Patent 6,873,134 B2. Filed 21 July 2003 and Issued March 29, 2005.

27 Landis, D.H. (2017). *Dead Bus Recovery Requirements for Earth Orbiting Spacecraft*. The Aerospace Corporation, TOR-2018-00316, 31 January 2017.

28 Lenertz, B.A. (2005). *Electrical Power Systems, Direct Current, Space Vehicle Design Requirements*. The Aerospace Corporation, TOR-2005(8583)-2 (11 May 2005).

29 Knight, F.L. (2009). *Space Vehicle Checklist for Assuring Adherence to "Test-Like-You-Fly" Principles*. The Aerospace Corporation, TOR-2009(8591)-15 (30 June 2009).

5

Earth-Orbiting Satellite Batteries

Penni J. Dalton, Eloi Klein, David Curzon, Samuel P. Russell, Keith Chin, David J. Reuter, and Thomas P. Barrera

5.1 Introduction

Beginning in the 1990s, the space industry recognized that significant battery mass and volume savings could be realized with the threefold increase in Li-ion cell specific energy when compared to heritage spacecraft nickel-hydrogen ($Ni\text{-}H_2$) battery cell technology. The transition from space $Ni\text{-}H_2$ to Li-ion cell technology has enabled significant lithium-ion battery (LIB) electrical power subsystem (EPS) mass savings, which has translated into increased satellite payload power capability, higher EPS reliability, and longer mission lifetimes. These system-level improvements have transitioned into favorable business case outcomes for the satellite supply chain, battery manufacturers, and user customers. Additional advantages of LIBs such as higher operating voltages, decreased thermal subsystem demands, and absence of on-orbit reconditioning requirements have provided spacecraft manufacturers with added mission design flexibility. Due to these increased capabilities, LIB-based EPS have proven to be a disruptive energy storage technology for spacecraft platforms operating in Earth-orbiting environments. Over the past 5 years, LIB technology has enabled the rapid growth of the communications broadband services industry via launches of Cubesat-sized internet satellite constellations. As of early 2022, there are over 4852 operational satellites orbiting the Earth, with more than 90% owned by the USA, China, and the UK combined [1].

The purpose of this chapter is to provide a general overview of LIB-based EPS characteristics in selected on-orbit Earth-orbiting spacecraft, with a special focus on LIB design and operation. Representative LIB power system examples ranging from the smallest pico-CubeSats to the largest NASA International Space Station (ISS) LIBs are described. Spacecraft missions spanning Earth-orbiting communications, space exploration, and scientific mission applications demonstrate the wide range of performance capabilities of space-qualified LIBs.

Spacecraft Lithium-Ion Battery Power Systems, First Edition. Edited by Thomas P. Barrera.

5.2 Earth Orbit Battery Requirements

An orbit is defined as the path of an object that is moving around a second object or point under the influence of gravity [2]. For the purpose of this chapter, the orbiting object is a satellite or spacecraft moving around the Earth as the second object. As discussed in Chapter 4, orbiting spacecraft utilize EPS architectures that use both solar array power generation and rechargeable batteries for energy storage. The solar arrays provide power to the spacecraft bus and payload subsystems and simultaneously recharge the batteries during insolation (sunlight) periods. During eclipse orbital periods, rechargeable batteries provide power to all spacecraft subsystem loads.

First-generation Earth-orbiting spacecraft used rechargeable nickel-cadmium (Ni-Cd) battery EPSs. Ni-Cd batteries were later replaced by higher specific energy density Ni-H_2 battery technologies. As battery technologies have continued to evolve and improve, LIBs have become the technology of choice because of their improved specific energy, energy density, and thermal characteristics, as compared to heritage space rechargeable battery power systems. The orbital inclination of a spacecraft depends on its planned usage (Figure 5.1). The main orbital paths around Earth are the low Earth orbit (LEO), geosynchronous orbit (GEO), medium Earth orbit (MEO), and high Earth or highly elliptical orbits (HEOs). As further described in Chapter 3, other orbits beyond HEO include the Lagrange points and Lunar orbits. Battery design features vary depending on key spacecraft requirements such as orbit type, mission lifetime, and environments. The majority of battery performance requirements are derived from the spacecraft mission-specific requirements such as the bus operating voltage, battery depth-of-discharge (DOD), and EPS load duty cycle. The orbit type also dictates the spacecraft environmental requirements (such as temperature variations and radiation dosage), which the batteries must survive during the entire mission lifetime. Battery environmental design and test requirements are also derived from the spacecraft launch vehicle type (random vibration and shock) as well as ground processing and operations (calendar life and storage characteristics) requirements. Finally, Earth-orbiting spacecraft may be either human-rated (crewed) or non-human-rated (unmanned). Crewed spacecraft batteries must comply with more stringent safety and reliability requirements to avoid impacting

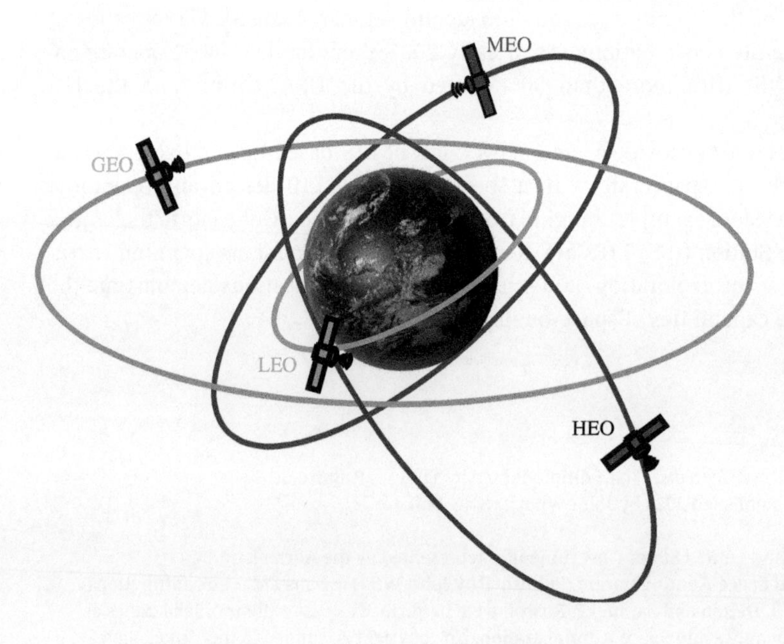

Figure 5.1 Earth-orbiting satellite representation. Credit: Julianna Calin.

the safety of spacecraft crew members, ground processing personnel, and spacecraft systems hardware. Batteries used in unmanned Earth-orbiting satellites must also meet Range Safety requirements due to the extensive ground handling, transportation, and storage activities that precede spacecraft launch.

5.3 NASA International Space Station - LEO

5.3.1 Introduction

The NASA ISS is a joint international collaboration, with US, Russia, Japan, Canada, and Europe providing hardware, personnel, and logistics and operational support. The ISS is the largest man-made spacecraft currently orbiting the Earth (Figure 5.2). The ISS provides a long-duration habitable laboratory and test bed for research and development activities. The ISS mission is to advance science and technology research and foster the commercial development of space and demonstrate capabilities to enable future exploration missions beyond LEO [3]. The ISS has been in continuous LEO since November 1999, with continuous crewed habitation since November 2000. It has the largest and most complex EPS ever flown in space. Due to the size of the EPS, it launched as four separate photovoltaic (PV) modules, each with an integrated equipment assembly (IEA) that contained batteries and power management and distribution boxes. The complete EPS was assembled in orbit over a period of nine years from first launch to assembly complete (AC). The first launch of the EPS, the Port 6 truss (P6), was in late November 2000. The fourth PV module, Starboard 6 truss (S6), was launched in March 2009 for AC.

5.3.2 Electrical Power System

The ISS has eight power channels that supply power to housekeeping and user electrical loads. Each channel consists of one solar array wing (SAW) with two panels, three batteries electrically

Figure 5.2 NASA International Space Station. Credit: NASA.

connected in parallel, and a direct current power management and distribution system. At AC, the ISS has 24 batteries. The Russian Functional Cargo Block segment provides its own 28 V power, using solar arrays and Ni-Cd batteries, but can also draw power from the main 160 V ISS bus through the Russian to American Convertor Units.

5.3.3 Ni-H_2 Battery Heritage

The ISS batteries launched with each PV module were Ni-H_2, selected for their long cycle life and high reliability. Since the ISS is a human-rated mission, the batteries needed to meet NASA's stringent two-fault tolerant design requirements for safety. Two-fault tolerance requires that no two failures can cause loss of mission or life. The Ni-H_2 batteries were designed for either robotic or crew installation. Each battery consisted of two orbital replacement units (ORUs), connected in series. Each battery ORU contained 38 series-connected 81 Ah individual pressure vessel (IPV) Ni-H_2 cells, for a total of 76 cells in series for a battery. Each ORU had approximately 4.4 kWh of energy. The ISS power system was the first on-orbit use of such a large quantity of series-connected IPV Ni-H_2 battery cells. At AC, there were a total of 24 batteries/48 Ni-H_2 battery ORUs in operation.

Each battery ORU was individually fused for fault propagation protection of the EPS in the event of a cell short. The 120A fuse block consisted of two parallel fuse strings, with one 60A string on each power cable. The fuses were constructed using high-voltage, high-reliability space-rated components.

The Ni-H_2 battery ORU contained a battery signal conditioning and control module (BSCCM). The BSCCM provided conditioned battery monitoring signals from the ORU to the local data interface (LDI) located within another ORU called the battery charge–discharge unit (BCDU). Available data included 38 cell voltages, four strain gauge (pressure) readings, and six cell and three baseplate temperatures. These data were provided as an analog multiplexed voltage. A separate signal provided ORU total voltage output. The BSCCM also accepted and executed commands from the BCDU/LDI to control the ORU cell heater and resistor letdown functions.

The batteries interfaced to the source bus through a BCDU and provided power for the ISS during its eclipse periods. For battery charging, the BCDU conditioned power from the 160 V source bus and charged the battery at pre-determined currents that are calculated based on the state of charge (SOC). The charging algorithm was based on a temperature and pressure SOC calculation, with current stepping down as batteries neared 100% SOC. During periods of eclipse, the BCDU extracted power from the battery, conditioned this power, and supplied power to the source bus.

The batteries were actively cooled using the PV thermal control system (TCS). The battery cells were assembled in an ORU box, using a unique finned radiant heat exchanger baseplate. The original battery ORU was then mounted on the IEA using mechanical fastening screws and mated to the PV TCS. The PV TCS was designed to maintain the Ni-H_2 Battery ORUs at a nominal operating temperature range of $5 \pm 5\,°C$ with minimum heater operation when run at a 35% DOD LEO regime.

The original Ni-H_2 battery ORUs life specification was for 6.5 years, cycling at 35% DOD. Since operational DOD was lower than specified (15–20% DOD), the ORUs were able to operate beyond their design life of 6.5 years. The original batteries did require reconditioning twice throughout their service in order to improve battery performance by maintaining cell balancing. Reconditioning consisted of a full discharge to 0% SOC, followed by a charge to approximately 104% SOC, and a discharge. New charging parameters were uploaded within a week of reconditioning, with a capacity test shortly after the parameter upload. The first battery replacements in 2009 and 2010 were with spare Ni-H_2 ORUs. However, in 2011, the ISS implemented a LIB project to replace the aging and obsolescent Ni-H_2 ORUs.

5.3.4 Transition to Lithium-Ion Battery Power Systems

The new LIB ORUs were designed to be direct drop-in replacements for the existing Ni-H_2 batteries. As such, the LIB interfaces were designed to use the existing ISS EPS infrastructure such as the heritage PV TCS and BCDUs. Two series connected Ni-H_2 ORUs were replaced with one LIB, with nearly twice the energy (15 kWh). The new LIB consists of 30 series-connected GS Yuasa LSE134 Li-ion cells (134 Ah) housed in one ORU enclosure with the same footprint as the Ni-H_2 ORU (Figure 5.3). Each LIB is approximately 0.9 (length) × 1.0 (width) × 0.5 m^3 (height, including the micro-square assembly) and has a mass of 195 kg. A second ORU, the adapter plate (AP), was built to replace the second Ni-H_2 ORU, maintain temperature during the transit to orbit, and complete the circuit (Figure 5.4). An adapter cable connects the AP to the LIB battery. The same mechanical mounting screws as those used in the Ni-H_2 attach the LIB to the IEA. One additional advantage

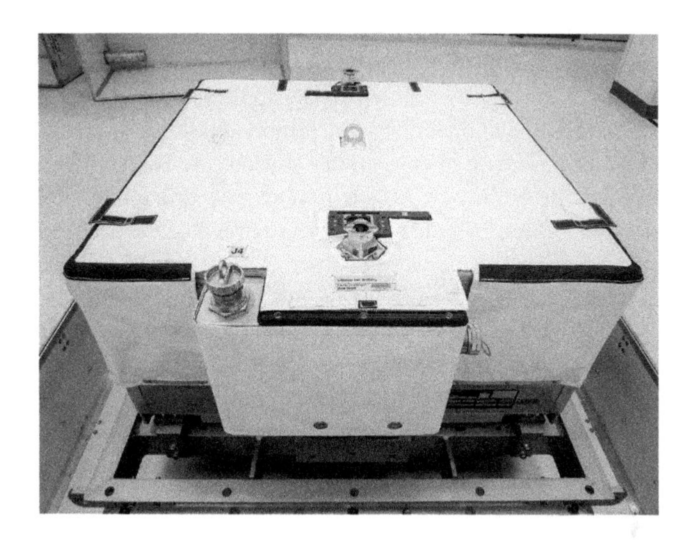

Figure 5.3 ISS 30s1p LIB ORU manufactured by Aerojet Rocketdyne. Credit: NASA.

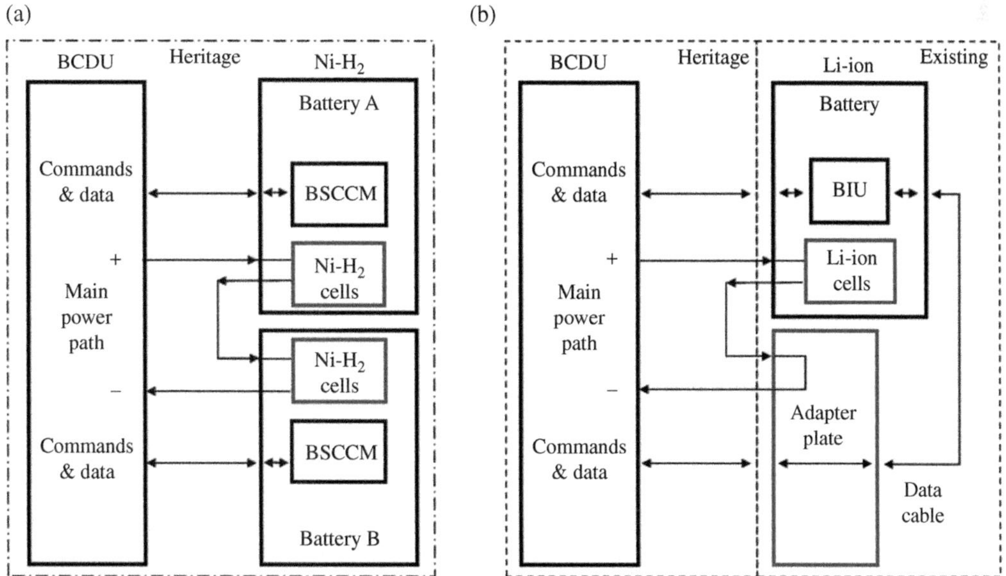

Figure 5.4 ISS EPS Battery block diagram: (a) Ni-H_2 76 series 81 Ah cells; (b) Li-ion 30 series 134 Ah cells.

with the LIB is that battery reconditioning is no longer necessary, thus preserving more battery operation time and saving many hours of ground controller planning and operational execution.

Inside the LIB, a battery interface unit (BIU) is responsible for the multiplexing of the battery data to the LDI in the BCDU. Individual cell voltage and temperature are monitored, along with battery baseplate temperatures. The same fuse used in the Ni-H_2 ORU is also in the LIB. The SOC calculation for Ni-H_2 charging was replaced with a look-up table, based on the cell voltage. Cells are charged at 50 A to an end-of-charge voltage (EOCV) of 3.95 V plus up to 19 mV. The maximum allowed mV offset value at each charge set-point in the look-up table was determined by ground testing for most efficient cell charging. The current is stepped down to 1 A as cells reach the maximum EOCV. At the lowest charge current, cells that reach EOCV prior to the end of insolation are bypassed to balance the cells. The EOCV can be changed via command by mission operation ground controllers.

Since Li-ion cells do not need to operate at the low temperatures of $5 \pm 5\,°C$ that the Ni-H_2 ORUs required, the battery enclosure baseplate fins were shortened. New algorithms for the PV TCS were developed to maintain the battery temperatures at $30 \pm 7\,°C$. In order to prevent catastrophic propagating cell-to-cell thermal runaway, additional safety features were added into the LIB design. Each cell temperature and voltage is individually monitored, with automatic and permanent isolation from the circuit of any cell that exceeds pre-set temperature or voltage limits.

Design life of the new LIBs is 10 years at approximately 18% DOD. The first replacement of six Ni-H_2 with three LIBs via robotics and extravehicular activities (EVA) occurred in December 2016, followed by replacements in September 2018, March 2019, September 2019, and July 2020. One LIB fuse was blown by a failed BCDU during installation in March 2019. The battery was subsequently replaced on January 28, 2021 via robotics, with an EVA on February 1, 2021 to connect the AP cable. With the successful replacement of the Ni-H_2 battery and start-up of the new LIB on February 1, 2021, all 48 Ni-H_2 batteries have been removed and replaced. All 24 LIBs are meeting ISS-specified system requirements. Representative on-orbit data are discussed in Chapter 9.

5.4 NASA Goddard Space Flight Center Spacecraft

5.4.1 Introduction

For over a period of two decades, EnerSys/ABSL has been supporting the NASA-Goddard Space Flight Center (GSFC) to design, manufacture, test, and qualify a series of LIB modules for various LEO and GEO mission applications. The early introduction of the small commercial-off-the-shelf (COTS) cell approach based on the Sony 18650HC Li-ion cell design was launched on the Space Technology-5 (ST-5) mission in 2006. This technology demonstrator consisted of three microsatellites in constellation format using 2s6p battery modules composed of COTS 18650 cells. The success of this first mission led to the adoption of the design concept for future use in a number of NASA-GSFC applications.

Following the mission success of ST-5, three additional NASA missions were baselined between 2005 and 2015. Although each NASA mission had a unique set of spacecraft mission objectives and environments, a shared set of LIB specification requirements enabled a common set of battery design solutions. All of the battery modules utilized a COTS 18650 cell design. The cells were configured in an s-p topology to provide the required voltage and capacity. On each battery, the cells were assembled into modular blocks comprising between 80 and 100 Li-ion cells. These blocks were then assembled into trays with aluminum sidewalls that were then stacked on top of each other to form the main battery chassis.

No cell-to-cell balancing electronics, integrated battery management system (BMS), or cell bypass systems were required to meet battery and EPS requirements. All of the battery chassis were rectangular in shape and manufactured from aluminum alloy. The mechanical interface to the spacecraft was achieved through mounting feet placed around the base unit perimeter. Battery telemetry included voltage sense lines on both the positive and negative rails, as well as integrated temperature sensors for monitoring battery thermal performance. Additionally, the battery electrical interface utilized circular, scoop-proof connectors for power, telemetry, ground charge, and relay control. All of the batteries had two orthogonally mounted relays installed within the battery.

5.4.2 Solar Dynamics Observatory – GEO

The Solar Dynamics Observatory (SDO) satellite was the first NASA mission to be launched in support of the Living with a Star Program, which focused on a better understanding of the effect of solar variability on the Earth's environment. SDO was also designed to further understand the Sun's influence on Earth and near-Earth space by studying the solar atmosphere on small scales of space and time (Figure 5.5). SDO was successfully launched on February 11, 2010, into an inclined GEO orbit. This orbit is located on the outer reaches of the Earth's extreme radiation environment. Additional shielding was added to the instruments, battery, and electronics to reduce the risk from hazards caused by extreme radiation exposure. SDO has now been in successful operation for over 10 years.

Figure 5.5 Solar Dynamics Observatory is readied for launch on its United Launch Alliance Atlas V rocket. Credit: Barbara Lambert/NASA.

Figure 5.6 NASA SDO flight 8s104p LIB manufactured by EnerSys/ABSL. Credit: EnerSys/ABSL.

The basic design concept of the SDO LIB has allowed an evolution of battery development at NASA-GSFC. The SDO battery used three decks of eight cell blocks and an integral relay unit. This modular approach has since been adapted and modified to produce a battery using two decks and extended in length, incorporating seven cell blocks (Figure 5.6). This concept was adapted for both the Lunar Reconnaissance Orbiter (LRO) and Global Precipitation Measurement (GPM) satellites. The SDO EPS is a direct energy transfer (DET) unregulated battery-voltage dominated bus system. The EPS was designed to support 1380 W of continuous load power and 1500 W of peak power. The SDO battery topology was an 8s104p configuration using Sony COTS 18650HC Li-ion cells. The battery consisted of eight 8s13p blocks arranged in three decks. There are three cell blocks on each of the two upper decks and two cell blocks on the lower deck. The third block volume at the lower deck houses two parallel battery disconnect 100 A relays, connectors, wiring, and power bus electronics with voltage sense resistors and relay drive decoupling diodes. The SDO battery is $52.5 \times 30 \times 24\,cm^3$ and has a mass of 43 kg.

The SDO battery temperature was monitored with six platinum resistor thermistors mounted on selected cells at the predicted hot and cold battery locations. NASA-provided heaters were mounted to the battery panel interface to maintain thermal control. The cell blocks have a thermal plate that facilitates heat transfer from the cell blocks to the battery aluminum chassis and through the spacecraft mounting interface. Cell temperatures within a battery were specified to be maintained within 3.6 °C when operated between 10 and 30 °C. The battery was qualified to survive from 0 to 40 °C. The nameplate beginning-of-life (BOL) battery capacity was 156 Ah and the end-of-life (EOL) capacity was 120 Ah when tested to an end-of-discharge voltage (EODV) of 24 V. The EOCV per cell was 4.2 V for a total battery voltage of 33.6 V. The charging profile charges to a voltage set point and the current is then tapered. There are two different set points: one for solstice and another during eclipse seasons. Battery EOCV set points are commandable by ground control operations.

5.4.3 Lunar Reconnaissance Orbiter – Lunar

LRO was the first mission of NASA's Robotic Lunar Exploration Program. LRO was designed to map the surface of the Moon and characterize future landing sites in terms of terrain roughness, usable resources, and radiation environment with the ultimate goal of facilitating the return of humans to the Moon. Originally planned as a one-year mission, the LRO mission was launched on June 18, 2009, and has been in operation in an eccentric lunar polar orbit (165 km altitude) for over 10 years. The battery supports an unregulated power bus in a DET EPS architecture with an on-orbit average load of 824 W.

The two-deck concept based on the SDO design was lower in height using seven cell blocks composed of SONY 18650HC Li-ion COTS cells. The battery topology was based on a smaller 8s84p configuration. The LRO battery was made up of seven identical 8s12p blocks arranged in two decks. There are a total of four cell blocks on the upper deck and three cell blocks on the lower deck. The lower deck houses two parallel-battery disconnect 100 A Hartman relays in a separate compartment. This also incorporated the three electrical connectors, power bus electronics boards with voltage sense resistors, and relay drive decoupling diodes. The LRO battery is $68 \times 28 \times 17\,cm^3$ and has a mass of 35 kg. The LRO battery decks are configured to draw cell-generated heat to a thicker mid-deck, which then provides an interface to the heat pipe and radiator. The heat pipe and radiator maintain the battery temperature and provide for low cell-to-cell temperature gradients. The nameplate BOL capacity was 126 Ah and the predicted EOL battery capacity was 80 Ah when operated to an EODV of 24 V. The battery was qualified to survive from 0 to 40 °C and to operate between 10 and 30 °C.

5.4.4 Global Precipitation Measurement – LEO

The GPM LEO spacecraft was jointly developed by the NASA-GSFC and the Japan Aerospace Exploration Agency (JAXA). Launched on February 27, 2014, GPM provided next-generation global observations of rain, snow, and ice to better understand climate, weather, and hydrometeorological processes from space, which improves forecasting of extreme weather events (Figure 5.7). The spacecraft was in a circular 400 km altitude, 65° inclination nadir-pointing orbit with a three-year baseline mission life and five-year extended mission life. The four-panel solar array consisted of two sun tracking wings populated with triple-junction solar cells of nominal 29.5% efficiency. One axis is canted by 52° to provide power to the spacecraft at high beta angles.

The GPM EPS is a DET system capable of supporting 1950 W on-orbit average power with a solar array and LIBs connected directly to the spacecraft's unregulated power bus. The GPM battery consisted of three 8s16p batteries electrically connected in parallel to form a single battery EPS. The battery-dominated bus voltage range for essential loads varies between 21 and 35 V, depending upon battery SOC. The power system electronics (PSE) contained the circuits needed for solar array power control, charging the LIBs, and the electronics for power distribution and current monitoring of spacecraft loads. The PSE controlled battery charging with current-limited selectable levels until the battery voltage reached a ground-selectable voltage limit.

The GPM batteries were manufactured using the same Li-ion cell design employed on SDO and LRO flight batteries. The GPM battery's nameplate BOL and EOL capacity was 126 Ah and 80 Ah, respectively, when measured to an EODV of 24 V. Each flight battery consisted of seven 8s12p blocks arranged in two decks; three blocks at the lower deck and four blocks on the upper deck. The GPM battery is $68 \times 31 \times 17\,cm^3$ and has a mass of 41 kg. All three LIBs were mounted on a single baseplate with heat pipes to maintain low battery temperature gradients. Non-flight test

Figure 5.7 NASA GPM LEO satellite system. Credit: NASA.

batteries, identical in form, fit, and function to flight units, but with cells from different manufacturing lots, were used during spacecraft assembly, integration, and test.

5.5 Van Allen Probes – HEO

5.5.1 Mission Objectives

The Van Allen Probes (formally known as the Radiation Belt Storm Probes) were twin NASA spacecraft managed, manufactured, and operated by The Johns Hopkins University Applied Physics Laboratory (APL) [4]. Launched together on August 30, 2012, on an Atlas V 401 launch vehicle, the two Van Allen Probes spacecraft operated for over seven years, investigating the hazardous Van Allen Radiation belts for a better understanding of extreme-space weather physics (Figure 5.8). Both spacecraft operated in similar-lapping HEO, ranging from an altitude of approximately 600 km perigee, to an altitude of approximately 30 500 km apogee at a 10° inclination.

5.5.2 Electrical Power System

Each independent spacecraft utilized an unregulated DET EPS, operating between 29 and 34 V. The spacecraft EPS components included PSE, BMS, a single 50 Ah LIB, and four solar array panels. The spacecraft EPS was designed to support an operational orbit average load power of 350 W. The BMS function provided cell voltage telemetry and shunt electronics for balancing each battery cell's EOCV [5]. The solar panels used triple junction cells with a minimum efficiency of 28.5%.

Figure 5.8 NASA's twin Van Allen Probes in orbit within Earth's magnetic field to explore the radiation belts. Credit: NASA.

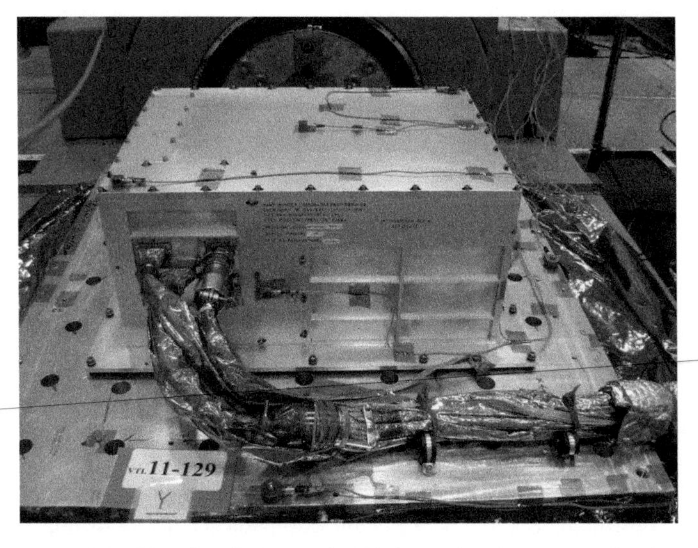

Figure 5.9 Van Allen Probes 8s1p LIB during *Y*-axis random vibration testing. Credit: The Johns Hopkins University APL.

5.5.3 LIB Architecture

Designed and manufactured by the APL, the spacecraft battery consisted of eight GS Yuasa LSE50 (50 Ah) Li-ion cells connected electrically in series (Figure 5.9). The LSE50 elliptic-cylindrical cells provided a specific energy and energy density of 136 Wh/kg and 277 Wh/l, respectively [6]. The electrical, mechanical, and thermal design of the battery was optimized to survive the extreme-energetic trapped-proton and electron-radiation environment of the Van Allen Belts. The design features included connector radiation shields, radiation-hardened electronic parts, and an increased battery chassis wall thickness [7].

The Van Allen Probes flew in a 10° inclined HEO, spending the majority of each orbit in the electron belt. The orbit precession provided a maximum eclipse of 115 minutes, with some periods of full sun for the nine-hour orbit period. The majority of the eclipses were less than 40 minutes, resulting in a DOD of less than 15%. The EPS battery management electronics contained an end-of-charge (EOC) bypass shunt cell balancing system to keep the cell's EOCV balanced within 5 mV. The shunt system was only capable of shunting a maximum C/66 battery charge current. Balancing the battery was planned once the minimum to maximum cell voltage difference was greater than 25 mV and the spacecraft was in full sun. Throughout the mission, the individual cell voltages were monitored and the cell differential voltage at EOC was reviewed. The need to balance the cells did not arise during the mission until one year prior to the planned spacecraft disposal. Due to the low EOCV dispersion rate, on-orbit battery cell balancing was waived for both Van Allen Probe spacecraft.

5.6 GOES Communication Satellites – GEO

5.6.1 Mission Objectives

In co-operation with NASA, the National Oceanic and Atmospheric Administration (NOAA) operates the Geostationary Operational Environmental Satellite (GOES) series of US geostationary weather satellites. Beginning in 1975 with the launch of GOES-A (later redesignated as GOES-1), the GOES family of satellites has supported weather forecasting, severe storm tracking, and meteorology research as a global source for weather information.

5.6.2 Battery Heritage

Early GOES satellites utilized first-generation 3 Ah prismatic Ni-Cd 28 V batteries manufactured by Eagle-Picher Industries, Inc. (now EaglePicher Technologies) [8]. In order to support increasing spacecraft power demands in the mid-1990s, the GOES program transitioned to higher-energy density 12 Ah capacity Ni-Cd 28 V batteries manufactured by Gates Aerospace Batteries (later acquired by Saft) [9]. Launched on May 24, 2006, GOES-N (GOES-13) was the first GOES spacecraft to utilize Ni-H_2 battery technology. Manufactured by The Boeing Company, the GOES N-O-P series Ni-H_2 123 Ah batteries supported a nominal 10-year GEO mission at 2.1 kW power capability at EOL [10]. GOES spacecraft have been manufactured by Ford Aerospace and Communications Corporation (GOES-A through -C), Space Systems/Loral (GOES-I through -M), Hughes Space and Communication Company (GOES-D through -H), and The Boeing Company (GOES-N through -P). The GOES-R series is currently being built by Lockheed Martin with the first and second in the series, GOES-16 and -17, launched in 2016 and 2018, respectively.

5.6.3 LIB and Power System Architecture

GOES-R (GOES-16) was the first GOES spacecraft to adopt a LIB-based EPS. GOES-16 was launched on November 19, 2016, into GEO and entered service in December 18, 2017, for a planned 15-year mission. The spacecraft EPS is supported by two 3p12s LIBs manufactured by Saft using high-energy VL48E (48 Ah) Li-ion cells. The Saft VL48E cell has a specific energy of 168 Wh/kg [11]. Each GOES-R LIB has a rated capacity of 136 Ah at BOL [12–14]. Using a constant-current constant-voltage (CC/CV) algorithm, the EPS BMS utilizes cell balancing circuits under flight

Figure 5.10 GOES-R undergoing final launch preparations prior to fueling inside the Astrotech payload processing facility (Titusville, FL, USA). Credit: NASA.

software control for managing individual cell bank voltages. Cell temperature is monitored to facilitate battery thermal control via heaters and radiators. Similar to previous GOES series spacecraft, the EPS utilizes a voltage-regulated 70 V power bus supporting approximately 3500 W. The GOES-S launched on March 1, 2018, and entered service on February 12, 2019 (Figure 5.10).

5.7 James Webb Space Telescope – Earth–Sun Lagrange Point 2

5.7.1 Mission Objectives

Launched on December 25, 2021, the James Webb Space Telescope (JWST) is the largest and most powerful telescope ever built and launched into space (Figure 5.11). It is not just the successor to the NASA Hubble Space Telescope (HST), but it also complements HST with new capabilities and unrivaled sensitivity that will enable scientific breakthroughs in a better understanding of the formation of the universe. The 6.6 m diameter segmented primary mirror is composed of 18 hexagonal, gold-coated beryllium mirrors that is 100 times more powerful than HST. The scientific instruments of JWST are infrared optimized over the wavelength range of $0.6 < \lambda < 28.5\,\mu m$. The five-layer sunshield is a key component of the observatory. It is approximately the size of a tennis court, and isolates the hot, sun-facing side of the spacecraft from the cold telescope side by over

Figure 5.11 James Webb Space telescope. Credit: Northrop Grumman/NASA.

260 °C to allow the high-sensitivity instruments to detect extremely faint signals. The first JWST full-color images and spectroscopic data were released on July 14, 2022.

5.7.2 Lagrange Orbit

While many different orbit locations were explored, JWST ultimately was located at the second Lagrange point in the Earth–Sun system (L2). The L2 location has the necessary benefit of allowing a constant cold thermal environment for the telescope to take advantage of the high-sensitivity science instruments having an unobstructed view of deep space. Another advantage of L2 is that the spacecraft will always appear in approximately the same location in the Earth sky at the same distance from the Earth throughout the mission, simplifying communications between the spacecraft and the ground station.

The JWST orbit about L2 is approximately 1.5 million km from Earth and lies in an orbital plane inclined with respect to the ecliptic plane. This halo orbit is sufficiently large to avoid the Earth and Moon eclipses of the Sun, ensuring continuous electrical power generation from the solar array, and no planned battery discharges during the six-year minimum mission. JWST will have a six-month halo orbit period about the L2 point in the rotating coordinate system, moving with the Earth around the Sun.

5.7.3 Electrical Power System

The JWST EPS bus utilizes an unregulated, battery-clamped voltage architecture (Figure 5.12). A body-fixed, five-panel solar array with advanced triple-junction gallium arsenide solar cells is the primary power source during the mission, providing over 2300 W at EOL. The LIB provides power during launch and ascent until solar array deployment and for any contingency power required for station-keeping maneuvers or any fault scenarios. The power control unit (PCU) provides power control, conversion, and distribution of primary power, plus battery charge/discharge

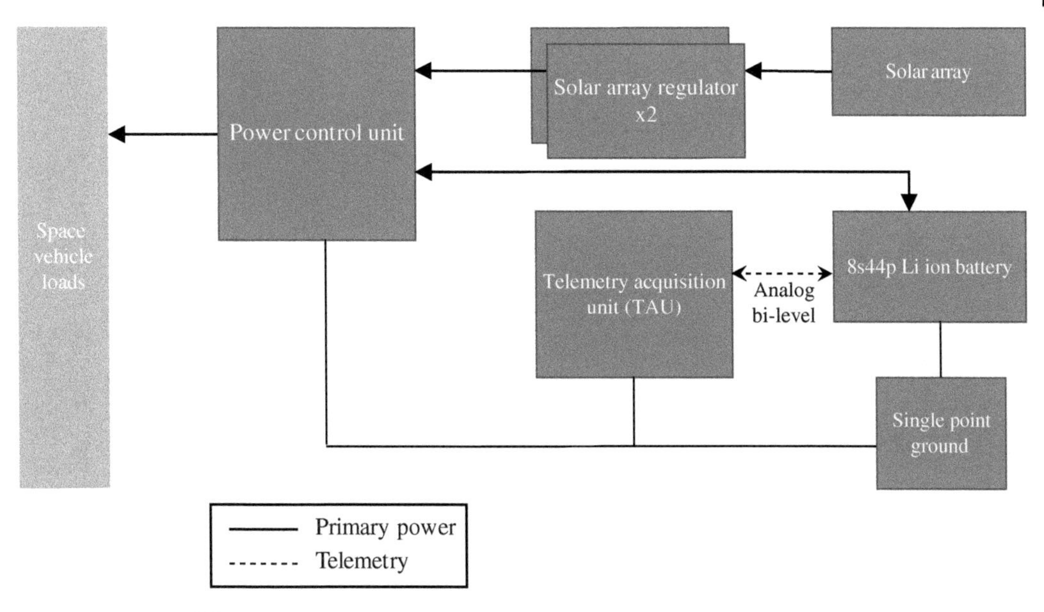

Figure 5.12 JWST EPS architecture.

control. The PCU provides CC/CV charge control of the battery via descending constant-current step values until the battery voltage reaches its charge voltage limit (CVL). The voltage is then held constant at the CVL while the charge current tapers asymptotically toward 0 V. Both the charge current steps and CVL can be changed throughout mission life via ground command. The telemetry acquisition unit (TAU) provides telemetry, precision actuator motor control, precision heater control, and both non-explosive actuator and traditional ordnance drivers. There are two solar array regulators (SARs) that provide conversion and control of the solar array power output into the PCU.

5.7.4 LIB Architecture

Manufactured by EnerSys/ABSL, the JWST LIB employs an 8s44p configuration rated at 105.6 Ah. The battery consists of four 8s11p blocks of 18650 type, Moli-M cells arranged in a dual-deck configuration (Figure 5.13). The JWST LIB is $43 \times 26 \times 16\,cm^3$ and has a mass of 21.3 kg. The electrical interface assembly (EIA) mounted to the end of one stack of cell blocks houses the electrical connectors, printed wiring boards, relays, and all associated voltage sense resistors, relay steering, and diodes, while providing physical protection and radiation shielding. All primary power is isolated by a minimum of two layers of electrical isolation. Primary and redundant heaters controlled by the TAU provide closed-loop software controlled thermal control. Thermistor temperature sensors mounted at strategic locations on the battery provide the temperature measurements needed for onboard thermal control and are down-linked for battery state-of-health telemetry monitoring and trending.

Relays are installed between the most positive internal terminal and the power terminals on the outside of the unit. These relays are integral to safety for both hardware and personnel. The relays are open during all handling, transportation, integration, and storage events, including spacecraft power off, in order to minimize any parasitic loads on the battery. The relays are in a dual series/parallel quad- arrangement in order to provide redundancy and fault tolerance to either failed

Figure 5.13 JWST 8s44p battery. *Source:* Northrop Grumman.

open or closed conditions. The relays are also used for EOL passivation/discharge. The JWST mission is ideal from a battery standpoint since battery usage is minimal. There are no eclipses in L2, and station keeping discharges result in low discharge currents and minimal DODs. The only significant discharge is during the launch and ascent phase, with the nominal launch and ascent resulting in approximately 25% DOD. The LIB is sized for launch and ascent, and to provide sufficient energy to execute all planned reasonable fault recovery scenarios.

5.8 CubeSats – LEO

5.8.1 Introduction

A new class of small satellites, called CubeSats, has been developed for LEO missions and is positioned to dominate future space exploration. Due to its inherent low cost, this class of satellite allows the participation of a substantially larger user-base in technology demonstrations, science, and education missions. CubeSats typically range in mass between 1 and 15 kg, and can be up to 12 U in volume ($1\,U = 10 \times 10 \times 10\,cm^3$), as summarized in Figure 5.14 [15–17]. Their low mass and volume enable CubeSats to ride along or "piggyback" on other launches or to be deployed from larger spacecraft at much lower costs or at no cost under NASA's CubeSat Launch Initiative (CSLI) Program [18, 19]. CubeSat missions are currently dominated by university and commercial programs, with more than 70% of ~650 CubeSats launched since 2000 provided by universities and commercial programs, with the balance coming from both military and civilian government applications [20].

Advances in attitude control, communications, and power subsystems technology have rapidly expanded the capabilities of today's CubeSats. Such advances have a profound impact on the implementation of more advanced science payload packages [21–23]. CubeSat instrumentation payload advancements have enabled scientific research ranging from astrobiology to earth science applications [24, 25]. While early CubeSats were typically 1 U tumblers (no attitude control), more recent CubeSats have access to both passive and active attitude determination and control systems (ADCS) [26, 27]. Larger 2 U and 3 U spacecraft can accommodate sophisticated communication systems such as high data-rate S-band transmitters for broadcasting voice data and images and

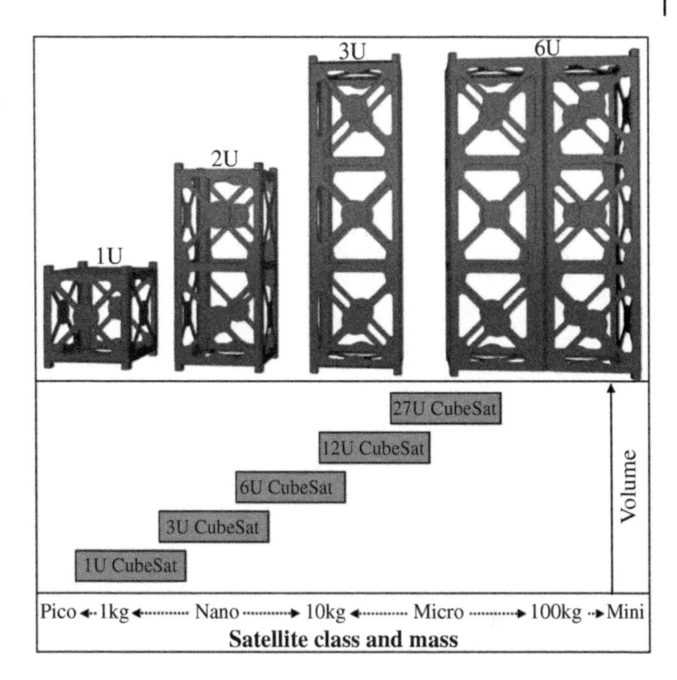

Figure 5.14 CubeSat volume and mass classifications. *Source:* Poghosyan and Golkar [15], ELSEVIER.

even laser-based communications [28, 29]. At the NASA Jet Propulsion Laboratory (JPL), CubeSat missions have significantly contributed to technology demonstration through flight demonstrations in space [30–33]. These technology demonstrations employ CubeSats ranging from 1 U up to 6 U. The Michigan Multipurpose Minisat (M-Cubed)/CubeSat On-board Processing Validation Experiment (COVE-2) is a 1 U CubeSat, launched in December 2013 to demonstrate advanced signal processing technologies [31]. The Integrated Solar Array and Reflectarray Antenna (ISARA), a 3 U CubeSat, was launched in November 2017 to demonstrate a high bandwidth Ka-band communications package. ISARA also highlighted advancements in deployable solar array architectures to enable higher power generation without increasing volume. In an effort to stay within the mass and volume limits while payload packages increase, JPL has adopted a standard size CubeSat of 6 U. Rain Cube, launched in May 2018 to study precipitation using a new cost-effective radar technology, is an example of this standard size [30].

5.8.2 Electrical Power System and Battery Architecture

Early 1 U CubeSat (approximately 1 kg mass) battery EPSs operated in the nominal range of 3–5 W [34]. CubeSat battery EPS technologies will likely remain the limiting factor as power demands continue to increase, particularly as deployable solar arrays with increasing energy generation and power distribution of up to 60 W or more are employed [35]. High specific energies of Li-ion cell technologies make them the most viable solution for CubeSat energy storage. As discussed in Chapter 2, several Li-ion cell chemistries are now available. Since lithium polymer (LiPo) and COTS Li-ion 18650 cylindrical cells exhibit the highest specific energies, they are the most common cell form factors used in CubeSat EPS designs.

Several commercial CubeSat vendors utilize COTS cells to provide fully integrated EPS designs. In order to utilize the full capacity of the LiPo cells employed by one vendor, the operating temperature was limited to ≥20 °C; other vendors predominantly use 18560 cell designs in their EPS

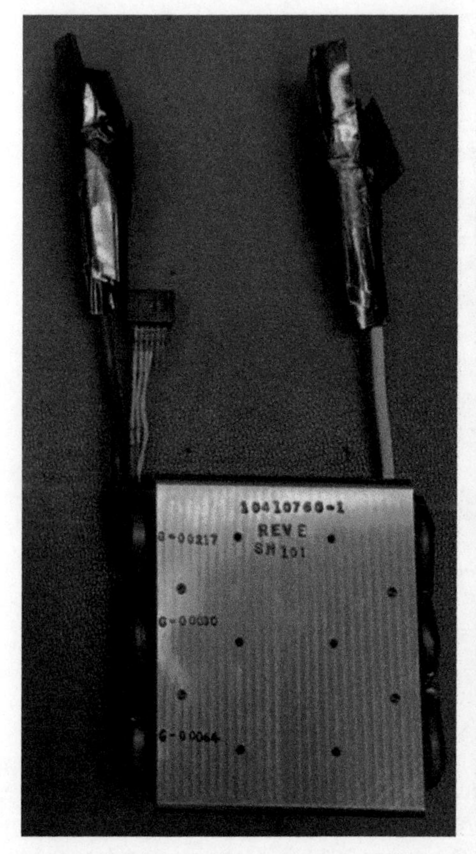
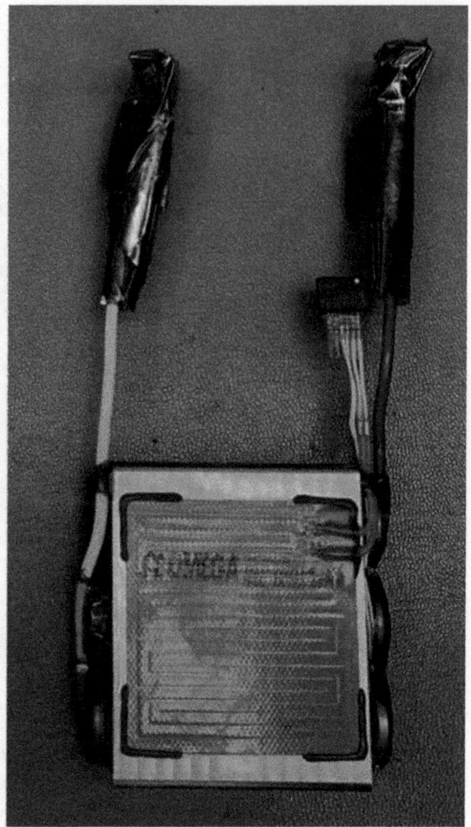

Figure 5.15 JPL 3s1p high power battery. Credit: NASA.

architectures [36]. Many EPS are available with space-qualified battery pack sizes ranging in operating bus voltages from 8.4 to 33 V and energy from 72 to 100 Wh with a wider temperature range than the LiPo-based EPS [37, 38]. Figure 5.15 is an example of a JPL-designed high-power (>250 W pulse) 3s1p payload LIB using Sony VTC4 18650 cells. Heating is provided by a small (<3 W) resistor heater. The 2.1 Ah battery is compatible for up to a 6 U CubeSat. This JPL high-power payload battery is $2.2 \times 6.6 \times 6.6\,cm^3$ and has a mass of 0.16 kg.

5.8.3 Advanced Hybrid EPS Systems

Hybrid energy storage systems (ESS) have advantages of both a high specific power and energy as well as a long cycle life by combining super-capacitor and Li-ion chemistries [39, 40]. Recently, JPL has successfully qualified a hybrid ESS for space applications that demonstrated lower resistance at low temperatures (−40 °C), suggesting a synergistic load sharing between the low temperature Li-ion cells and super-capacitors during discharge [41].

Based on these promising test results, JPL demonstrated a hybrid ESS onboard a 2 U CubeSat called CSUNSat1, in collaboration with California State University, Northridge (CSUN). In this flight demonstration, the CubeSat hybrid ESS assembly had a volume of ½ U including the electronics, with the battery cells consisting of a single Li-ion cell and two COTS 310F-rated electrochemical double-layer (EDL) super-capacitor cells. Funded by the CSLI program, CSUNSat1 was launched to the NASA ISS onboard the OA7 resupply vehicle in 2017. CSUNSat1 was deployed from ISS and was successful in achieving all of its primary and extended mission objectives in testing the hybrid ESS in the space environment. More importantly, the CSUNSat1 project underscored the benefits of CubeSats for cost-effective technology demonstration along with significant educational impact via academic collaboration [42].

5.9 European Space Agency Spacecraft

5.9.1 Introduction

European Space Agency (ESA) spacecraft are developed and launched through the joint cooperation of the 22 member states, two associate member states, and six co-operative states, of the ESA. ESA's purpose is to provide for and promote cooperation among European states in space research and technology and their space applications, with a focus on scientific and operational space mission applications [43].

5.9.2 Sentinel-1 Mission Objectives

Sentinel-1 is part of the Copernicus European Union's Earth observation program, coordinated and managed by the European Commission in partnership with the ESA, the EU Member States, and EU agencies. The Sentinel-1 mission is composed of a constellation of two identical satellites, Sentinel-1A and Sentinel-1B. These are maintained in the same orbit, phased 180° apart, providing a revisit time of six days at the equator [44]. The constellation is on a sun-synchronous, near-polar (98.18° inclination), 693 km altitude orbit. The orbit has a 12-day repeat cycle and completes 175 orbits per cycle. The first satellite, Sentinel-1A, was launched on April 3, 2014, and Sentinel-1B was launched on April 25, 2016.

The Sentinel spacecraft payloads include a C-band Synthetic-Aperture Radar (C-SAR) instrument, which provides a collection of data in all weather, day and night. This instrument has a spatial resolution of down to 5 m and a swath of up to 400 km.

The ESA and European Commission's policies make Sentinel-1's data easily accessible. Various users can acquire the data and use it for public, scientific, or commercial purposes at no cost. There is a wide range of applications for the data collected via the Sentinel-1 mission. A few of these uses include sea and land monitoring, emergency response due to environmental disasters, and economic applications. The spacecraft is based on the PRIMA (Piattaforma Riconfigurabile Italiana Multi Applicativa) bus of Thales Alenia Space, of RADARSAT-2 and COSMO-SkyMed heritage, with a mission-specific payload module.

The two SAWs are organized in 24 sections, with a PV assembly design based upon AZUR 3G28/150-8040 solar cells. Each SAW interfaces with the spacecraft by means of an SAW Reel Mechanism, which is able to rotate the SAW within a fixed angular region of about 350°, and to transfer SAW power and signals to the spacecraft via dedicated flex cables. Due to the particular orbit characteristics (sun synchronous), no SAW rotation is foreseen during the nominal mission lifetime; consequently, the power lost due to missing sun tracking had to be taken into account in the EPS performance evaluation [45]. The C-SAR requires more than 4 kW of peak radio frequency power. To limit the current and consequently minimize the corresponding harness size, it was decided to power it at a relatively high voltage of about 65 V, provided by an unregulated EPS power bus. A more traditional regulated 28 V bus powers the rest of the satellite, through a power control and distribution unit.

Figure 5.16 Sentinel-1 16s16p LIB flight battery modules (during ground processing) manufactured by EnerSys/ABSL. Credit: EnerSys/ABSL.

The LIB strings are manufactured from 16 Sony COTS 18650HC Li-ion cells electrically connected in series, so that battery voltage matches the unregulated bus voltage feeding the radar antenna. Each battery module has a 16s16p topology [46]. A series–parallel architecture was selected for the battery due to its narrow voltage range compared to a parallel–series architecture (which has to take into account the possible loss of one cell in a series string). The battery (Figure 5.16) consists of 10 modules mounted on a honeycomb baseplate panel. The modules are connected in parallel with a single junction box, allowing the battery to have only one electrical interface with the satellite. This battery is among the largest batteries in LEO, both in terms of nominal energy (14 kWh) and mass (130 kg). The dimensions of a single Sentinel-1 battery module are $37.9 \times 20.4 \times 17.7\,\text{cm}^3$.

5.9.3 Galileo Mission Objectives – MEO

Galileo is a global navigation satellite system (GNSS) created by the European Union through the ESA, operated by the European GNSS Agency (GSA). The Galileo satellites compose a constellation in MEO, on a circular 23 200 km altitude, 56° inclination orbit. This orbit has a period of 14.1 hours with two Earth eclipse seasons per year. The eclipses have a maximum duration of 60 minutes [47]. The constellation is composed of 24 reference satellites on three orbital planes separated by 120° right ascension of the ascending node. The satellite's design lifetime is 12 years. There are two types of satellites: the first four, called in-orbit validation (IOV), were manufactured by Astrium GmbH and Thales Alenia Space, while the remaining satellites, called full operational capability (FOC), were built by Otto Hydraulic Bremen (OHB) System and Surrey Satellite Technology Limited. The four IOV spacecraft were launched between December 2005 and October 2011, while the first two FOC spacecraft were launched in August 2014.

Both of the IOV and FOC EPSs are based on a 50 V fully-regulated bus [48]. The nominal spacecraft power is approximately 1550 W, with a constant payload power consumption. Both spacecraft designs use the same solar array, provided by Airbus Netherlands, capable of delivering more than 1904 W at EOL. The solar array design, based on GaAs triple junction 28% cells, is made of 29 cells per string with a total of 88 strings. Each spacecraft uses sequential switching shunt regulators to manage the solar array sections. The LIBs were supplied by two different battery manufacturers and use four different types of cells: the earlier IOV satellites were equipped with Saft VES180SA (50 Ah) and VES140SA (39 Ah) Li-ion cells. Later FOC satellites are equipped with EnerSys/ABSL-supplied Sony 1.5 Ah COTS 18650HC and 18650HCM Li-ion cells.

The IOV Galileo LIB is constructed from nine identical cell packages electrically connected in series, each cell package being composed of three high-capacity Saft Li-ion cells connected in parallel (3p9s) for a total nameplate energy between 3700–4800 Wh depending on which cell was used. Once assembled, each cell package behaves like a single cell, having three times the capacity of a single cell. Battery hazard tolerance to cell failure is achieved by nine electromechanical NEA8023 bypass switches. Two cell voltage balancing functions operate in cold redundancy and are individually on/off switchable via an MIL-STD 1553 bus protocol. Each function operates fully autonomously without any need of external commanding. The IOV battery is $68.5 \times 26.5 \times 23.5\,\text{cm}^3$ and has a mass of 37 kg.

A single cell balancing function consists of nine closed-loop, parallel-load regulators operating individually on its associated cell package, leveling out the cell package voltage mismatch to better than 2.1 mV. The balancing function reacts as a pure analog regulation. The conditioned cell voltages are mutually compared to identify the lowest value, which will be used as reference for other cell voltages. The difference between each conditioned cell voltage and the reference minimum voltage proportionally drives a load circuit that differentially acts on the associated cell package. The load capability is 130 mA at 4.1 V per cell.

Each FOC LIB is comprised of one module, which contains 64 strings in parallel, each string consisting of 11 Li-ion cells in series (11s64p). The module is physically composed of two decks, each deck containing four cell bricks made of eight strings (Figure 5.17). The FOC battery is $69.5 \times 26.9 \times 18.0\,\text{cm}^3$ and has a mass of 37 kg.

The nameplate energy is 3800 Wh and the power interface is made by four D-sub 5W5 connectors. The main design difference between IOV and FOC LIBs is that the EnerSys/ABSL LIB designs do not contain any battery module voltage balancing system electronics. Elimination of the module balancing system is feasible due to the production quality of the COTS cells and implementation of a rigorous cell screening and matching process.

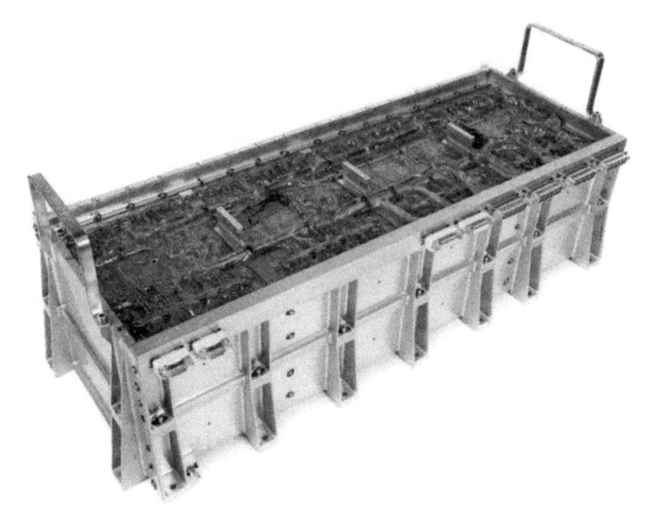

Figure 5.17 Galileo FOC 11s64p HCM LIB manufactured by EnerSys/ABSL. Credit: EnerSys/ABSL.

5.10 NASA Astronaut Battery Systems

5.10.1 Introduction

NASA's astronaut spacesuit, the Extravehicular Mobility Unit (EMU), is designed to provide crew life support while outside the spacecraft in the extreme environment of space. The EMU has expanded to include five separate batteries that operate three independent subsystems: the Portable Life Support System (PLSS), the EMU Accessory System (including the Pistol Grip Tool), and the Simplified Aid for EVA Rescue (SAFER) [49–51]. Over the last 40 years, NASA's EVA missions have evolved from single-flight experiments and repair to ISS assembly, sustained operation, and maintenance. The introduction of LIB technology, in place of heritage silver-zinc (Ag-Zn) chemistry, has allowed the consolidation of EMU batteries and chargers and expansion of existing power systems. For example, the PLSS LIB eliminated the need for frequent logistics flights with battery replacements. Use of a common Li-ion technology enabled consolidation of chargers and enabled the addition of new powered loads for improved operational performance and safety.

The rechargeable EVA LIBs use COTS 18650 Li-ion cells and a dedicated charger that performs the necessary battery management functions. A standard battery 20 V servicing interface simplifies charger design, which increases operational reliability during EVA. In order to ensure maximum operational life, several protocols are followed. These protocols include low-rate charge–discharge cycling, limiting the time above 50% SOC to less than two years, maintaining battery storage temperature conditions to 10–25 °C (at <50% SOC) during ground storage, and limiting charge–discharge cycles to less than 25.

The EVA battery operations terminal (EBOT) provides battery maintenance and storage aboard the ISS [52]. EBOT and the portable LIB charger provide redundant overcharge protections, overdischarge detection and isolation, and maintenance cycling and data reporting for ground crew analysis. The LIB charger operates on 120 V power from ISS or 28 V power from the Servicing, Performance, and Checkout Equipment (SPCE). Each charger reports faults and signifies successful completion of pre-EVA charge or preparation for extended storage.

Since EVA batteries are in close contact with crew, each LIB design complies with industry and government specifications, which includes design features that prevent single-cell catastrophic failures (such as thermal runaway) from propagating through the battery assembly [53]. Each battery includes redundant electrical insulation as protection from external short abuse conditions. In addition, each EVA battery must demonstrate tolerance to worst-case failure conditions without the aid of circuit-interrupt devices or a properly functioning charging system.

The SAFER system uses a non-rechargeable lithium battery to power the emergency device and provide consumable reporting to the user. An adaptor is under development for deployment in 2022 that accepts a rechargeable EVA battery, enabling on-orbit recharge, and further consolidating the EVA energy storage system. Table 5.1 lists the US astronaut EMU LIB performance characteristics.

5.10.2 EMU Long-Life Battery

The Long-Life Battery (LLB) fits within the EMU and provides up to eight hours of EMU PLSS operation with a minimum of five years of useful life. The first-generation battery was launched in 2010 and powered 39 EVAs, including the record-breaking US Space Transportation System (STS)-102 EVA lasting 8 hours and 56 minutes. Replaced in 2017, the second-generation LLB2 includes provisions to mitigate the risk of a single-cell catastrophic failure from propagating through the battery assembly and potentially damaging the spacesuit, the ISS, or harming a crew member.

Table 5.1 US astronaut EMU lithium battery performance characteristics.

Battery characteristic	LLB2	LREBA	LPGT	SAFER
Cell design type	18650	18650	18650	Duracell Ultra CR123
Cell rated capacity (Ah)	2.60	2.60	2.60	1.50
Topology	14p5s	9p5s	2p5s	14s3p
Operating voltage range (V)	15.5–20.5	15.5–20.5	15.5–20.5	28–44
Capacity (Ah)	36.4	23.4	5.20	3.75
Specific energy (Wh/kg)	98.9	96.2	85.1	75.4
Energy density (Wh/l)	183	112	79.5	78.3
Operating temperature (°C)	0 to +45	+3 to +56	0 to +60	−20 to +75

The LLB2 is stored on ISS in a soft-stowed launch configuration or in the EVA EBOT when not in use. During maintenance or EVA charge operations, the battery remains attached to the portable charger or installed in the EMU PLSS. During EVA, the LLB2 is an integral part of the PLSS system and remains inaccessible by the crew. The battery contains two electrical interfaces: one for the PLSS and one for connecting to the dedicated charger. During EVA, the EMU can receive power from either the LLB2 or the umbilical connection to SPCE [54]. Selected by a user-actuated switch located on the displays and control module on the front of the EMU, power is supplied to the PLSS, including the caution and warning system and fan/separator/pump assembly [55]. Both the SPCE and the LLB2 support operation of the bends treatment adaptor (BTA) should a crew member require treatment for decompression sickness.

A safe and reliable LIB for human spaceflight application requires addressing the entire system: the cell, the battery, and the battery charger system. Cell lot testing provides a measure of performance uniformity and allows identification and removal of non-conforming cells. Samples of conforming cells undergo destructive testing to demonstrate performance of inherent protective features, and characterize the effect of cell failure. The battery design incorporates multiple protection features to protect against known and perceived hazards, and must demonstrate tolerance to known and perceived scenarios during testing. The battery assembly complies with industry and aerospace workmanship and performance standards.

The second generation LLB2 employs a 14p5s assembly configuration using replaceable cell bundles. Designed for EOL performance, the LLB2 stores 673 Wh with a specific energy of 99 Wh/kg and an energy density of 183 Wh/l. Both generations of LLB include a user-accessible fuse holder allowing on-orbit recovery (Figure 5.18). The LLB2 is $29 \times 10 \times 13\,cm^3$ and has a mass of 6.8 kg.

5.10.3 Lithium-Ion Rechargeable EVA Battery Assembly

The Li-Ion Rechargeable EVA Battery Assembly (LREBA) fits underneath the EMU thermal garment and provides eight hours of regulated power for operating EMU accessories with a seven-year useful life. Use of higher-capacity 18650 COTS Li-ion cells allowed LREBA to replace two nickel metal-hydride (Ni-MH) batteries with a single, regulated power source configured to use a common approach to EVA battery management. In addition, LREBA allowed the expansion of the Accessory System to include a high-definition camera system and wireless data recorder. The LREBA arrived on the ISS in 2019.

Figure 5.18 LLBs during thermal vacuum testing. Credit: NASA.

The LREBA employs a 9p5s configuration of replaceable cell bundles to ensure charging consistency. A voltage regulator provides a 13.5±0.5V output for powering the accessory systems. A manually-actuated toggle switch provides a secondary isolation feature during use and storage, and an internal low-voltage cut-off circuit prevents overdischarge during extended storage in the event the switch remains in the closed position. Designed for EOL performance, the LREBA stores 433 Wh with a specific energy of 96 Wh/kg and an energy density of 112 Wh/l. The regulation circuitry in LREBA reduces the available battery energy to 303 Wh. The LREBA is $52 \times 27 \times 3\,cm^3$ and has a mass of 5.4 kg.

The LREBA powers the glove tip heaters, a helmet-mounted wireless camera and light assembly, and a wireless data recorder. The glove tip heaters regulate supply voltage to power crew-controlled resistive heaters that reduce the risk of finger damage when in contact with cold materials. The wireless camera system provides a video feed to ground operators during EVA. The light assembly provides either low-or high-intensity light during periods of shadow or darkness. The data recorder collects and transmits unique EMU telemetry for ground analysis. During EVA, the astronaut turns devices on or off as needed to perform the EVA mission. In the event one of the user accessible power interrupt features fails during EVA, the LREBA includes a toggle switch that interrupts power to the regulation circuit. The toggle switch is accessible to the second EVA astronaut. The LREBA battery contains a non-resettable current interrupt and a low-voltage cut-off feature on the regulated output. The service connection bypasses the regulation function, allowing direct access to battery power and telemetry circuits. The low-voltage interrupt prevents over-discharge during extended storage if the toggle switch were to remain in the closed position, thus preserving the battery for future use.

5.10.4 Lithium-Ion Pistol-Grip Tool Battery

The Li-ion Pistol-Grip Tool (LPGT) battery provides operational voltage and current to power the motorized ratchet. The Pistol-Grip Tool (PGT) was originally operated with a configurable rechargeable Ni-MH battery. Relocating the configuration function into an adaptor installed in the PGT reduced cycle wear of the aging electrical connector and aided in mitigating the consequences of catastrophic battery failure. The configurable nature of the battery enables use by a range of devices including the PGT, EBOT, EVA infrared (IR) camera, and, most recently, SAFER.

The propagation-resistant LPGT battery contains 10 Li-ion cells, two temperature sensors, and retains the mechanical interface of the previous Ni-MH battery design in a shorter package. When installed in the PGT, a permanently mounted adaptor fills the previously occupied region, configures the battery into 1p10s, and allows communication with one temperature sensor. When installed in the EBOT, a permanently-mounted adaptor engages the battery and arranges the cells into a 2p5s configuration for maintenance. Two other devices have taken a similar approach: the EVA IR camera (5p2s) and a SAFER adaptor (1p10s). A non-resettable fuse protects each cell from external abuse. Designed for EOL performance, the LPGT battery stores 96 Wh with a specific energy of 85 Wh/kg and an energy density of 79 Wh/l. The LPGT battery is $11 \times 10 \times 6\,cm^3$ and has a mass of 1.4 kg. LPGT battery was the second pathfinder for LLB (after LREBA) for mitigating the hazards of a propagating thermal runaway condition in the suite of Li-ion EVA batteries.

The principal function of the LPGT battery is to power the hand-held PGT during EVA. Designed for worksite replacement, a spare LPGT battery remains in the airlock unless needed at the worksite. The EVA IR camera and SAFER application do not accommodate replacement during EVA. Designed for use during high-altitude EVA, without the protection of the EMU, the PGT absorbs solar radiation. This presents a potential overtemperature risk during ISS EVA, which is not shared by the other, protected EVA batteries. Like the EMU and other temperature-sensitive equipment, worksite analysis and mission planning mitigates the risk to thermal overstress conditions. During EVA, the PGT remains in standby mode until needed. Unlike the other EVA devices, the PGT can operate unpowered if necessary. The LPGT battery has maintained PGT functionality while providing a modular 4, 8, 20, or 40 V battery on ISS. The flight LPGT was certified in 2017 and entered into service on ISS in 2019.

5.10.5 Simplified Aid for EVA Rescue

First flown in 1994 on STS-64, the ISS SAFER is a self-contained, 24-jet, free-flyer jet pack that provides adequate propellant and control capability to allow an EVA crew member separated from the ISS to perform a self-rescue back to the ISS (Figure 5.19). Derived from an earlier model called the Manned Maneuvering Unit (MMU), the SAFER system unit is used to arrest relative rate and rotation, which allows the EVA astronaut to fly back to the ISS should the astronaut become separated from the ISS structure. The Russian Orlan spacesuit includes a SAFER unit interface, but the SAFER has not yet been used with the Orlan on the ISS since it was deployed in 1994. The integrated SAFER is always installed when a US astronaut conducts an EVA.

The SAFER consists of the main unit, tower latches, hinges, avionics unit, and three hardware modules: propulsion, hand controller module (HCM), and intravehicular activity (IVA) replaceable battery pack. The SAFER fits around the EMU PLSS without limiting suit mobility. The propulsion system provides stored nitrogen gas to provide up to 10 ft/s of delta velocity with 6° of freedom for a period of approximately 13 minutes. The propulsion system also contains motion sensors and when activated by deployment of the hand controller, autonomously stabilizes the detached crew member. Control is provided through crew member inputs from a single HCM. The HCM is stowed in a cavity on the right side of the SAFER propulsion module when not in use and is activated when needed. To deploy the HCM, the crew member pulls up the deployment handle mounted on the front, right side of the propulsion module. The crew member then grabs the HCM from the tray, holds the module in the left hand, and turns on the power switch. This switch fires a pyrotechnic device that pressurizes the propulsion system. The HCM can then be used to perform a self-rescue. The SAFER battery assembly is launched unattached and is soft-stowed in the ISS airlock or other pressurized locations.

Figure 5.19 USA astronaut on EVA with SAFER attached to EMU. Credit: NASA.

The SAFER battery is a human-rated non-rechargeable lithium battery ORU designed for on-orbit installation, removal, and ground refurbishment. The battery is located underneath a hook-and-loop secured insulation barrier on the bottom of the SAFER unit and is mechanically secured with captive fasteners. The SAFER battery includes an electrical cable that mates to the SAFER and resides underneath the mechanical battery structure. The electrical connection between the SAFER unit and the SAFER battery allows for transmission of power and data during checkout and on-orbit use. The battery assembly is composed of a gauge board, an electrical cable, and 42 non-rechargeable lithium cells. The cells are COTS Duracell® Ultra® CR123 lithium manganese dioxide (Li-MnO_2), designs used in various commercial high-power electronic devices [51].

The batteries provide energy for the avionics subsystem to perform 52 one-minute on-orbit IVA checks and one EVA self-rescue of at least 13 minutes in duration, within an operating voltage range of 28–44 V. To meet SAFER system voltage and capacity mission requirements, the battery cells are electrically connected into a 14s3p battery topology. The SAFER battery contains individual four- and ten-cell bundles with their cells connected electrically in series. Each cell bundle is protected by a positive temperature coefficient (PTC) thermal fuse and a dedicated Schottky blocking diode. Individual 4 s- and 10 s-cell bundles are connected electrically in series to form a 14-cell series string. Finally, three 14-cell series strings are electrically connected in parallel to form the 14s3p battery architecture. Battery cell bundles and gauge board are packaged in an aluminum case lined with foam. The SAFER battery is $51 \times 8 \times 5\,\text{cm}^3$ and has a mass of 1.99 kg. The SAFER battery gauge board is continuously powered to measure battery capacity during storage and when connected to the SAFER unit. An integrated harness cable assembly utilizes an RS232 communication link to communicate with the SAFER unit avionics subsystem.

5.11 Summary

This chapter provides a general overview of operational Earth orbiting spacecraft, with emphasis on their LIB EPS characteristics. LIBs have successfully powered Earth-orbiting spacecraft for over two decades and are now the preferred rechargeable energy storage technology for all spacecraft EPS. Representative examples from Earth orbiting communications, space exploration, and scientific mission applications, as well as astronaut battery systems, demonstrate today's wide range of LIB design and performance characteristics. LIB designs incorporating safety features which mitigate cell-to-cell propagation of thermal runaway will become more common on high-reliability and human-rated spacecraft systems. As LIB designs continue to realize increased specific energy and improved safety characteristics, they will remain the energy storage solution for all future spacecraft systems.

Acknowledgment

Some of the work described here was carried out at the Jet Propulsion Laboratory, California Institute of Technology, under contract with the National Aeronautics and Space Administration (NASA) (80NM0018D0004).

References

1 UCS Satellite Database, Union of Concerned Scientists (August 2020). www.ucsusa.org/resources/satellite-database (accessed 21 July 2022).

2 Wertz, J.R. (2011). Space mission geometry. In: *Space Mission Engineering: The New SMAD*, Chapter 8 (ed. J. Wertz, D. Everett and J. Puschell), 149–196. Microcosm Press.

3 International Space Station (ISS) (2013). Program Plan, Revision A. *NASA ISS Program*, Johnson Space Center, Houston, TX.

4 Butler, M. and Laughery, S. (2012). The RBSP spacecraft power system design and development. AIAA 2012–4059. In: *10th International Energy Conversion Engineering Conference*, Atlanta, GA (30 July–1 August 2012). https://doi.org/10.2514/6.2012-4059.

5 Smith, E., Fretz, K., Butler, M., and Newlander, K. (2013). Lithium ion battery fault management design on the Van Allen Probes. In: *AIAA SPACE 2013 Conference and Exposition*, San Diego, CA (September 10–12, 2013). https://doi.org/10.2514/6.2013-5526.

6 Pusateri, T. and Aldrich, C. (2014). State of GS Yuasa Lithium-ion Space Products and introduction to life and thermal modeling capabilities. In: *The Aerospace Corporation Space Power Workshop*, Manhattan Beach, CA (May 6–8, 2014).

7 Kirby, K., Bushman, S., Butler, M. et al. (2012). Radiation Belt Storm Probe Spacecraft and impact of environment on spacecraft design. In: *2012 IEEE Aerospace Conference*, Big Sky, MT, 1–20. https://doi.org/10.1109/AERO.2012.6187020.

8 Armantrout, J.D. (1977). *SMS/GOES Cell and Battery Data Analysis*, NASA CR-156739

9 Singhal, S.P., Alsback, W.G., and Rao, G.M. (1997). Performance of nickel-cadmium batteries on the GOES I-K Series of Weather Satellites, NASA CP-1998-208536. In: *Proceedings of the NASA Aerospace Battery Workshop*, Huntsville, AL (November 18–20, 1997).

10 GOES *NO/P/Q The Next Generation*, NP-2001-7-324-GSFC.

11 GOES I-M Databook, DRL 101–08, *Space Systems Loral*, Rev. 1, 31 August 1996.

12 GOES *R Series Databook*, CDRL PM-14, Rev. A, May 2019.

13 Borthomieu, Y. and Reulier, D. (2012). VES16 cell and battery design. In: *Proceedings of the NASA Aerospace Battery Workshop*, Huntsville, AL (November 6–8, 2012).

14 Tucker, J., Bauer, R., Springer, J. et al. (2018). GOES-R lithium-ion battery life test & workhorse battery performance. In: *Proceedings of the NASA Aerospace Battery Workshop*, Huntsville, AL (November 26–27, 2018).

15 Poghosyan, A. and Golkar, A. (2017). CubeSat evolution: Analyzing CubeSat capabilities for conducting science missions. *Prog. Aerosp. Sci.* 88: 59–83.

16 Bouwmeester, J. and Guo, J. (2010). Survey of worldwide pico- and nanosatellite missions, distributions and subsystem technology. *Acta Astronaut.* 67: 854–862.

17 Buchen, E. (2015). Small satellite market observations. *AIAA/USU Conference on Small Satellites*, Logan, UT

18 Woellert, K., Ehrenfreund, P., Ricco, A.J., and Hertzfeld, H. (2011). CubeSats: Cost-effective science and technology platforms for emerging developing nations. *Adv. Space Res.* 47: 663–684.

19 NASA CubeSat Launch Initiative, www.nasa.gov/directorates/heo/home/CubeSats_initiative (accessed 27 July 2022).

20 Swartwout, M.A. (2020). Associate Professor at St. Louis University, CubeSat Database website: www.sites.google.com/a/slu.edu/swartwout/home/cubesat-database (accessed 27 July 2022).

21 Kopacz, J.R., Herschitz, R., and Roney, J. (2020). Small satellites and overview and assessment. *Acta Astronaut.* 170: 93–105.

22 Selva, D. and Krejci, D. (2012). A survey and assessment of the capabilities of CubeSats for Earth observation. *Acta Astronaut.* 74: 50–68.

23 Welle, R.P., Janson, S., Rowen, D., and Rose, T. (2015). CubeSat-Scale laser communications. In: *31st Space Symposium*, Colorado Springs, CO (April 2015).

24 Archer, M.O., Horbury, T.S., Brown, P. et al. (2015). The MAGIC of CINEMA: First in-flight science results from a miniaturized anisotropic magneto-resistive magnetometer. *Ann. Geophys.* 33 (6): 725–735.

25 Woods, T.N., Caspi, A., and Chamberlin, P.C. et al. (2017). New solar irradiance measurements from the miniature X-ray solar spectrometer CubeSat. *Astrophys. J.* 835 (122): 1–6.

26 Gerhardt, D.T. and Palo, S.E. (2016). Volume magnetization for system-level testing of magnetic materials within small satellites. *Acta Astronaut.* 127: 1–12.

27 Lee, D.Y., Park, H., Romano, M., and Cutler, J. (2018). Development and experimental validation of a multi-algorithmic hybrid attitude determination and control system for a small satellite. *Aerosp. Sci. Technol.* 78: 494–509.

28 Palo, S., O'Conner, D., DeVito, E. et al. (2014). Expanding CubeSat capabilities with a low cost transceiver. In: *28th Annual AIAA/USC Conference on Small Satellites*. Logan, UT.

29 Welle R.P., Janson S., Rowen D., Rose T. (2015) Cubesat-scale laser communications. *31st Space Symposium, Technical Track*. Colorado Springs, Colorado (April 13–14, 2015).

30 Peral, E., Imken, T., Sauder, J. et al. (2017). RainCube, a Ka-band precipitation radar in a 6U CubeSat *5th Workshop on Advanced RF Sensors and Remote Sensing Instruments & 3rd Ka-band Earth Observation Radar Missions Workshop*, Noordwijk, Netherlands (September 12–14, 2017).

31 Pingree, P.A. (2014). Look Up: The MCubed/COVE Mission. *CubeSat Developer's Workshop.*

32 Reising, S.C., Kummerow, C.D., Chandrasekar, V. et al. (2017). Temperal experiment for storms and tropical systems technology demonstration (TEMPEST-D) mission: Enabling time-resolved cloud and precipitation observation from 6U-class satellite constellations. In: *31st Annual AIAA/USU Conference on Small Satellites*, Logan, UT.

33 Klesh, A. and Krajewski, J. (2015). MarCO: CubeSats to Mars in 2016. In: *29th Annual AIAA/USC Conference on Small Satellites*, Logan, UT.

34 Elbrecht, A., Dech, S., and Gattscheber, A. (2011). 1U CubeSat Design for increased power generation. In: *1st IAA Conference on University Satellite Missions and CubeSat Workshop*, Rome, Italy.

35 Clark, C. (2010). Huge power demand. . .Itsy-Bitsy Satellite: Solving the CubeSat Power Paradox. In: *24th Annual AIAA/USU Conference on Small Satellites*, Logan, UT.

36 Clark, C.S. and Simon, E. (2007). Evaluation of lithium polymer technology for small satellite applications, SSC07-V-9. In: *Annual AIAA/USA Conference on Small Satellites.*

37 Clark, C. (2008). The Space E-Commerce Revolution, SSC08-I-4. In: *22nd Annual AIAA/USU Conference on Small Satellites.*

38 Edpuganti, A., El Moursi, M.S., and Al-Sayari, N. (2022). A comprehensive review on CubeSat electrical power system architectures. *IEEE Trans. Power Electron.* 37 (3): 3161–3177.

39 Steffan, S. and Semrau, G. (2014). High power density modular electric power system for aerospace applications. In: *AIAA Propulsion and Energy Forum, 12th International Energy Conversion Engineering Conference*, Cleveland, OH (July 28–30, 2014).

40 Hu, X., Deng, S., Suo, J., and Pan, Z. (2009). A high rate, high capacity and long life ($LiMn_2O_4$+AC)/$Li_4Ti_5O_{12}$ hybrid battery-supercapacitor. *J. Power Sources* 187: 635–639.

41 Chin, K.C., Smart, M.C., Brandon, E.J. et al. (2014). Lithium-ion battery and super-capacitor hybrid energy system for low temperature SmallSat applications. In: *28th Annual AIAA/USU Conference on Small Satellites*, Logan, UT (August 2014).

42 Chin, K.B., Bolotin, G.S., Smart, M.C. et al. (2021). Flight demonstration of a hybrid battery/supercapacitor energy storage system in an earth orbiting CubeSat. *IEEE Aerosp. Electron. Syst. Mag.* 36: 24–36.

43 The European Space Agency (2019). *ESA Convention and Council Rules of Procedure*, ESA SP-1337, 8e. ESA Publications.

44 European Space Agency (2012). *Sentinel-1: ESA's Radar Observatory Mission for GMES Operational Services*, ESA SP-1322/1. ESA Publications.

45 Catalano, T.F., Costantini, S., and Daprati, G. (2011). Sentinel-1 EPS architecture and power conversion trade-off. In: *9th European Space Power Conference*, Saint-Raphaël, France.

46 Curzon, D. and Schrantz, K. (2015). Large volume production of lithium-ion battery units for the space industry. In: *NASA Aerospace Battery Workshop*, Huntsville, AL.

47 Bard, F., Carré, A., Fernandez, P. et al. (2016). In-orbit trend analysis of Galileo satellites for power sources degradation estimation. In: *11th European Space Power Conference*, Porto Palace, Thessaloniki, Greece (October 3–7, 2016).

48 Douay, N. (2011). Galileo IOV electrical power subsystem relies on Li-ion battery charge management controlled by hardware. In: *9th European Space Power Conference*, Saint-Raphaël, France (June 6–10, 2011).

49 UTC Aerospace Systems (2017). *NASA Extravehicular Mobility Unit (EMU) LSS/SSA Data Book.* Houston, TX: NASA Johnson Space Center.

50 Richards, P.W., Wagner, K., King, R. et al. (1999). Pistol-grip torque-measuring power tool. *NASA Tech Briefs* 23 (7): 36.

51 Iannello, C.J., Barrera, T.P., Doughty, D. et al. (2017). *Simplified Aid for Extra-Vehicular Activity Rescue (SAFER) Battery Assessment*, Hampton, VA: NASA Engineering Safety Center (NESC), March 30, 2017. NASA/TM-2018-219818.

52 Marmolejo, J., Landis, P. and Sommers, M. (2002). Delivery of servicing & performance checkout equipment to the International Space Station Joint Airlock to Support Extravehicular Activity: SAE Technical Paper 2002-01-2366, 2002. In: *International Conference on Environmental Systems.*

53 Russell, S., Delafuente, D., and Darcy, E. (2017). *JSC 20793, Crewed Space Vehicle Battery Safety Requirements.* Houston: NASA Johnson Space Center.

54 Russell, S.P., Elder, M.A., Williams, A.G. et al. (2010). The Extravehicular Mobility Unit's new long life battery and lithium ion battery charger. In: *AIAA Space 2010 Conference and Exposition.* AIAA. https://doi.org/10.2514/6.2010-8917.

55 National Aeronautics and Space Administration (1998). *The Space Shuttle Extravehicular Mobility Unit (EMU), an Activity Guide for Technology and Education.* Washington DC: NASA.

6

Planetary Spacecraft Batteries

Marshall C. Smart and Ratnakumar V. Bugga

6.1 Introduction

Planetary missions use both rechargeable and non-rechargeable batteries as part of their electric power subsystem (EPS). While lithium primary batteries are preferentially used as non-rechargeable energy storage devices, rechargeable Li-ion batteries (LIB)s are the preferred energy storage option for numerous planetary exploration missions due to their superior performance compared to aqueous systems. With the wide variety of electrode and electrolyte materials and cell designs available, LIB technology can be specifically tailored to mission-specific requirements of energy, power, and low temperature operation. Not surprisingly, aided by their superior performance characteristics, LIBs have successfully supported numerous planetary exploration missions in the past two decades, operating in conjunction with either photovoltaic or nuclear radioisotope power sources in robotic missions, which include planetary orbiters, landers, rovers, and sample return capsules (SRCs) to different destinations. These destinations include the Moon, Mars, Jupiter, Venus, and various asteroids, such as Itokawa and 162173 Ryugu. As the use of LIBs continues to expand, the technology has been rapidly improving, with higher specific energy, energy density, and cycle life characteristics. This chapter describes the various planetary missions that have utilized LIB technologies and the prospects for LIBs for future planetary missions [1].

6.2 Planetary Mission Battery Requirements

In contrast to commercial LIBs, space LIBs have unique design and performance characteristics, many of which arise due to the unique environments encountered. Some of the key characteristics are that they must be highly reliable, robust, and safe. Planetary exploration missions have great scientific, national, and global importance, and typically have a high-budget value. Therefore, batteries, like other critical items, are designed with extreme care and caution and are expected to function well without any single-point failures.

Space LIB design characteristics are driven by stressful electrical, mechanical, thermal, and environmental requirements. LIB specification requirements are specific to particular planetary mission objectives since the environments to which the batteries are exposed can be distinctly different and the configuration on the spacecraft can vary. However, there are a number of requirements that are common to various classes of planetary missions (ranging from "Discovery" to

Spacecraft Lithium-Ion Battery Power Systems, First Edition. Edited by Thomas P. Barrera.

"Flag-Ship" missions), depending upon the final destination and the operational service life of the battery. Beyond any common generic requirements, it is imperative that space batteries meet the form, fit, and functionality needs of a particular mission.

6.2.1 Service Life and Reliability

Since batteries cannot be replaced on the spacecraft after it is launched, they need to demonstrate long service life characteristics. Besides the operational lifetime after launch, the lifetime of the battery must account for pre-launch acceptance testing, spacecraft integration and testing, and pre-launch storage. The operational lifetime can be very long for some planetary missions (more than 10 years), owing in part to long cruise periods to reach their destinations, including the outer planets and their moons. To meet these lifetime requirements, the battery design and sizing most often accounts for the possibility that one or more battery cell strings fails at any point in the mission, such that the remaining cell strings can effectively support the primary mission.

Several steps are involved in achieving a high-reliability battery for a space mission: (i) the chemistry is first validated for mission needs through extensive testing and demonstration of the performance in the mission environments, with adequate margins and uncertainties, (ii) the manufacturing methods for the cells should provide good cell-to-cell consistency, (iii) the cells and batteries are tested in the flight environments to qualify the battery, and (iv) the cells are subjected to a rigorous screening process as part of the acceptance procedure, to exclude any cells that fall outside the acceptable performance levels that could compromise the successful operation, performance, and service life of the battery, as described in Chapter 2.

6.2.2 Radiation Tolerance

Some of the outer planetary missions require batteries to survive high-intensity radiation. For example, Jupiter is surrounded by an enormous magnetic field and charged particles are trapped in the magnetosphere that forms intense radiation belts 10 times stronger than Earth's Van Allen belts. Inherent tolerance of the LIB to high total ionizing doses of radiation is preferred compared to radiation shielding, which increases spacecraft mass. The potential risks associated with high radiation exposure include decomposition of the organic electrolyte, loss of separator integrity, electrode delamination, and possible loss of functionality of electrochemically inert materials, all culminating in decreased capacity, increased impedance, loss of performance, and possibly cell failure. A number of studies at the NASA Jet Propulsion Laboratory (JPL) and elsewhere have tested and analyzed the tolerance of Li-ion cells to radiation. These studies involved irradiation of cells with ^{60}Co gamma rays, which were determined to effectively simulate the high-energy electrons and ions in the environment around Jupiter. A number of cell chemistries and cell designs have been evaluated, including: (i) Yardney 7 Ah NCO-based cells [2], (ii) Saft 9 Ah NCA-based cells [2], (iii) Sony HC 18650 LCO-based cells [3], (iv) Panasonic NCRA and NCRB cells [4, 5], E-One Moli ICRM cells [4, 5], and LG Chem MJ1 cells [6]. The results of these studies demonstrated good tolerance to total ionizing doses of up to 20 Mrad with minimal capacity loss and no failures.

6.2.3 Extreme Temperature

LIBs for planetary missions should ideally have the ability to operate under extreme temperatures. Missions exploring inner planets commonly require high temperature batteries, especially for aerial and surface missions. For example, missions that land spacecraft on the surface of Venus would need batteries to survive 465 °C, whereas missions to Mars and the outer planets benefit from low temperature batteries, since the environment can be as low as −120 °C. Although these spacecraft do possess complex thermal management systems to maintain the batteries in the required operating temperature ranges, they consume part of the battery's energy, which otherwise could be allocated to other spacecraft loads. Therefore, Li-ion cell technologies capable of operating over wide operating temperature ranges are preferred due to the benefits of mass and energy savings.

6.2.4 Low Magnetic Signature

Due to sensitive on-board instrumentation used for science operations, some missions require the battery to exhibit a very low magnetic signature. For example, the NASA Magnetospheric Multiscale (MMS) Mission, which consists of four separate spacecraft, studies how magnetic fields around Earth connect and disconnect. Each MMS observatory possesses 11 scientific instruments that contain 25 different sensors (such as magnetometers), some of which are highly sensitive to any magnetic field that might be generated by the battery. Depending upon the Li-ion cell chemistry, the electroactive materials can be ferromagnetic (such as certain cathode materials) and may require demagnetization. Additional measures can be implemented to the battery design with respect to the wire routing scheme to cancel out the magnetic properties of current flowing through the battery wire harness.

6.2.5 Mechanical Environments

As described in Chapter 3, LIBs are designed to survive high levels of vibration and acceleration environments during launch and pyrotechnic events. However, planetary mission batteries also experience shock levels from various pyro events during planetary entry, descent, and landing. Battery vibration and shock level requirements depend on spacecraft location, mechanical interface, and structural dampening measures. In the case of planetary surface operation, besides the normal launch stresses, compliance to mechanical environments associated with the entry, decent, and landing (EDL) operations process is required.

In addition to shock and vibration requirements, the LIBs must possess the ability to operate under space vacuum conditions ($<1\times10^{-6}$ Torr) without the possibility of the outgassing of any undesirable materials that may contaminate surrounding spacecraft components. Another requirement for the batteries will be orientation insensitivity, since they will need to perform in any orientation.

6.2.6 Planetary Protection

Unlike Earth-orbiting satellites, LIBs that are developed for planetary missions need to comply with planetary protection requirements. The purpose of planetary protection is to prevent harmful contamination of pristine planetary bodies that could potentially confuse scientific measurements related to detecting the presence of biosignatures or extant life. The closely related discipline of contamination control is intended to avoid damaging the sensitivity or operations of on-board scientific instruments. Depending upon the levels of bio-reduction required for the specific mission, it can be quite challenging to effectively sterilize the battery. Typically, most spacecraft components are subjected to DHMR (dry heat microbial reduction) procedures, which involve exposing components to high temperatures for several hours. In the case of the Mars Viking mission, nickel-cadmium (Ni-Cd) batteries were sterilized by exposure to temperatures up to 135 °C for 40 hours [7]. LIBs cannot be exposed to such high temperatures without permanent performance loss or failure. In lieu of DHMR procedures, irradiation to approximately 10 Mrad of Υ-radiation combined with surface sterilization (such as through the use of vapor hydrogen peroxide treatment) is one proposed option to sterilize space LIBs. However, this sterilization protocol has not yet been established by international space agencies and relevant governing bodies. Currently, LIBs are built with hermetically sealed cells using rigorous battery manufacturing protocols and are thoroughly cleaned externally after assembly with isopropanol to comply with planetary protection requirements and contamination control requirements.

6.3 Planetary and Space Exploration Missions

Planetary space exploration and Earth-orbiting spacecraft missions can be classified into robotic (unmanned) and human-rated (crewed) exploration mission categories. The robotic missions include: (i) orbiters, (ii) fly-by and sample return missions, (iii) landers, (iv) rovers, (v) probes, (vi) penetrators and impactors, (vii) sample return capsules, and (viii) miscellaneous science missions. Given the challenges associated with human exploration from the extreme environments and the distances involved, robotic exploration is often preferred for planetary exploration missions. Due to rapid advances in robotics, deep space communication, and advanced power generation and storage technologies, NASA has successfully completed a number of robotic missions exploring various planetary bodies in the solar system.

6.3.1 Earth Orbiters

In addition to numerous Earth-orbiting commercial and government satellite systems, many terrestrial satellites have been launched to perform various science functions. For example, in support of NASA's Earth Science Division, there have been many orbiting spacecraft launched in order to further understand Earth's interconnected systems. LIB technologies have supported numerous missions focused on science related to the Earth, including studying the atmosphere, the climate, the ocean, land and vegetation, etc.

An example of the transition from nickel-hydrogen (Ni-H_2) to LIB technology was the Gravity Recovery and Climate Experiment (GRACE) mission, which was a joint NASA-German Aerospace Center program and focused upon obtaining detailed measurements of Earth's gravity field anomalies. The GRACE mission consisted of two satellites utilizing 16 Ah Ni-H_2 batteries operating in low Earth orbit (LEO). Launched in 2018, the GRACE Follow-On (or GRACE FO) spacecraft (built by Airbus Defense and Space) [8] utilized 78 Ah LIBs (manufactured by EnerSys/ABSL), which provided an average of 355 W of electrical power during eclipse.

Another example of a NASA Earth-orbiting satellite utilizing LIBs is the Soil Moisture Active–Passive (SMAP) mission, which was launched in January of 2015 to measure the soil moisture content and its freeze/thaw state [9]. This spacecraft utilizes LIBs that consist of four batteries manufactured by EnerSys/ABSL (one 8s52p module and three 8s10p modules) comprised of Sony HCM 18650 cells. The primary mission requirements included successful operation for at least 40 months and completion of 4000 cycles at a DOD of 25%. After completing the primary mission in 2018, the mission was extended through 2023 owing to nominal LIB and spacecraft health.

6.3.2 Lunar Missions

After a very prolific period of lunar exploration in the 1960s and 1970s, primarily by the United States and the Soviet Union, who were engaged in a "Space Race," the number of missions to the Moon steadily declined in the ensuing decades. However, in recent years, there has been considerable interest in revisiting the Moon with more sophisticated instrumentation employed on unmanned spacecraft. These missions, which have been launched by several nations, have primarily consisted of orbiting spacecraft performing scientific measurements and reconnaissance. However, in recent years there have been some examples of surface missions landing on the lunar surface. Although not exhaustive, the discussion highlights some of these more recent missions that have utilized LIBs with an emphasis on the performance requirements and the on-orbit performance.

6.3.2.1 Gravity Recovery and Interior Laboratory

On September 10, 2011, NASA launched the Gravity Recovery and Interior Laboratory (GRAIL) mission as part of its Discovery Program to perform high-quality gravitational field mapping of the Moon, with the goal of determining its interior structure [10]. The mission consisted of two separate spacecraft (GRAIL A-Ebb and GRAIL B-Flow) using the same launcher.

The twin GRAIL spacecraft were based upon the heritage Experimental Satellite System-11 (or XSS-11) that was manufactured by Lockheed Martin Astronautics. The EPS consisted of two non-articulated solar arrays used in conjunction with a LIB manufactured by Yardney Technical Products, Inc. (now EaglePicher Technologies). The battery consisted of a 28 V, 30 Ah design containing eight NCP 25-1 prismatic cells electrically connected in series. The NCP 25-1 cells contained a heritage chemistry (MCMB-NCO) that was used on many other planetary missions. In terms of the mission, the battery was required to provide at least 1500 cycles at 40% DOD with excursions to 70% DOD, to operate within a temperature range of 0 to +30 °C, and to support an on-orbit operational life of five years. In practice, the on-orbit power duty cycle was less severe than expected and the actual mission was only one year in duration.

6.3.2.2 Lunar Crater Observation and Sensing Satellite

Launched in 2009 aboard an Atlas V launch vehicle, the Lunar Crater Observation and Sensing Satellite (LCROSS) was a NASA robotic mission that has a secondary payload along with the Lunar Reconnaissance Orbiter (LCO) [11]. The mission was a Lunar impactor with the objective of taking measurements of water ice present in the permanently shadowed craters at the poles. LCROSS benefited from the use of an EELV Secondary Payload Adaptor (ESPA) ring as the primary structure, which greatly reduced the cost of the mission. LCROSS was a NASA Class D mission attempting to usher in a new approach for NASA missions to use commercial off-the-shelf (COTS) parts to reduce schedule impact and cost. The mission utilized four 8s16p LIBs manufactured by EnerSys/ABSL, which contained the Sony HCM 18650 cells with a battery nameplate capacity of 24 Ah and a nominal bus voltage of 28 V. This 8s16p module was used on a number of other missions, including NASA-JPL Kepler, NASA-JPL NuSTAR, NASA-Ames LADEE, and NASA-MIT TESS. The battery was qualified for an operational temperature range of −5 to +40 °C, pyro-shock load of 1907 g, and an overall 10.8 G_{RMS} during random vibration testing. Due to the COTS 18650 cell-based battery 8s16p topology, cell consistency, and uniformity in performance, the flight battery does not require cell-to-cell balancing.

6.3.3 Mars Missions

Over the last two decades NASA has successfully utilized space LIBs in several Mars missions, which include orbiters and surface missions. Because of the proximity of Mars, the cruise time is short, ranging from 7 to 11 months, and the surface conditions of Mars are relatively benign with no issues from pressure or radiation. The surface temperature can be, however, low (about −120 °C), requiring a combination of low-temperature LIBs and appropriate thermal management.

6.3.3.1 Mars Orbiters

Mars orbiter missions are primarily utilized to map planetary surface and geological formations, to assist in the understanding of the atmosphere, and to monitor daily global weather. In addition, orbiting spacecraft can serve as a communication link between the Earth and other planetary surface missions, such as a lander or rover. Due to the proximity to the Sun, Mars orbiters have traditionally used solar arrays as the primary power source, which is augmented by rechargeable batteries. Since these spacecraft orbit Mars approximately 12 times per day, rechargeable batteries used for these applications are required to meet unique cycle life characteristics (approximately 4380 cycles per year) and provide a lifetime of at least 5–10 years after initial fabrication. Until recently, Mars orbiters have been dominated by the use of rechargeable Ni-H_2 batteries, since most missions were launched prior to the maturation of space Li-ion technologies. Launched in 1996, the Mars Global Surveyor (MGS) mission incorporated an Ni-H_2 battery-based EPS. Due to a series of events involving human ground operator error, the spacecraft was inadvertently reoriented such that one of the two common pressure vessel (CPV) Ni-H_2 batteries [12] was exposed to direct sunlight, causing it to overheat, leading to the depletion of capacity from both batteries. Prior to the failure, the MGS mission had been the longest operating spacecraft at Mars. More recently, other Mars orbiters, such as the Mars Reconnaissance Orbiter and the Mars Odyssey missions, utilize Ni-H_2 batteries. These two orbiters are still operational and supporting the surface assets, Mars Curiosity and Mars Perseverance rovers, which are being powered by LIBs. A summary of the successful Mars Orbiter missions that have launched since 2000 is shown in Table 6.1.

Mars Express In contrast to the NASA approach of using custom cells for Mars missions, the European Space Agency (ESA) Mars Express used COTS cells for many of their planetary missions, starting with the Mars Express, which included a lander named Beagle 2. The LIB used on the orbiting spacecraft was manufactured by EnerSys/ABSL UK and consisted of Sony HCM 18650 cells in three 6s16p modules. The three modules were electrically connected in parallel to provide a battery nameplate capacity of 67.5 Ah [13]. After 18 years and 9 months, the battery is still operational. As of 2022, the battery had entered the 21st eclipse season and has completed over 47 000 cycles, with the majority being of very shallow DOD (<5%). Thus, the Mars Express mission

Table 6.1 Summary of Mars Orbiter missions since 2000.

Mission	Launch date	Battery design	Battery chemistry	Capacity (Ah) rated	In-orbit life (Years)
2001 Mars Odyssey	April 2001	11s	Ni-H_2	16 Ah	>19
Mars Express	June 2003	6s16p	Li-ion	67.5 Ah	>18
Mars Reconnaissance Orbiter	August 2005	2 × 11 s	Ni-H_2	2 × 50 Ah	>15.5
Mars Orbiter Mission (Mangalyaan)	November 2013	Unknown	Li-ion	36 Ah	6.0
MAVEN	November 2013	8s2p	Li-ion	2 × 55 Ah	7.44
ExoMars Trace Gas Orbiter	March 2016	Unknown	Li-ion	Unknown	4.37
Emirates Mars Mission (Hope)	July 2020	Unknown	Li-ion	Unknown	0.625 (on-going)

represents the second longest surviving active spacecraft orbiting around another planetary body other than Earth, only being surpassed by the 2001 Mars Odyssey. However, it holds the record for the longest surviving planetary mission with LIBs.

MAVEN In June of 2013, NASA launched the Mars Atmosphere and Volatile EvolutioN (MAVEN) spacecraft as part of its "Mars Scout" program, with the objective of investigating the nature of the upper atmosphere and ionosphere of Mars and how it interacts with the solar wind [14, 15]. The orbiting spacecraft is powered by solar arrays that were used in conjunction with two 28 V, 55 Ah LIBs manufactured by Yardney Technical Products, Inc. This technology consists of MCMB anodes, $LiNiCoO_2$ cathodes, and a low-temperature electrolyte that was originally developed at JPL in the late 1990s. The same cell chemistry was also utilized for the Juno mission. The battery telemetry includes temperature and battery voltage measurements as well as internal cell balancing and cell monitoring electronics. The cell balancing circuitry consisted of a custom Yardney design that was also used in the Grail mission. In addition to cell balancing capability, the electronics provided an overvoltage indication and individual cell voltage analog outputs.

The primary science mission of MAVEN was only one year in duration, which started in November of 2014. After performing productive science investigations, including measuring how volatile gases are swept away by the solar wind and characterizing the entire upper atmosphere, the science phase was extended through September 2016. After completing one full Martian year of scientific investigations, the mission was approved again for another mission extension through September of 2018. To serve as a communication relay for landers and future rovers (such as the Perseverance rover, which successfully landed in February of 2021), the MAVEN spacecraft entered a lower orbit around Mars in April of 2019 by implementing a successful aerobraking maneuver.

6.3.3.2 Mars Landers

Missions in which an immobile probe is landed on the surface of the planet is referred to as a lander spacecraft. In general, a number of scientific objectives can be pursued with the use of Mars landers, including investigations on the chemical composition of the surface, and any potential biology, meteorology, seismology, and magnetic properties. The landers are generally exposed to extreme environmental temperature ranges as the Mars surface temperature can vary widely between −120 and +30 °C in a single Martian day (or sol). Due to the design complexity and the cost of landing on Mars, it is highly desirable to limit the mass and volume of the lander. Furthermore, to allow for sufficient science equipment and adequate landing systems, reducing the mass and volume of the battery translates into more spacecraft capabilities.

The first successful US missions to land on the surface of Mars were the Viking 1 and Viking 2 landers, which reached the surface in July 1976 and September 1976, respectively [16]. One of the key objectives of the Viking landers was to perform life detection experiments in an attempt to determine the existence of present or extant life in the Martian soil. The power generation sources for the landers were radioisotope thermoelectric generator (RTG) units that contained ^{238}Pu. The energy storage devices utilized were rechargeable Ni-Cd batteries, which were designed to meet lander peak power demands. Viking 1 lander operated on the surface of Mars for over six years but the mission ended due to a human error that resulted in the antenna retracting, thus ceasing communication and power. Viking 2 lander operated for over three years and seven months prior to a battery failure that ended the mission. Since this early success, there was a considerable delay until the next successful lander reached Mars in the late 1990s. A summary of various Mars Lander missions is shown in Table 6.2.

Table 6.2 Summary of Mars Lander missions.

Mission	Launch date	Battery design	Battery chemistry	Capacity (Ah) rated	Surface life (Years)
Viking 1	August 1975	4×12s1p	Ni-Cd	4×8 Ah	6.25
Viking 2	September 1975	4×12s1p	Ni-Cd	4×8 Ah	3.50
Mars Pathfinder	December 1996	18s1p	Ag-Zn	50 Ah	0.23
Phoenix Lander	May 2008	8s2p	Li-ion	2×25 Ah	0.44
InSight Mars Lander	May 2018	8s2p	Li-ion	2×25 Ah	2.27 (on-going)

Mars Pathfinder The next successful Mars lander mission was the Mars Pathfinder that landed on Mars in July of 1997. This mission consisted of a lander that served as a base station and a robotic rover named Sojourner, which became the first rover to ever operate beyond the Earth and Moon. At the time, the Mars Pathfinder mission represented one of the most successful Mars surface exploration missions for NASA, by providing a significant amount of the data, over 16 500 images from the lander and 550 images from the rover, and conducted a number of studies on the chemical analysis of the rocks and soils. The energy storage device on the Pathfinder lander consisted of a silver-zinc (Ag-Zn) rechargeable battery, which was used in conjunction with solar arrays [17]. Although Ag-Zn batteries generally exhibit moderately high specific energy and power density characteristics, the cycle life is poor compared to other space rechargeable battery chemistries. For the Mars Pathfinder mission, the battery was required to provide 40 operational cycles, to survive 14 months of total wet stand life, and to have the ability to be launched in an inverted orientation. Upon landing, the Ag-Zn battery was managed successfully such that it survived its design life and provided nearly three times beyond the cycle life requirement [18]. The small Mars Sojourner six-wheel rover that was carried aboard the Pathfinder lander was equipped with a solar array that was able to generate up to 16 W of power. This energy generation source was augmented by three non-rechargeable lithium thionyl chloride (Li-$SOCl_2$) batteries connected in parallel to form a 3s3p configuration capable of providing 9 V and 12 Ah of capacity designed to power the rover during the night-time and to support communications.

Mars Surveyor Program After the success of the Mars Pathfinder mission, NASA intended to launch the Mars Surveyor Program 2001 (MSP'01) lander with an LIB. This would have represented the first use of Li-ion technology in planetary exploration. Unfortunately, the mission was canceled for programmatic reasons. However, the MSP'01 LIBs were fully developed, designed, and qualified prior to mission cancelation. These batteries, which were fabricated by Yardney Technical Products, Inc., consisted of two strings of eight cells connected in series to provide an operational voltage of 24–32.8 V [19], as shown in Figure 6.1. The cells were of 25 Ah nameplate capacity and consisted of MCMB anodes and Li-$NiCoO_2$ cathodes, with a JPL-developed low temperature electrolyte. The LIBs were required to complete at least 90 cycles on the surface of Mars and operate over a wide temperature range of −20 to +40 °C, with tolerance to non-operational excursions of −30 to +50 °C. Each of the batteries possessed a dedicated charge-control unit, with individual cell bypass features that enabled cell balancing. A unique mission requirement was that the battery was

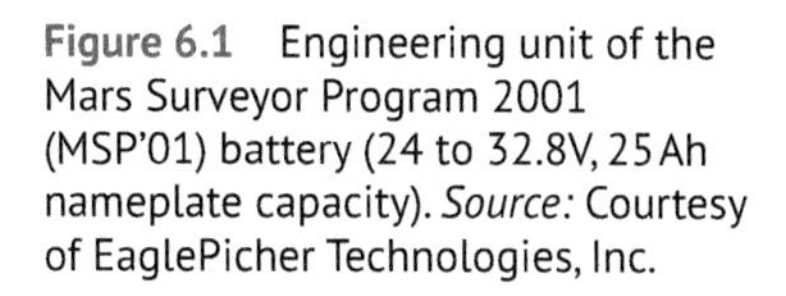

Figure 6.1 Engineering unit of the Mars Surveyor Program 2001 (MSP'01) battery (24 to 32.8V, 25 Ah nameplate capacity). *Source:* Courtesy of EaglePicher Technologies, Inc.

required to be both charged and discharged continuously at C/5 rates at low temperatures down to −20 °C and provide a minimum beginning-of-life (BOL) capacity of 25 Ah. Given that the typical DOD on the surface of Mars was anticipated to be 50% based on the available capacity, a single battery was expected to fulfill most of the needs of the entire mission, so the second battery served as a redundant back-up. Although the mission was canceled, the battery was qualified through ground testing, which resulted in the generation of relevant engineering data that enabled the adoption of this same cell chemistry for a number of subsequent missions.

Mars Phoenix Lander After the success of the Mars Pathfinder and Mars Exploration Rover missions, NASA launched the Phoenix Mars Mission, which landed in the Martian north polar region (Figure 6.2). The Phoenix Mars Mission was the first of the missions in NASA's Scout Program, which were relatively low-cost innovative missions that relied on legacy hardware [20]. Toward this end, the Phoenix Mission utilized the Mars Surveyor 2001 Lander built in 2000 by Lockheed Martin that included the LIB designed and qualified by Yardney Technical Products, Inc., as shown in Figure 6.3. The goal of this mission was to study the history of water in the Martian arctic ice-rich soil and the potential for the habitability of life. The Phoenix Lander completed its three-month mission and met the primary objectives and operated for an additional two months prior to becoming inoperable due to the reduced sunlight associated with the winter, which led to insufficient energy to power the spacecraft.

Mars InSight NASA's Mars InSight mission, which is an acronym for "Interior Exploration using Seismic Investigations, Geodesy, and Heat Transport," is a Discovery mission that placed a single lander, based on the previous Phoenix lander design, to perform geophysical measurements and study its deep interior [21, 22]. After a delay to repair the prime instrument in the science payload, the InSight mission was launched on May 5, 2018, and landed on November 26, 2018, in the Elysium Planitia region of Mars. The InSight mission required a higher specific energy battery (approximately 15% more energy) than in the Phoenix mission, that could operate over a wider temperature range, with both charging and discharging between −30 and +35 °C. A calendar life of four years and operation on the surface of Mars for 709 sols (a duration of over one Martian year) were also required.

Figure 6.2 The Mars Phoenix Lander on the surface of Mars. *Source:* Courtesy of NASA-JPL.

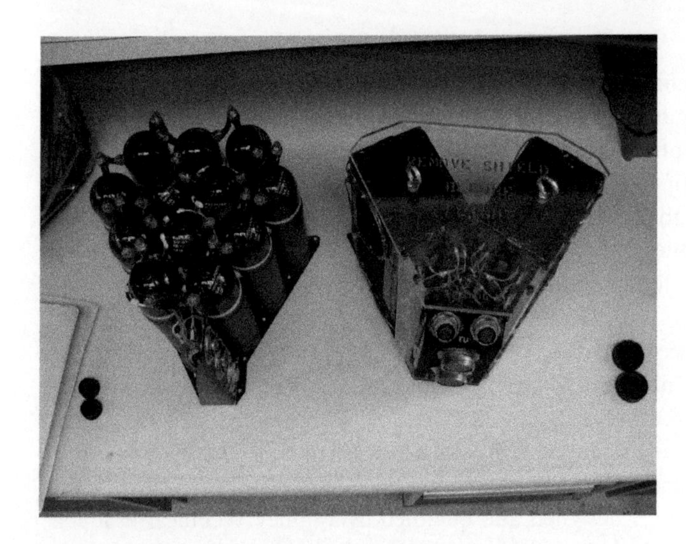

Figure 6.3 The Mars Phoenix Lander battery manufactured by Yardney Technical Products, Inc. compared to a mock-up of a single Ni-H2 battery. *Source:* Courtesy of EaglePicher Technologies, Inc.

To facilitate this, a lower temperature electrolyte consisting of 1.0 M $LiPF_6$ in ethylene carbonate (EC) + ethyl methyl carbonate (EMC) + methyl propionate (MP) (20 : 60 : 20 vol%) was used [23–25]. The most challenging battery requirement was the capability to both charge and discharge the 25 Ah battery at very low temperatures (−30 °C) using C/5 rates. Besides the challenge of supporting the power and energy requirements, there was a risk that lithium plating on the anode could occur during low-temperature charging, which may lead to performance degradation [26]. Furthermore, there was some concern that the low-temperature capability of the battery could be compromised by being subjected to long, high-temperature operation, making end of mission (EOM) requirements difficult to meet [27].

As a result of this ground qualification test campaign, it was determined that at the BOL the NCA-based cells delivered >15% improvement in delivered capacity and energy at an ambient temperature compared to the heritage NCO chemistry. When cells were both charged and discharged at −25 °C, the InSight NCA-based cells with the MP-containing electrolyte delivered 31% more capacity at −25 °C (>28.7 Ah corresponding to >108 Wh/kg) compared with the heritage NCO-based chemistry (19.6 Ah corresponding to approximately 75 Wh/kg) (Figure 6.4). As the project matured

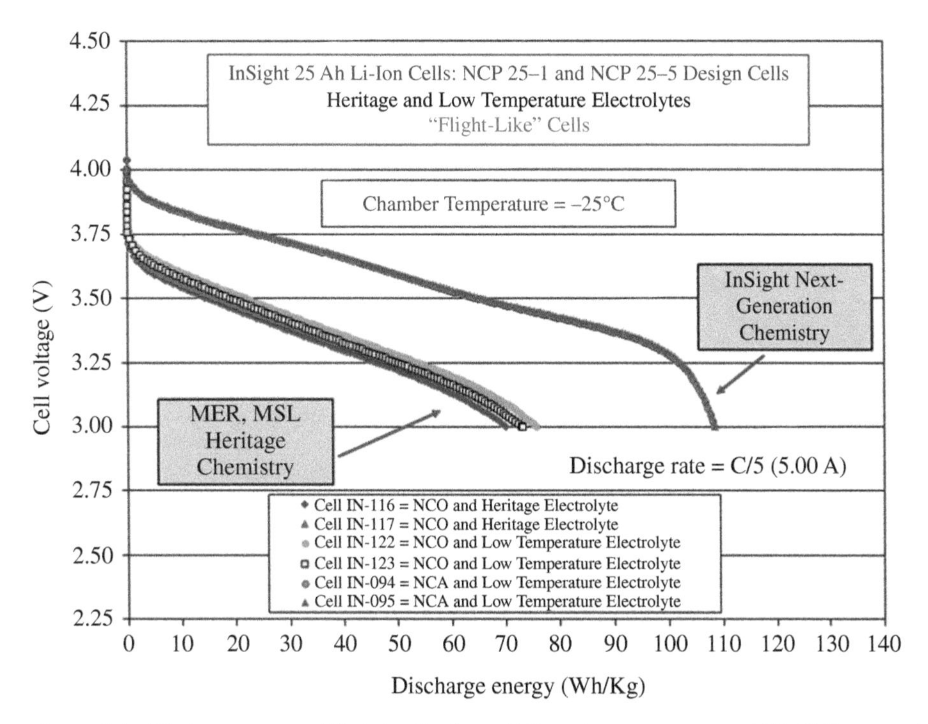

Figure 6.4 Discharge energy (Wh/kg) of heritage NCO and next-generation NCA 25 Ah nameplate Li-ion cells delivered at −25 °C using 5.00 A charge to 4.10 V (with a C/50 taper current cut-off) and a 5.00 A discharge to 3.00 V. Cells were both charged and discharged at −25 °C [27].

during the course of this test campaign, mission projections suggested the possibility of the battery being exposed to very cold environments under certain scenarios, so there was a strong desire to determine the capability of the cells to operating continuously at −30 °C at C/5 rates.

As shown in Figure 6.5, a significant observation under these cycling conditions was that there was no evidence of lithium plating on the negative electrode, as ascertained by the absence of any pronounced higher voltage plateau at the beginning of the subsequent discharge profiles at low temperature. The cycling data at −30 °C presented in Figure 6.5 was performed on the cells after they had completed 234 cycles under a variable temperature cycling regime (alternating the cycling at +30, +20, and −25 °C).

In summary, the qualification test program demonstrated the capability of the cells to meet cycling requirements over the required −30 to +35 °C operating temperature range. Furthermore, the cells were demonstrated to provide over 100 Wh/kg at −40 °C and good performance to temperatures as low as −60 °C when charged at ambient temperatures (over 25 Ah and 76 Wh/kg delivered at a C/10 rate when discharged to 2.0 V). Since the InSight required excellent low-temperature capability throughout the mission, the InSight project adopted the NCA and the MP-containing electrolyte chemistry for the mission.

After landing in November of 2018, the InSight Lander successfully completed over 680 sols of operation, which met the primary mission duration requirements. Due to the success of the mission, it received a mission extension of another two years into 2022. As of this writing, one challenge for the Lander in the months to come is to survive the impeding winter when the energy generation from the solar arrays will be decreased when Mars moves toward aphelion. To exacerbate this further, the power output in February 2021 was only approximately 27% of the initial

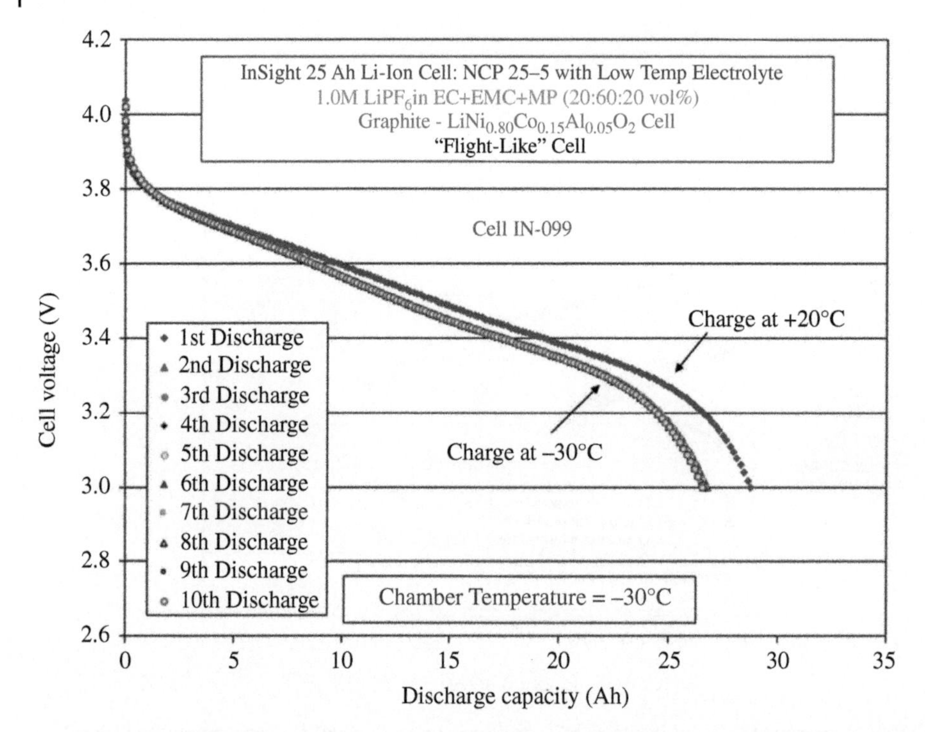

Figure 6.5 Discharge capacity (Ah) of a next-generation NCA 25 Ah nameplate Li-ion cell delivered at −30 °C. Cell was charged at 5.00 A (C/5 rate) to 4.10 V (with a C/50 taper current cut-off) and discharged (C/5 rate) at 5.00 A to 3.00 V. The first discharge displayed was following a charge at +20 °C and all subsequent cycles charged at −30 °C [27].

capacity, due to dust collecting on the solar panels. This will force the ground mission operation planners to reduce the amount of science that is performed during the winter to preserve a positive energy balance of the spacecraft.

6.3.3.3 Mars Rovers

A list of the operational Mars rovers launched to date is summarized in Table 6.3.

Mars Exploration Rovers In 2003, the NASA launched two separate rovers, named "Spirit" and "Opportunity," to explore the surface of Mars (Figure 6.6). Designed to meet the demanding requirements of the MER mission, the LIBs were required to (i) provide energy during launch,

Table 6.3 Summary of Mars Rover missions.

Mission	Launch date	Battery design	Battery chemistry	Capacity (Ah) rated	Surface life (Years)
Mars Sojourner	December 1996	3s3p	Li-$SOCl_2$	36 Ah	0.23
Spirit (MER-A)	June 2003	8s2p	Li-ion	2×8 Ah	6.05
Opportunity (MER-A)	June 2003	8s2p	Li-ion	2×8 Ah	14.4
Curiosity (MSL)	November 2011	8s2p	Li-ion	2×43 Ah	9.70 (on-going)
Perseverance (M2020)	July 2020	8s2p	Li-ion	2×43 Ah	1.08 (on-going)

Figure 6.6 Artist's portrayal of Opportunity on the surface of Mars. *Source:* Courtesy of NASA-JPL.

(ii) support attitude adjustments during a spacecraft cruise, (iii) provide sufficient energy to support surface operations under worst-case conditions, (iv) provide adequate cycle life and calendar life characteristics, and (v) possess the ability to be cycled over a wide operating temperature range (−20 to +30 °C) [27–31].

Upon landing in January of 2004, the two rovers completed the primary mission of operating for at least 90 sols (or 90 Martian solar days), with the objective of at least one rover traveling 600 m on the surface. MER-A, or Spirit, landed on January 4, 2004, and operated for over six years until its final transmission on March 22, 2010. The mission ended for mechanical reasons that were unrelated to the battery, which was doing well. The second rover, MER-B or Opportunity, landed on January 25, 2004, and was operational until June 10, 2018, representing over 14 years of operation and 5111 sols after landing. By this time, the rover had traversed the surface of Mars 45.16 km (28.06 miles). After successful completion of the primary mission objectives, each rover mission was extended five times during their lifetimes. Much of this success can be attributed to the excellent life characteristics that were displayed by the EPS comprising solar arrays and the LIBs.

The LIB technology that was first utilized on the 2003 MER project and was later adopted for a number of NASA-JPL missions to Mars was originally developed under a NASA-DoD consortium that included Yardney Technical Products, the Jet Propulsion Laboratory, USAF-WPAFB, and NASA-Glenn Research Center (GRC) [32, 33]. The chemistry developed consisted of meso-carbon microbeads (MCMB) anode, NCO cathode, and a low temperature ternary all-carbonate-based electrolyte developed at JPL, namely 1.0 M $LiPF_6$ in ethylene carbonate (EC) + diethyl carbonate (DEC) + dimethyl carbonate (DMC) (1 : 1 : 1 vol %) [19, 34–36].

The MER spacecraft were powered by deployable solar arrays that utilize triple-junction GaInP/GaAs/Ge cells [37]. The BOL energy of the solar array was approximately 1000 Wh per sol, but over

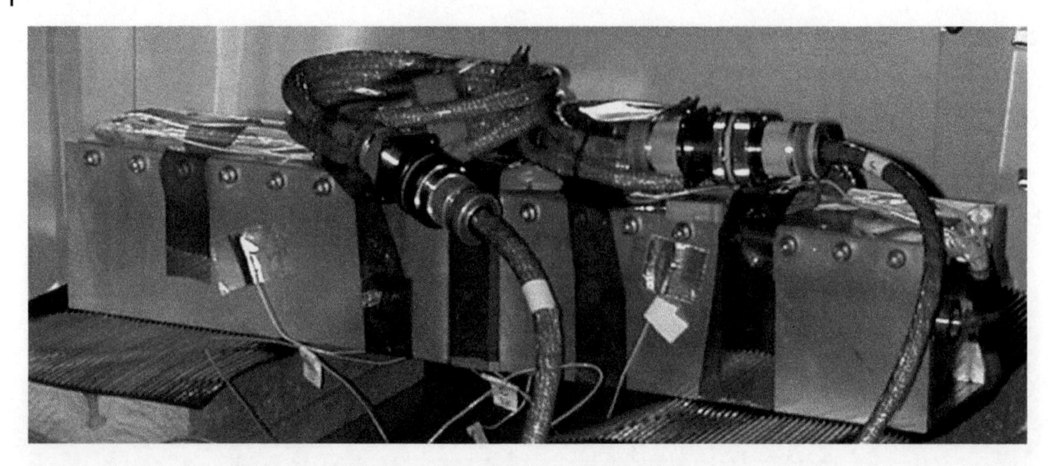

Figure 6.7 Picture of the LIB (24–32.8V, 20 Ah capacity) that is similar to that used on Spirit and Opportunity Mars Exploration Rovers. *Source:* Courtesy of NASA-JPL.

the course of the mission the energy generation was significantly decreased mainly due to the accumulation of dust on the array. Fortunately, the arrays of the Rovers were partially cleaned of the accumulated dust by localized wind, which restored much of the energy generation capability at different times throughout the mission [38]. The energy storage for the Rovers consisted of two strings of eight cells (8s2p) electrically connected in series (16 cells total) to provide an operating voltage range of 24–32.8V and a capacity of 20Ah at BOL (Figure 6.7). During operation, each of these batteries were controlled independently with individual battery control boards (BCBs) that were designed and fabricated at JPL to control the battery charge–discharge characteristics. The BCBs monitored and controlled individual battery cell voltages to prevent cell overcharge and overdischarge. To accomplish cell balancing during charge, the BCBs contained individual cell bypass shunt capabilities through 120-Ω resistors. The batteries were designed to meet a number of requirements, including: (i) providing 200Wh during launch, (ii) providing 160Wh during cruise for supporting anomalies during trajectory control maneuvers (TCMs), (iii) providing 280Wh for surface operations, and (iv) providing energy to fire three simultaneous pyros (each with a load of 7A) multiple times during the EDL sequence.

Spirit operated on the surface of Mars for over six years until its final transmission on March 22, 2010, representing over 2200 sols of operation. Given that the rover was designed to meet a mission requirement of 90 sols of operation, Spirit exceeded the prime mission by over 24 times. Due to variable environmental conditions, there were a number of times during the mission that operational protocols were modified to preserve the health of the rover. For example, in June of 2007, a series of dust storms resulted in a dramatic reduction in the solar energy available due to the atmosphere blocking a very large proportion of the direct sunlight. This resulted in the energy generation being reduced to approximately 128Wh per sol, whereas the arrays would normally produce up to 700Wh at the BOL. To avoid a negative energy balance at a low power output, the rover was placed into the lowest power setting until the storms passed, resulting in a steady decrease of the battery SOC. After the storms had passed there was sufficient energy generation to effectively charge the battery. During the course of the mission, the battery was subjected to 40–45% DOD cycles each sol over a temperature range of 0 to −20 °C, as shown in Figure 6.8. Ultimately, in May 2011, NASA lost contact with Spirit and it speculated that the rover was subjected to excessively cold internal temperatures and that there was inadequate energy to properly power the

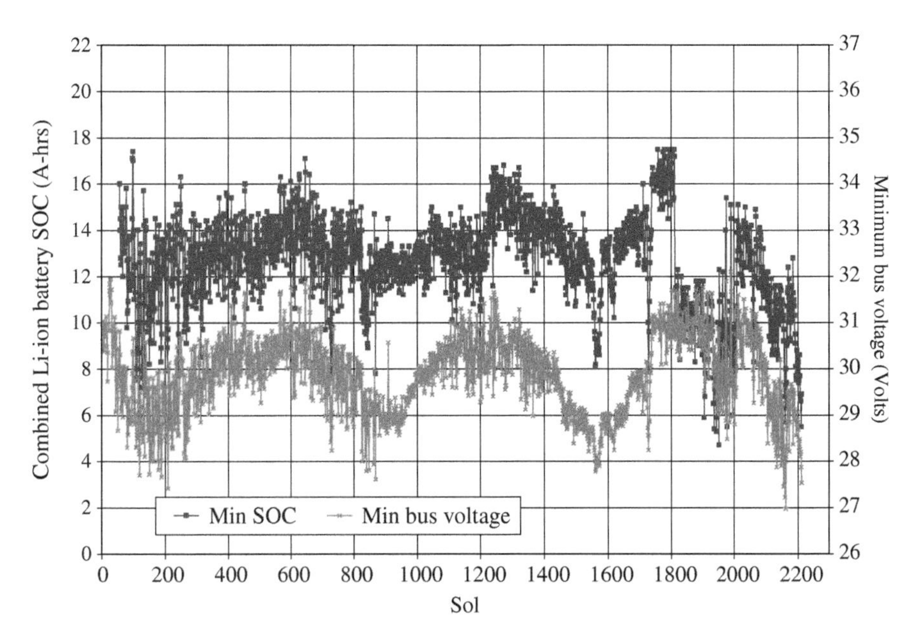

Figure 6.8 Performance of the Spirit battery on the surface of Mars. *Source:* Courtesy of NASA-JPL.

survival heaters. Of the two rovers, Spirit experienced harsher environmental conditions and also became stuck in soft sand that was difficult to extricate itself from.

On the other side of the planet, Opportunity operated on the surface of Mars for over 14 years until it stopped communicating on June 10, 2018. By this time, the rover had been operational for 5111 sols, representing 57 times longer than the design requirement of 90 sols of operation. In early June of 2018, a dust storm engulfed the planet which led to very low power generation by the solar panels. During this period the rover entered a hibernation state, with mission planners hoping to regain contact after the storm had passed. However, NASA was never able to regain contact and speculated that either a catastrophic failure occurred or a thick layer of dust on the solar panels prevented sufficient energy generation.

During its mission, Opportunity supported a similar power profile as Spirit, with approximately a 44–50% DOD duty cycle. In the earlier stages of the mission, the thermal profile was somewhat milder than that of Spirit, but later in its life colder temperatures were experienced, as illustrated in Figure 6.9. In total, from the initial acceptance testing of the battery, which occurred in 2002, to the mission ending in 2018, the wet life of the battery exceeded 16 years and completed over 5000 cycles over a wide temperature range, as shown in Figure 6.10, setting a record for the longest operation on the surface of Mars.

Mars Science Laboratory Curiosity Rover Since the batteries used for the MER Rovers met and far exceeded requirements, the same chemistry was adopted for use on the larger 2011 MSL Curiosity Rover, which successfully landed on the planet Mars on August 6, 2012 [39]. The Curiosity Rover is still in operation, and as of July 2022 had completed over 3540 sols on the surface of Mars. The primary science goals of this mission are to investigate the Martian climate and geology, assess the habitability of the planet, and determine if the Gale crater (where it landed) had previously possessed the environmental conditions favorable for microbial life. By December of 2012, NASA extended the initial two-year mission indefinitely.

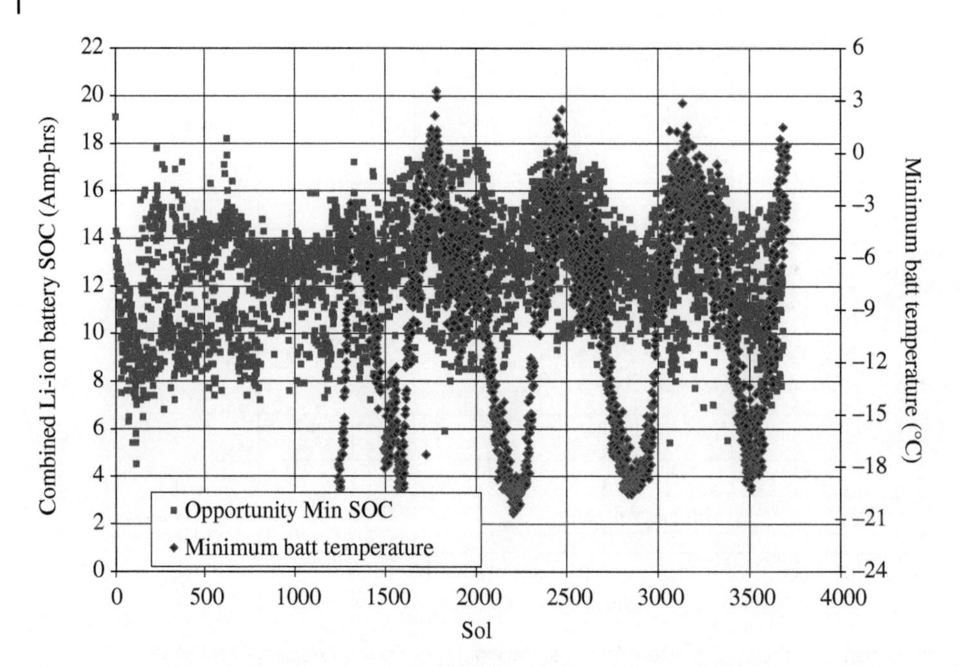

Figure 6.9 Performance of the Opportunity battery on the surface of Mars. *Source:* Courtesy of NASA-JPL.

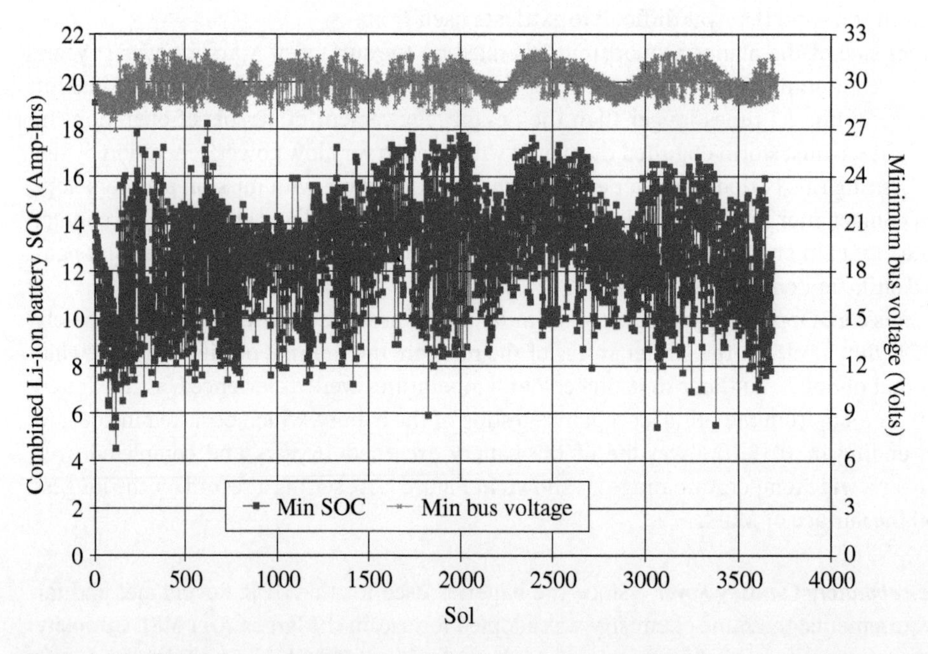

Figure 6.10 Performance of the Opportunity battery on the surface of Mars. *Source:* Courtesy of NASA-JPL.

The MSL Curiosity Rover operates the most advanced suite of instruments ever sent to the Martian surface, which includes the capability to analyze samples scooped from the soil and cored from rocks. Although there are some similarities, the MSL mission has more demanding battery performance requirements when compared to the MER mission, including a longer mission

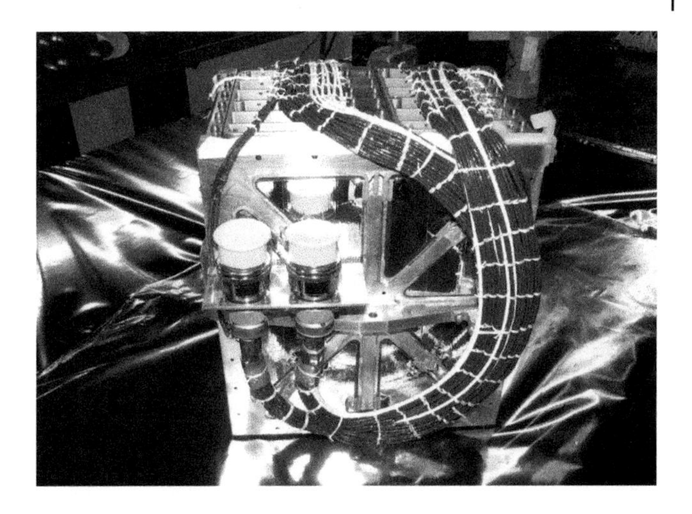

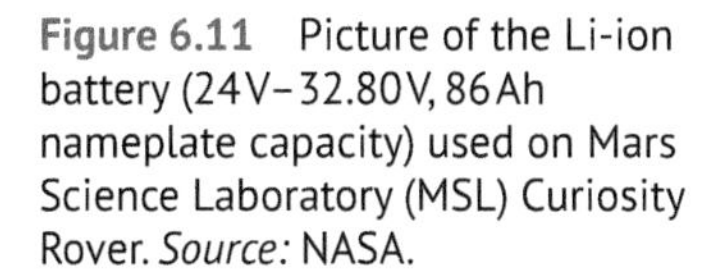

Figure 6.11 Picture of the Li-ion battery (24V–32.80V, 86Ah nameplate capacity) used on Mars Science Laboratory (MSL) Curiosity Rover. *Source:* NASA.

duration (approximately 687 sols vs. 90 sols), increased power demands, and the need to withstand higher temperature excursions. Furthermore, the larger Curiosity rover necessitated the use of a much larger battery, which led to the adoption of two eight-cell strings of 43Ah prismatic cells manufactured by EaglePicher Technologies/Yardney Division (Figure 6.11). Unlike the MER mission in which power generation was accomplished by solar arrays, the MSL mission is powered by a Multi-Mission Radioisotope Thermoelectric Generator (MMRTG), which has little sensitivity to cold environments and to dust storms and expectedly a much longer lifetime [40]. The use of the MMRTG also aids in the thermal stability of the Rover, by using waste heat generated to keep science instruments and battery within the required operating temperature range. In addition, the MMRTG power generation provides a very predictable output since the amount of heat produced is directly linked to the decay of the radioisotope ^{238}Pu.

To augment the EPS MMRTG, the LIB functions to provide power during the launch phase of the mission, to assist the thermal batteries during the EDL phase, and to support the power loads on the Martian surface that exceed the output of the MMRTG. To meet these operational requirements, the battery was developed to meet the following design requirements: (i) operate for more than 40months after launch including a calendar life of greater than 4years, (ii) consist of two redundant LIBs, each with a nameplate capacity rating of 43Ah, (iii) provide up to 1300Wh during launch at a temperature of +20°C to +30°C with a 34A maximum discharge current, (iv) provide the capability of supporting a 21A for 18ms (each battery) and support sequential (grouped) pyro events as close together as 120ms in duration (such as during EDL), (v) support 670 sols of surface operation with two discharge cycles per sol, with one cycle being 1000Wh at 0°C and 17.2A, (vi) possess the capability of meeting the performance requirements with an average battery temperature of +15°C and an absolute maximum of +30°C on the surface of Mars, (vii) provide a capacity of 59Ah at 0°C using a C/5 discharge rate at EOM [41]. In contrast to the MER mission, the battery temperature experienced throughout the mission is less severe for the MSL rover, with an average temperature of approximately +17°C, although the battery is capable of operating over a much wider range of temperatures (−20 to +30°C).

Mars 2020 Perseverance Rover Following the success of the Mars Curiosity Rover, NASA-JPL designed, built, and launched the Mars 2020 Perseverance Rover, which was a build-to-print design based on the Curiosity platform. However, the Perseverance Rover contains new instruments: (i) a sample caching instrument designed to collect and prepare Martian soil samples that will be retrieved at a later date by a potential sample return mission, (ii) a Mars Oxygen In-situ Resource Utilization Experiment (MOXIE) as a technology demonstration of producing oxygen for future human missions, and (iii) the Mars Helicopter Ingenuity to demonstrate the first controlled aerial platform for planetary exploration. Similar to the previous mission, the Perseverance Rover uses a radioisotope thermoelectric generator for power generation that charges an LIB utilized for peak power and science operations.

The LIB was also a build-to-print design based on the MSL battery, with a notable exception that the cell consists of graphite-NCA rather than the previous MCMB-NCO chemistry. In the same prismatic cell envelope, the NCA chemistry provides a higher capacity (50 vs. 46 Ah) and increased cell-level specific energy (157 vs. 145 Wh/kg). For this mission, since very low temperatures were not anticipated, the previous all-carbonate electrolyte was used. Based on JPL data, the NCA-based chemistry was shown to provide improved cycle life, lower cell impedance, and improved tolerance to high temperature exposure. The battery configuration was the same as before with two eight-cell strings electrically connected in parallel to comprise the Rover Battery Assembly Unit (RBAU).

The RBAU was required to meet the following requirements: (i) support 1003 sols of Mars surface operations, with two discharge cycles per sol, (ii) during cruise and surface operations, support continuous discharge current of up to 22 A, with a maximum DOD of 45% (800 Wh), over the temperature range of 0 to +30 °C, (iii) during launch operations, support up to 1300 Wh and a 34 A maximum discharge current, and (iv) support a maximum continuous charge current of 8 A in the temperature range of −20 to +30 °C. Two different cell chemistries (the heritage NCO-based and the newer NCA-based chemistry) were evaluated in the same cell prismatic format with brief cycling at various temperatures. As illustrated in Figure 6.12, a number of cells were continuously cycled at alternating temperatures of +30, +20, and −25 °C. To characterize cell capacity loss with cycling, diagnostic health checks were performed periodically at +20 °C to assess the health. After 313 cycles, the NCO-based cells exhibited 8.0% capacity loss, whereas cells with the NCA-based cells lost approximately 3.4% capacity. The NCA-based cells also displayed approximately 40% lower initial DC resistance and displayed much less impedance growth during cycling.

To date, the Perseverance Rover has completed 394 sols of operation on the surface of Mars. Since the MMRTG is projected to provide a lifetime of at least 14 years, it is anticipated that the M2020 Perseverance Rover LIB will also support the mission this long, based on the excellent performance of LIBs on previous Mars missions.

Planned ExoMars Rover The ESA, in collaboration with the Russian Roscosmos State Corporation, were planning to launch the Rosalind Franklin rover, which was previously known as the ExoMars rover. This mission inherits much of its design from the ExoMars 2016 Schiaparelli mission, which crashed upon landing on Mars. The Rosalind Franklin rover was originally planned to launch in 2020 but was delayed to September 2022 to address parachute testing issues. The design of the rover consists of using solar arrays for energy generation and an LIB for energy storage and is equipped with radioisotope heater units (RHUs) to maintain the internal temperature of the rover above −40 °C. The mission battery would employ a wide operating temperature range using Saft MP176065 Xtd 5.6 Ah cells, with a total battery nameplate energy of 1140 Wh [42, 43]. For this mission, the required operating temperature range on the surface of Mars ranges between −40 and +50 °C. The primary mission requirement is to support 220 sols of operation on the surface of Mars by providing energy for peak power events during the day and supporting night-time operations when the solar arrays are not

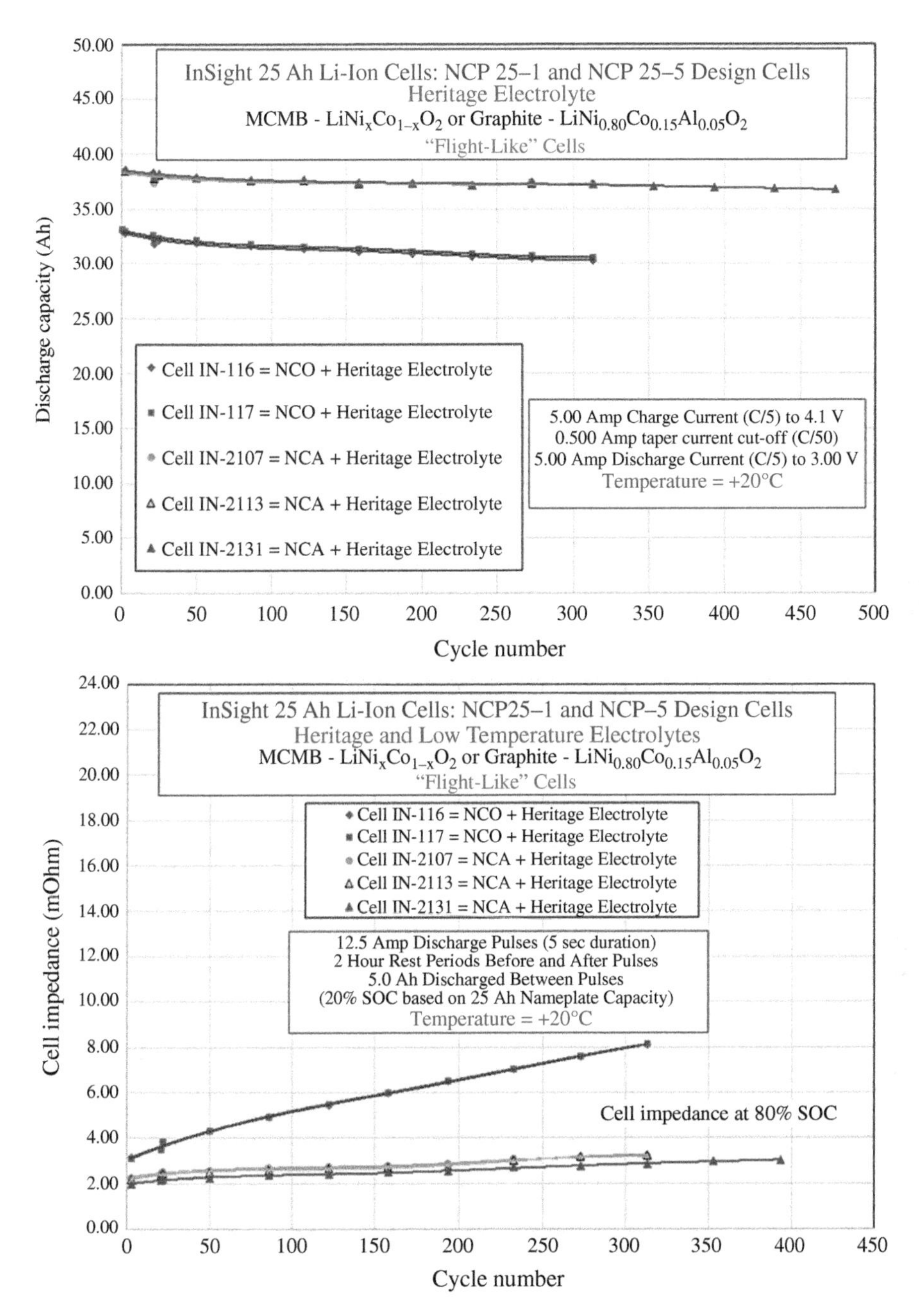

Figure 6.12 Yardney 25 Ah cells tested under the InSight program with heritage electrolyte (MSL, M2020) and either NCO/MCMB or NCA/graphite electrodes; cycles were performed at +30 °C, +20 °C, and −25 °C. Upper: capacity at 20 °C, C/5, 4.10 V–3.00 V. Lower: impedance growth (measured at 80% SOC). *Source:* Courtesy of NASA-JPL.

illuminated. For the cold cases, the battery is anticipated to support a 40% DOD cycle at temperatures ranging between −4 and −20 °C, with charging occurring at >−10 °C. However, the fate of this mission is uncertain since ESA suspended their co-operation with Roscosmos in March 2022.

6.3.3.4 Mars Helicopters, CubeSats, and Penetrators

Mars 2020 Mars Helicopter Ingenuity On-board the Perseverance Rover was a small helicopter that is intended to serve as a technology demonstration to illustrate that the aerial exploration of Mars is possible. The Mars helicopter, named Ingenuity, was developed by the JPL in collaboration with AeroVironment Inc. and other NASA centers [44]. It is more difficult to fly an aircraft on the surface of Mars, since the atmosphere is only about 1% as dense as it is on Earth at sea level, making it harder to generate the sufficient lift. This aspect is partially offset by the fact that the gravity is much lower on Mars (3.71 m/s^2 compared to 9.81 m/s^2). To achieve flight, the team developed a very light (1.8 kg) co-axial rotor design helicopter that is fully autonomous (Figure 6.13). The helicopter is powered by an LIB that is recharged daily by a solar panel manufactured by SolAero Technologies using Inverted Metamorphic (IMM4J) cells. The LIB consists of six COTS Sony US18650 VTC4 high-power cells connected electrically in series to deliver a nameplate capacity of 2.0 Ah. The battery is contained within the fuselage that is built around a central mast, which is a hollow structural tube that runs from the top to the bottom of the helicopter. The battery assembly was designed and fully fabricated at the JPL (Figure 6.14). In addition, all of the acceptance testing of the cells and the battery modules were performed at JPL, which includes the cell selection process from the flight lot batch of cells.

The Mars helicopter battery was designed to meet the following requirements: (i) support continuous power loads of 360 W, (ii) support peak power loads of 510 W, (iii) support high power operation over a temperature range of +10 to +25 °C, and (iv) support low power night-time operations over a temperature range of −15 to +25 °C. In addition, the battery was required to meet all requirements after being subjected to a seven-month cruise period, as well as the shock and vibration associated with the EDL upon arrival at Mars. After which, the helicopter had a technology demonstration mission to complete five flights of increasing complexity over the course of roughly 30 sols. To prevent any possible harm to the Perseverance Rover, the helicopter battery was maintained at SOCs below 35% during cruise and EDL in the unlikely event that an external short was introduced. The comparative performance test studies showed that the Sony US18650 VTC4 cells have a superior power capability among the COTS 18650 cell technologies. The cell is capable of supporting up to 30 A continuous discharge currents and displayed the

Figure 6.13 Artist's portrayal of the Mars helicopter Ingenuity on the surface of Mars. *Source:* Courtesy of NASA-JPL.

Figure 6.14 Pictures of the Mars helicopter Ingenuity Li-ion battery. *Source:* Courtesy of NASA-JPL.

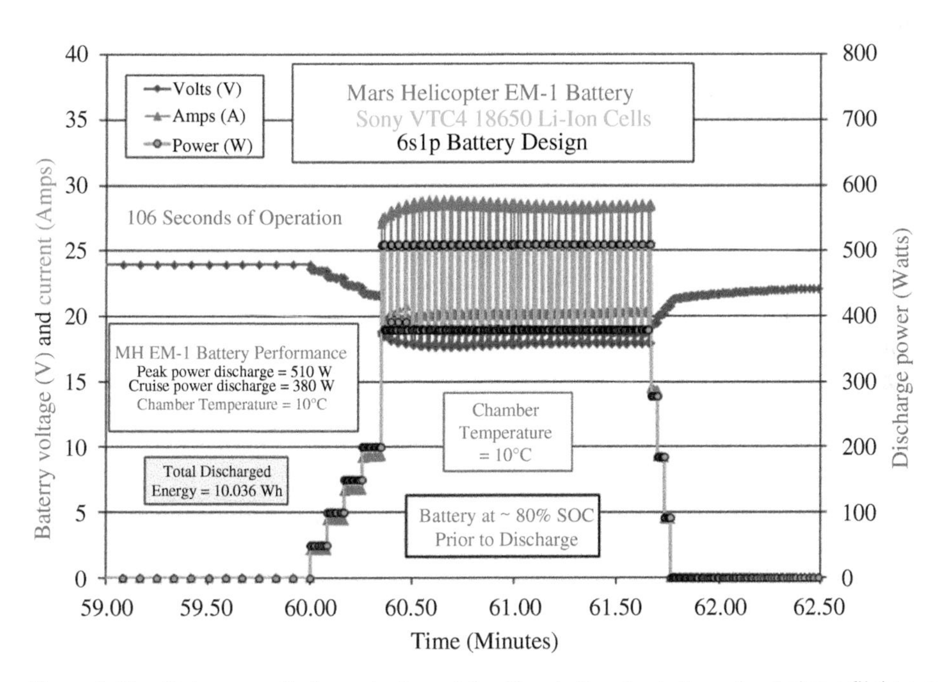

Figure 6.15 High power discharge testing of the Mars helicopter battery simulating a flight event. *Source:* Courtesy of NASA-JPL.

highest operating voltage, especially at low temperatures when subjected to mission relevant load profiles. As shown in Figure 6.15, an engineering model of the six-cell battery successfully supported a load profile simulating the anticipated power draw during a flight event, which consisted of a continuous power load of 380 W and peak power loads of 510 W [45]. In practice, the continuous and peak power loads were lower during preliminary integrated testing and on the surface on Mars.

After successfully landing on Mars on February 18, 2021, and subsequently deployed from the underbelly of the rover on April 3, 2021, the Ingenuity helicopter went on to successfully complete all five of the planned flights, thereby meeting the primary mission objectives. This represents the first time that powered flight has been achieved on another planetary body other than Earth. One of the challenges of the Mars helicopter is that it must survive when separated from the rover and operate autonomously, which means that it must possess all of the functionality of a complete spacecraft system. This includes the EPS, thermal management, guidance navigation and control (GNC), avionics, computing, and communication subsystems. Related to this, given the small size of the helicopter, thermal management was especially challenging, with roughly two-thirds of the stored available energy (not counting the reserve) being allocated to maintaining battery and electronics warm during the very cold nights. Each flight was implemented with increasingly more complexity, with the fourth flight consisting of rising to a height of 5 m and traversing approximately 133 m south before returning to the same spot for a round trip distance of 266 m over the course of 117 seconds. After completing the fifth flight, in which the helicopter flew to another landing spot 129 m away, Ingenuity had traveled a total of approximately 0.5 km. In practice, the battery performed very well, supporting all flights with no issues and displaying an operating voltage of >21 V with minimal string-level cell-to-cell voltage dispersion. Prior to each flight, the battery was heated to a temperature of +20 °C with the SOC being greater than 80%. To date, Ingenuity has completed over 29 flights on the surface of Mars, corresponding to over 7.0 km and over 55 minutes airborne.

Mars Cube One (MarCo) CubeSats The first CubeSats to operate beyond Earth's orbit are the two nano-spacecraft MarCO-A and MarCO-B, which were both 6 U CubeSats that were launched along with NASA's InSight Mars Lander. The objective of the MarCO CubeSats was to serve as a technology demonstration of new miniaturized communication and navigation capabilities and to provide real-time communications relay of telemetry during the Mars InSight EDL phase while the spacecraft was out of the line of sight of Earth [46]. These small CubeSats contained the following components: (i) an ultra-high frequency (UHF) antenna, (ii) three sets of X-band antennas, (ii) a small radio for communication, (iv) a star-tracker for attitude determination, (v) a cold-gas propulsion system for attitude control, (vi) miniature wide-angle and narrow-angle cameras to verify deployments, (vii) a pair of deployable solar panels, and (viii) a rechargeable LIB for energy storage. By all accounts, the two MarCO CubeSats completed their mission successfully in serving as communication relays during the EDL phase of the InSight spacecraft on November 26, 2018.

The EPS of the MarCO CubeSats consisted of two solar arrays containing triple-junction GaAs solar cells from MMC, LLC, an electrical power supply provided by AstroDev, and a LIB pack designed and built at JPL. The battery design consisted of using Panasonic NCR-18650B Li-ion cells in a 3s4p configuration (Figure 6.16). The requirements for the designed battery include the following: (i) provide an operating voltage of 9.0–12.3 V, (ii) deliver a BOL capacity of 11.8 Ah and energy of 126 Wh at +25 °C, (iii) support a maximum discharge of 4.2 A and charge current of 1.42 A, and (iv) operate between a temperature range of 0 to +30 °C.

The primary mission of MarCO was to test new miniaturized communication and navigation technologies. They were able to provide a real-time communication relay while the InSight lander was in the EDL phase. The MarCO spacecrafts were launched as a pair for redundancy and flew at either side of InSight with independent navigation. While there are a large number of CubeSats around Earth, the MarCO project is the first CubeSat mission to go beyond Earth's orbit. This allowed for collection of unique data outside of the Earth's atmosphere and orbit. In addition to serving as communication relays, they also tested the CubeSat component's endurance and navigation capabilities in deep space. Instead of waiting several hours for the information to relay back to

Figure 6.16 The final battery design utilized for the MarCO CubeSats. *Source:* Courtesy of NASA-JPL.

Earth directly from the InSight lander, MarCO thus relayed EDL-critical data immediately (subject only to the 8-minute Mars–Earth transmission time) after completion of the landing. The information sent to Earth included an image from InSight of the Martian surface right after the lander touched down.

6.3.4 Missions to Jupiter

Exploration of the outer planets is especially challenging, primarily due to the distance that the spacecraft must travel to reach its destination. Depending upon the launch vehicle employed and the orientation of the Earth relative to Jupiter, the cruise period can take anywhere from five to seven years to complete. Furthermore, since Jupiter is roughly five astronomical units (AUs) away from the Sun, the mean solar incidence ($50.5\,W/m^2$) is only approximately 8.6% of that observed at the Earth. This makes solar-powered missions more challenging and radioisotope-powered spacecraft more attractive. For example, the NASA Galileo spacecraft, which arrived at Jupiter on December 7, 1995, after roughly a six-year cruise period, utilized two RTGs that produced about 570 W of power at launch and 493 W by the time it reached Jupiter. In a similar fashion, the NASA Cassini mission to Saturn utilized three RTGs that were still able to produce 600–700 W of electrical power after completing the nominal 11-year mission. In fact, one of the spare RTGs for the Cassini mission was used to power the New Horizons mission to Pluto and the Kuiper belt. It was not until the Juno mission (launched in 2011) to Jupiter that solar-powered options were feasible and rechargeable LIBs were used as energy storage devices.

6.3.4.1 NASA Juno Mission

On August 5, 2011, NASA launched the Juno spacecraft to orbit Jupiter with the intention of measuring the planet's composition, gravitational field, magnetic field, and polar magnetosphere. One primary objective is to determine how much water is in Jupiter's atmosphere, which should help to validate current planet formation theories. Juno is the farthest deep space mission powered by solar arrays. After a roughly five-year cruise period and traveling almost two billion miles, the spacecraft entered into an elliptical, polar orbit approximately 53.5 days in duration. In contrast to the earlier RTG-powered Galileo probe, Juno is powered by three large solar arrays consisting of three panels that provide approximately 500 W of electrical power in orbit around Jupiter [47, 48]. Juno represents the mission with the three largest solar array wings ever deployed on a planetary probe (Figure 6.17).

For energy storage, the spacecraft employs two separate LIBs manufactured by Eagle-Picher Technologies/Yardney Division, consisting of 55 Ah NCO chemistry cells electrically connected in series to comprise eight-cell strings (8s2p) (Figure 6.18) [49]. The batteries are located outside the spacecraft electronics radiation enclosure, which required tolerance to the high radiation environment of Jupiter. With respect to the cell chemistry, the radiation tolerance was previously

Figure 6.17 Artist's portrayal of the Juno spacecraft at Jupiter. *Source:* Courtesy of NASA-JPL.

Figure 6.18 The Juno spacecraft 55 Ah, 28 V Li-ion battery. *Source:* Courtesy of EaglePicher Technologies, Inc.

demonstrated to be resilient up to 20 Mrad of exposure [2]. However, the battery management electronics (BME) are housed separately in a radiation enclosure to provide additional shielding from Jupiter's radiation. To maintain an acceptable operating temperature range, the batteries utilize heaters and are thermally isolated.

One of the challenges for the battery design was to provide good performance after being subjected to the long cruise period and have the ability to meet all other requirements at the end of the mission. Prior to this mission, aerospace Li-ion cells and batteries had not typically been required to provide such long service lifetimes. During the mission planning phase, the team relied upon a systematic study of the calendar life of various prototype cells that was performed at JPL, including 7 Ah Yardney prismatic and 9 Ah Saft DD-size Li-ion cells, in order to validate end-of-life performance projections [31, 50]. This study was performed at six different storage temperatures, while maintaining the cells at 50% SOC at a fixed float voltage to mimic flight-like conditions. In total, nine years of storage data was generated, with periodic full capacity and DC resistance measurements being performed to trend the rate of capacity loss. As a result of this study, the Juno program adopted the recommendation to maintain the batteries at 0 °C and 50% SOC during the duration of the cruise period, with periodic warming of the battery to support trajectory maneuvers. With the planned completion of the primary mission set to occur in July of 2021, in January 2021, the mission was extended through September 2025 to support an additional 42 orbits, including flybys of Ganymede, Europa, and Io. Given that the spacecraft was launched in 2006, this would represent a 19-year mission life (not including the additional wet-life associated with the acceptance and prelaunch activities). To date, the DOD observed on the battery has been relatively low (<10% DOD), so the battery is expected to meet mission requirements.

6.3.5 Missions to Comets and Asteroids

6.3.5.1 Hayabusa (MUSES-C)

The Hayabusa-1 mission (previously called MUSES-C) was launched in May of 2003, by the Japan Aerospace Exploration Agency (JAXA), to explore the asteroid Itokawa by touching down on the surface [51]. Hayabusa represented one of the first uses of large capacity Li-ion cells specifically developed for space [52]. The Li-ion cells used were prismatic 13.2 Ah capacity cells (manufactured by the Furakawa Battery Co. Ltd) and consisted of graphite and LCO anode and cathode, respectively. The battery design consisted of 11 cells electrically connected in series with charge control being achieved with a battery charge regulator (BCR). Independent of the BCR, cell voltage was monitored, and each cell was equipped with a bypass shunting circuit to protect against overcharge conditions and to balance the cell voltages at 4.1 V. After touching down of the surface of Itokawa in 2005, the spacecraft experienced loss of attitude control in two separate events, which led to the overdischarging of 4 of the 11 cells within the battery. Despite the damage to the overdischarged cells leading to copper substrate dissolution and possible internal shorts, the remaining cells were successfully used to complete the preparation of the asteroid sample container.

Building from this previous mission, JAXA launched another spacecraft in 2014, Hayabusa-2, to study the near-Earth asteroid (162173) 1999 JU3 (now designated 162173 Ryugu). On board, the spacecraft contained a small asteroid lander MASCOT, developed by the German Aerospace Center (DLR) that was designed to perform science operations and was powered by primary lithium batteries [53]. The Hayabusa-2 spacecraft was based largely upon the technological heritage of the Hayabusa-1, including the heritage 13.2 Ah LIB [54]. The spacecraft rendezvoused with the asteroid in June 2018 and took samples that were successfully returned to Earth in December 2020.

6.3.5.2 ESA Rosetta Lander Philae

As part of the Horizon 2000 program, the ESA launched the Rosetta spacecraft along with its lander module, Philae, in March of 2004 to study the comet 67P/Churyumov-Gerasimenko. Similar to MASCOT that accompanied the Hayabusa mission, the Philae lander was developed by an international consortium led by the DLR [55]. The Rosetta lander contained two different battery systems: (i) a non-rechargeable Saft LS 20 Li-$SOCl_2$ battery to support the lander release phase of the mission and (ii) a LIB manufactured by AEA containing Sony 18650 HC cells [56].

6.3.5.3 NASA OSIRIS-REx Mission

Following the Juno mission to Jupiter and the New Horizons mission to Pluto, the NASA Origins, Spectral Interpretation, Resource Identification, and Security Regolith Explorer (or OSIRIS-REx), was the third planetary science mission selected in the New Frontiers program [57]. The main objective of the mission is to return carbonaceous regolith from the near-Earth asteroid Bennu. With respect to the EPS, the spacecraft included two solar arrays and two LIBs (for redundancy) manufactured by Yardney Technical Products, Inc. [58]. Depending upon the range of the spacecraft from the sun, the solar arrays generated between approximately 1200 and 2800 W. The batteries were designed to provide power when the solar arrays are not pointed at the sun, which includes the sampling event at Bennu, and consist of two 28 V, 30 Ah modules similar to what was used on Grail. On October 20, 2020, OSIRIS-REx successfully touched down on Bennu and a 60 g sample was collected. It is currently returning back to Earth and is scheduled to arrive in September 2023. In total, the mission duration will be over seven years after launch.

6.3.6 Missions to Deep Space and Outer Planets

Besides the Juno mission, which is currently orbiting Jupiter, no other spacecraft has yet utilized LIB technology to explore deep space or the other outer planets, including Saturn, Uranus, and Neptune. Long before the advent of LIB technology, the two Voyager spacecraft that were launched in 1977 are still in operation and have moved past the outer boundary of the heliosphere into interstellar space. Remarkably, Voyager 1 and Voyager 2 have reached distances of ~153 and ~127 AU from Earth, respectively, and are still transmitting data. Electrical power for these spacecrafts is provided by three RTGs, containing radioactive ^{238}Pu, which have a steady power degradation rate of approximately 0.79% per year from an initial 470 W at launch. In general, for very long-duration missions to distant destinations where the solar irradiance is very low, radioisotope-powered options are often more operable than solar-powered designs. For these spacecraft, energy storage devices are primarily used for peak power events and load leveling, which was achieved by dielectric capacitors as in the case on the Cassini mission to Saturn. However, as evidenced by Juno, solar-powered spacecraft that utilize LIBs for energy storage can be viable for many mission concepts bound for the vicinity of Jupiter and Saturn and their moons. Since non-radioisotope spacecraft designs can be considerably less expensive and solar array technology continues to advance, it appears to be very probable that the number of mission concepts aimed at exploring Jupiter and Saturn, as well as their moons, will increasingly utilize LIB technologies.

6.4 Future Missions

There is increasing astrobiology interest to further explore the outer planets Jupiter and Saturn, with an emphasis upon their moons. As communicated by the NASA Planetary Science Division of NASA-SMD in a recent decadal survey [59], the exploration of ocean worlds and ice giants, such as Jupiter's moon Europa and Saturn's moons Enceladus and Titan, are especially attractive since these bodies hold the prospect for harboring life below their ice shells. The exploration of Europa is especially attractive, since there is now strong evidence that beneath the thick ice crust (10–15 miles thick) is an ocean of liquid water that contains the key chemical elements that would make life possible. For this reason, there is planned orbital mission to this icy moon called the Europa Clipper mission, which is anticipated to launch in 2024. Ultimately, this may be a precursor mission to a future Europa Lander mission. There is also a planned NASA mission to Saturn's moon Titan, scheduled to launch in 2027, which consists of landing a dual-quadcopter named "Dragonfly" that engages in the aerial exploration of a variety of locations on Titan [60]. The baseline EPS for the aerial spacecraft is an MMRTG, similar to what was used for the Curiosity and Perseverance rovers and will utilize LIBs to support EPS requirements. The current baseline for this mission is to utilize a battery constructed with large-capacity GS Yuasa cells. For all of these future missions, LIBs that might be used will have to provide long life, tolerance to high levels of radiation, and preferably good performance over a wide operating temperature range. For future aerial missions, rechargeable batteries with high specific energy, high energy density, and high-power capability will be required.

6.4.1 The Planned NASA Europa Clipper Mission

NASA-JPL is planning to launch a mission (scheduled to launch in October of 2024) to perform a detailed reconnaissance of the Jovian icy moon Europa, entitled the Europa Clipper Mission [61]. The planned mission would involve a Jupiter-orbiting spacecraft that performs a science investigation of the surface during multiple flybys of the moon by utilizing a number of instruments, including an ice-penetrating radar that is intended to measure the thickness of the icy shell and the possible presence of a subsurface ocean. The mission may also serve as a precursor to future landed missions focused upon determining the planet's potential habitability for life. For this mission, the baseline power subsystem consists of solar arrays, power management electronics, and a rechargeable LIB [62].

The current battery design involves the use of high specific energy COTS Li-ion 18650 cells that are assembled into large capacity multi-string batteries. Key cell-level requirements for this mission include tolerance to high levels of radiation and high specific energy over a wide range of temperatures. In addition, long storage life characteristics are required since the long cruise period to Europa could potentially be over 6.5 years in duration. To meet the challenging mission requirements, a number of potential cell chemistries were evaluated. In addition to evaluating the charge/discharge rate performance and the long-term storage and cycle life behavior, considerable effort was devoted to assessing the radiation tolerance of candidate cells, including the E-One Moli ICRM and the LG Chem MJ1 Li-ion cells [63]. Based on a number of performance and energy-related factors, the LG Chem MJ1 cell has been baselined for use on the Europa Clipper Mission, which consists of high nickel-content $LiNi_{0.80}Mn_{0.10}Co_{0.10}O_2$ (NMC 811) cathodes coupled with graphite containing a small percentage of silicon [64]. One of the attractive features of this cell is the very high specific energy, with >220 Wh/kg being delivered over a conservative voltage range of 3.0–4.10 V. To preserve the life and provide for additional safety margins, once incorporated into multi-cell modules for space applications, the batteries are typically cycled over conservative voltage ranges (nominal operation involves only charging to 32.8 V for the eight-cell string architecture, corresponding to 4.10 V at the cell level, and discharging to 26 V).

To meet requirements for the Europa Clipper and future missions to the outer planets, a study was enacted to demonstrate the cell and battery compatibility with high levels of radiation. Jupiter is surrounded by an enormous magnetic field and charged particles are trapped in the magnetosphere and form intense radiation belts 10 times stronger than Earth's Van Allen belts. Inherent resilience of the LIB is preferred compared to radiation shielding. Furthermore, due the inability to subject LIBs to DHMR, which involves unacceptably high temperatures, exposure to gamma-rays (irradiation with ^{60}Co γ-rays) was considered as a technique to reduce the overall bio-burden of the battery. However, due to the recent relaxing of planetary protection requirements for the future Europa Clipper mission, irradiation of the battery is not currently required. As shown in Figure 6.19, when LG Chem MJ1 18650 Li-ion cells were subjected to 20 Mrad of ^{60}Co γ-rays, no significant impact of radiation was observed upon the 100% DOD cycle life performance at +30 °C. Initially, a modest increase in the cell DC resistance is observed with the cells that were subjected to 20 Mrad of γ-ray irradiation. However, the growth in resistance of the irradiated cells with cycling is comparable (if not better) to that of the baseline cells that were not irradiated (Figure 6.20). In general, the LG Chem MJ1 cell displayed impressive impedance growth characteristics during cycling, with less than a 16% increase observed (at 80% SOC) for all cells after completing 900 cycles (at 100% DOD), even after being irradiated with 20 Mrad of γ-ray irradiation.

Besides verifying the resilience of individual cell chemistries to the effects of radiation, as part of a risk-reduction effort, a flight-like 8s16p module manufactured by EnerSys/ABSL was also exposed to 20 Mrad of γ-ray irradiation and then exposed to a full qualification program, including random vibration testing, pyro-shock testing, and thermal vacuum testing, which it successfully passed [65]. In addition to having to be radiation tolerant, the Europa Clipper battery design must also have a low magnetic signature due to the proximity of sensitive on-board instrumentation, such as the Plasma Instrument for Magnetic Sounding (PIMS) [66]. Furthermore, due to the radiation environment near Jupiter, the battery design must also mitigate against the possibility of charge build-up on dielectric materials due to induced electrostatic discharge.

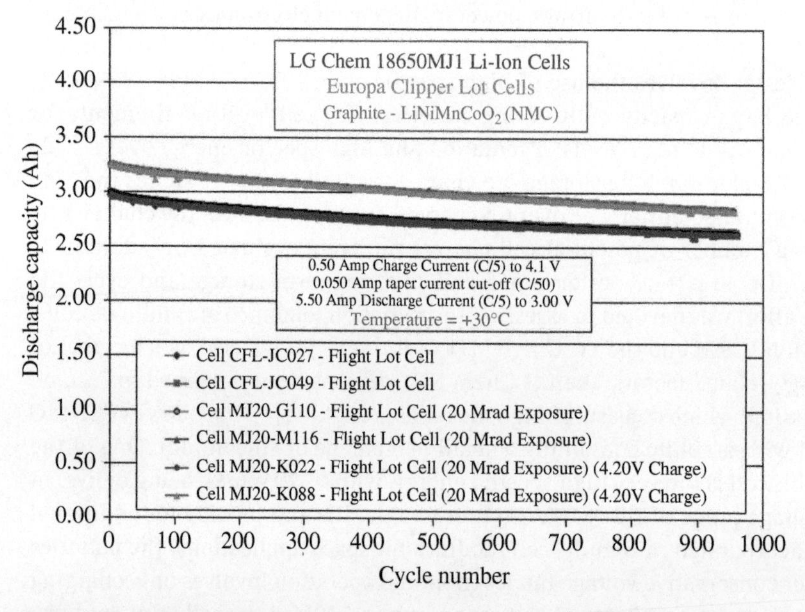

Figure 6.19 100% DOD cycle life performance of LG Chem MJ1 cells at +30 °C. *Source:* Courtesy of NASA-JPL.

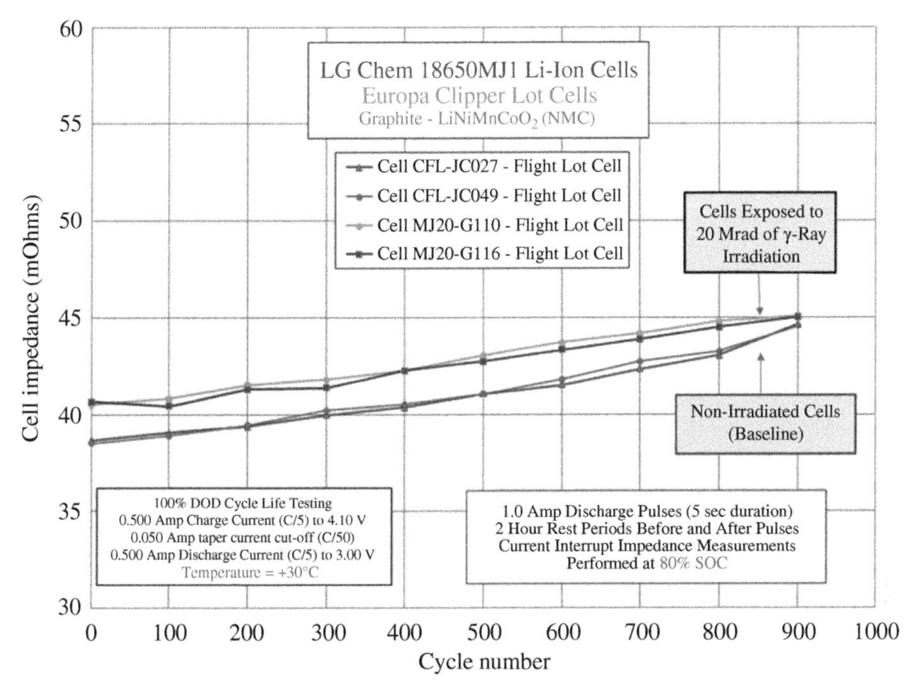

Figure 6.20 The cell impedance growth (in mΩ) of LG Chem MJ1 cells subjected to 100% DOD cycle life performance testing at +30 °C. Impedance measurements were performed using DC current-interrupt techniques. *Source:* Courtesy of NASA-JPL.

6.4.2 ESA JUICE Mission

Similar to the Europa Clipper, the ESA is planning to launch the Jupiter Icy Moons Explorer (JUICE) mission (tentatively scheduled to launch in August of 2023) to investigate three of Jupiter's Galilean satellites (Ganymede, Europa, and Callisto) [67]. The spacecraft is being built by Airbus Defense and Space as the main contractor. After completing a 7.6-year cruise period, the spacecraft is planned to orbit abound Ganymede. The power generation for the spacecraft will come from the use of an 85 m^2 solar array that will provide an estimated 820 W at Jupiter and will be augmented with the use of an LIB. For this mission, five 8s24p LIBs (manufactured by EnerSys/ABSL) that are connected in parallel will be used, with a total nameplate capacity of 288 Ah. Six thermistors are included within the battery together with main and redundant heaters.

6.5 Mars Sample Return Missions

There is continued interest in the aerospace community in conducting a Mars sample return mission to bring rocks and soil samples back to Earth. Toward this end, NASA is currently working on mission concepts, in collaboration with ESA, to retrieve samples that have been collected by the Mars 2020 Perseverance rover. Such a mission would potentially consist of a lander, rover, and an ascent vehicle, which would rendezvous with an orbiter. With respect to mission usage of battery technologies, similar requirements and restraints would be present as compared to previous missions to Mars. However, although past missions to Mars have primarily utilized custom large Li-ion cell formats optimized for low temperature and containing cell-balancing electronics, future missions may incorporate small COTS cell battery designs owing to their very high specific energy. This would simplify the power subsystem and lower costs, since complex electronic systems such as the BCB would not be required. The current Mars Sample Return mission is likely to use the Europa Clipper heritage battery chemistry and design.

6.6 Summary

In summary, LIBs have successfully supported numerous planetary exploration missions in the past two decades, operating in conjunction with either photovoltaic or nuclear radioisotope power sources in robotic missions. The robotic missions include various spacecraft such as planetary orbiters, landers, and rovers and to different destinations such as the Moon, Mars, Jupiter, Venus, and various asteroids. Based on the on-going improvements in the materials, it appears very likely that LIBs will be the rechargeable energy storage technology of choice for many years. Some advanced technologies hold some promise for the future, such as solid-state Li-NMC and lithium metal cells with conversion cathodes, offering further improvements in specific energy. However, it is unlikely that these next-generation technologies will achieve maturity in the next decade and surpass Li-ion in performance characteristics.

Acknowledgment

Some of the work described here was carried out at the Jet Propulsion Laboratory, California Institute of Technology, under contract with the National Aeronautics and Space Administration (NASA) (80NM0018D0004).

References

1 Ratnakumar, B.V. and Smart, M.C. (2007). Aerospace applications – planetary exploration missions (orbiters, landers, rovers and probes). In: *Industrial Applications of Batteries. From Electric Vehicles to Energy Storage and Toll Collection* (ed. G. Pistoia and M. Broussely), 327–387. Amsterdam: Elsevier B. V.

2 Ratnakumar, B.V., Smart, M.C., Whitcanack, L.D. et al. (2004). Behavior of Li-ion cells in high-intensity radiation environments. *J. Electrochem. Soc.* 151 (4): A652–A659.

3 Ratnakumar, B.V., Smart, M.C., Whitcanack, L.D. et al. (2005). Behavior of Li-ion cells in high-intensity radiation environments: II. Sony/AEA/ComDEV cells. *J. Electrochem. Soc.* 152 (2): A357–A363.

4 Krause, F.C., Lawrence, A., Smart, M.C. et al. (2015). Evaluation of commercial high energy lithium-ion cells for aerospace applications. In: *227th Meeting of the Electrochemical Society (ECS)*. Chicago, IL.

5 Smart, M.C., Krause, F.C., Lawrence, A. et al. (2017). The performance evaluation of high specific energy 18650-size lithium-ion cell for the planned Europa Clipper Mission. In: *232nd Meeting of the Electrochemical Society (ECS)*. National Harbor, MD.

6 Smart, M.C., Krause, F.C., Ratnakumarm, B.V. et al. (2019). The impact of radiation exposure upon lithium-ion batteries for future planned NASA missions to Europa. In: *235th Meeting of the Electrochemical Society (ECS)*. Dallas, TX.

7 Britting, A.O. Jr. (1984). Design, development, performance, and reconditioning of Ni-Cd batteries using polypropylene separators. *J. Power Sources* 12 (3–4): 305–316.

8 Kayali, S., Morton, P., and Gross, M. (2017). International challenges of GRACE follow-on. In: *2017 IEEE Aerospace Conference*, 1–8. Big Sky, MT. https://doi.org/10.1109/ AERO.2017.7943615.

9 Entekhabi, D., Yueh, S.H., O'Neill, P.E. et al. SMAP mission status, new products and extended-phase goals. In: *IGARSS 2018 – IEEE International Geoscience and Remote Sensing Symposium*, 3747–3750. Valencia, Spain. https://doi.org/10.1109/IGARSS.2018.8518011.

10 Lehman, D.H., Hoffman, T.L., and Havens, G.G. (2013). The gravity recovery and interior laboratory mission. In: *2013 IEEE Aerospace Conference*, 1–11. Big Sky, MT, USA. https://doi.org/10.1109/AERO.2013.6496866.

11 Ennico, K., Shirley, M., Colaprete, A., and Osetinsky, L. (2012). The Lunar Crater Observation and Sensing Satellite (LCROSS) payload development and performance in flight. *Space Sci. Rev.* 167: 23–69.

12 Di Stefano, S., Perrone, D., and Ratnakumar, B.V. (1997). Characterization of nickel-hydrogen 2-cell common pressure vessels for NASA missions. In: *IECEC-97 Proceedings of the Thirty-Second Intersociety Energy Conversion Engineering Conference* (Cat. No.97CH6203), vol. 1, 154–158. Honolulu, HI, USA. https://doi.org/10.1109/IECEC.1997.659177.

13 Genc, D.Z. and Thwaite, C. (2011). Proba-1 and Mars Express: An ABSL Lithium-Ion Legacy. In: *Proceedings of the 9th European Space Power Conference*. Noordwijk, Netherlands: European Southern Observatory, id.85. ESA SP, vol. 690. Saint Raphael, France. ISBN: 978-92-9092-257-5.

14 Jakosky, B.M. (2015). MAVEN explores the Martian upper atmosphere. *Science* 350 (6261): 643.

15 Jakosky, B.M., Grebowsky, J.M., Luhmann, J.B. et al. (2015). MAVEN observations of the response to Mars to an interplanetary coronal mass ejection. *Science* 350 (6261): 0210.

16 Siddiqi, A.A. (2018). *Beyond Earth: A Chronicle of Deep Space Exploration, 1958–2016*. NASA History Program Office.

17 Shirbacheh, M. (1997). Power and pyro subsystems for Mars Pathfinder. In: *IECEC-97, Proceedings of the Thirty-Second Intersociety Energy Conversion Engineering Conference* (Cat. No.97CH6203), vol. 4, 2231–2236. Honolulu, HI, USA. https://doi.org/10.1109/IECEC.1997.658215.

18 Surampudi, S., Otzinger, B., Perrone, D. et al. (1998). Mars Pathfinder Lander battery and solar array performance. In: *Proceedings of the 36th Aerospace Sciences Meeting and Exhibit*. Reno, NV.

19 Smart, M.C., Ratnakumar, B.V., Whitcanack, L. et al. (1999). Performance characteristics of lithium-ion cells for the NASA's Mars 2001 Lander application. *IEEE Aerosp. Electron. Syst. Mag.* 14 (11): 36–42.

20 Goldstein, B. and Shotwell, R. (2006). Phoenix – the first Mars scout mission (a mid-term report). In: *2006 IEEE Aerospace Conference*, 18. Big Sky, MT, USA. https://doi.org/10.1109/AERO.2006.1655749.

21 Golombek, M., Kipp, D., Warner, N. et al. (2017). Selection of the InSight landing site. *Space Sci. Rev.* 211: 5–95.

22 Lisano, M.E. and Kallemeyn, P.H. (2017). Energy management operations for the InSight solar-powered mission at Mars. *IEEE Aerosp. Conf.* 2017: 1–11.

23 Smart, M.C., Ratnakumar, B.V., Whitcanack, L.D. et al. (2008). Li-ion electrolytes containing ester co-solvents for wide operating temperature range. *ECS Trans.* 11 (29): 99.

24 Smart, M.C., Ratnakumar, B.V., Chin, K.B., and Whitcanack, L.D. (2010). Lithium-ion electrolytes containing ester co-solvents for improved low temperature performance. *J. Electrochem. Soc.* 157 (12): A1361–A1374.

25 Smart, M.C., Dawson, S.F., Shaw, R.B. et al. (2014). Performance validation of Yardney low temperature NCA-Based Li-ion cells for the NASA Mars InSight Mission. In: *NASA Aerospace Battery Workshop*. Huntsville, AL.

26 Smart, M.C. and Ratnakumar, B.V. (2011). Effect of electrolyte composition on lithium plating in lithium-ion cells. *J. Electrochem. Soc.* 158 (4): A379–A389.

27 Smart, M.C., Ratnakumar, B.V., Ewell, R.C. et al. (2018). The use of lithium-ion batteries for JPL's Mars missions. *Electrochim. Acta* 268: 27–40.

28 Ratnakumar, B.V., Smart, M.C., Kindler, A. et al. (2003). Lithium batteries for aerospace applications: 2003 Mars exploration rover. *J. Power Sources* 119–121: 906–910.

29 Smart, M.C., Ratnakumar, B.V., Ewell, R.C. et al. (2007). An update on the ground testing of the Li-ion batteries in support of JPL's 2003 Mars Exploration Rover Mission. In: *5th International Energy Conversion Engineering Conference (IECEC)*. St. Louis, MI.

30 Ratnakumar, B.V., Smart, M.C., Whitcanack, L.D. et al. (2007). An update on the performance of lithium-ion rechargeable batteries on Mars Rovers. In: *5th International Energy Conversion Engineering Conference (IECEC)*. St. Louis, MI.

31 Smart, M.C., Ratnakumar, B.V., Whitcanack, L.D. et al. (2010). Life verification of large capacity Yardney Li-ion cells and batteries in support of NASA missions. *Int. J. Energy Res.* 34: 116–134.

32 Ratnakumar, B.V., Smart, M.C., Ewell, R.C. et al. (2004). Lithium-ion rechargeable batteries on Mars Rovers. In: *2nd International Energy Conversion Engineering Conference (IECEC)*, 1–8. Providence, Rhode Island.

33 Marsh, R.A., Vukson, S., Surampudi, S. et al. (2001). Li-ion batteries for aerospace applications. *J. Power Sources* 97–98: 25–27.

34 Smart, M.C., Ratnakumar, B.V., and Surampudi, S. (1999). Electrolytes for low temperature lithium-ion batteries based on mixtures of aliphatic carbonates. *J. Electrochem. Soc.* 146: 486.

35 Smart MC, Ratnakumar BV, Surampudi S, Huang C-K. Organic solvents, electrolytes, and lithium ion cells with good low temperature performance. US Patent 6,492,064 (December 10, 2002).

36 Smart, M.C., Ratnakumar, B.V., Whitcanack, L.D. et al. (2004). Lithium-ion batteries for aerospace. *IEEE Aerosp. Electron. Syst. Mag.* 19 (1): 18–25.

37 Stella, P.M., Ewell, R.C., and Hoskin, J.J. (2005). Design and performance of the MER (Mars Exploration Rovers) solar arrays. In: *Conference Record of the Thirty-First IEEE Photovoltaic Specialists Conference*, 626–630. Lake Buena Vista, FL, USA. https://doi.org/10.1109/PVSC.2005.1488209.

38 Staab, M.S., Herman, J.A., Reich, K. et al. (2020). MER Opportunity dust-storm recovery operations and implications for future Mars surface missions. In: *2020 IEEE Aerospace Conference*, 1–20. Big Sky, MT, USA. https://doi.org/10.1109/AERO47225.2020.9172528.

39 Ratnakumar, B.V., Smart, M.C., West, W.C. et al. (2008). Li-ion Rover batteries and descent stage lithium thermal batteries on the Mars Science Laboratory. In: *6th International Energy Conversion Engineering Conference (IECEC)*. Cleveland, OH.

40 Woerner, D., Moreno, V., Jones, L. et al. (2013). The Mars Science Laboratory (MSL) MMRTG in-flight: A power update. In: *Proceedings of Nuclear and Emerging Technologies for Space 2013*. Albuquerque, NM.

41 Smart, M.C., Ratnakumar, B.V., Whitcanack, L.D. et al. (2014). Performance testing of Yardney's Li-ion cells and batteries in support of JPL's Mars Science Laboratory Curiosity Rover. In: *NASA Aerospace Battery Workshop*. Huntsville, AL.

42 Amos, S. and Brochard, P. (2017). Battery for extended temperature range Exomars Rover Mission. In: *E3S Web of Conferences, 2016 European Space Power Conference (ESPC 2016)*, vol. 16, 06001.

43 Tricot, H. and Brochard, P. (2019). EXOMARS 2020 mission Descent module & Rover batteries. In: *2019 European Space Power Conference (ESPC)*, 1–4. https://doi.org/10.1109/ESPC.2019.8931994.

44 Balaram, J., Canham, T., Duncan, C. et al. (2018). Mars Helicopter Technology Demonstrator. In: *2018 AIAA Atmospheric Flight Mechanics Conference*. Kissimmee, FL.

45 Smart, M.C., Krause, F.C., Jones, J.-P. et al. (2019). The performance evaluation of 18650-size lithium-ion cells and batteries for future NASA Missions. In: *Pacific Power Source Symposium 2019*. Waikoloa, HI.

46 Krause, F.C., Loveland, J.A., Steinkraus, J.M. et al. (2020). Implementation of commercial Li-ion cells on the MarCO Deep Space CubeSats. *J. Power Sources* 449: 227544.

47 Bolton, S.J., Lunine, J., Stevenson, D. et al. (2017). The Juno mission. *Space Sci. Rev.* 213: 5–37.

48 Matousek, S. (2007). The Juno New Frontiers mission. *Acta Astronaut.* 61: 932–939.

49 Gitzendanner, R., Byers, J., Buonanno, A., and Deroy, C. (2016). Juno lithium-ion batteries: Design and performance demonstrated for Mission operations 5 years after launch. In: *Space Power Workshop*. Manhattan Beach, CA.

50 Ratnakumar, B.V., Smart, M.C., and Whitcanack, L.D. (2010). Storage characteristics of lithium-ion cells. *ECS Trans.* 25 (36): 297–306.

51 Kawaguchi, J., Fujiwara, A., and Uesugi, T. (2008). Hayabusa – Its technology and science accomplishment summary and Hayabusa-2. *Acta Astronaut.* 62: 639–647.

52 Sone, Y., Ooto, H., Eguro, T. et al. (2009). Charge and discharge performance of over-discharged Lithium-ion secondary battery-lessons learned from the operation of the interplanetary spacecraft HAYABUSA. *Electrochemistry* 75 (12): 750–757.

53 Grimm, C.D., Grundmann, J.T., and Hendrikse, J. (2015). System design of the Hayabusa-2-asteroid sample mission to 199 JU3. In: *4th IAA Planetary Defence Conference*. Rome, Italy.

54 Tsuda, Y., Yoshikawa, M., Abe, M. et al. (2015). On time, on target – How the small asteroid lander MASCOT caught a ride aboard Hayabusa-2 in 3 years, 1 week, and 48 hours. *Acta Astronaut.* 91: 356–362.

55 Ulamec, S., Biele, J., Bousquet, P.-W. et al. (2014). Landing on small bodies: From the Rosetta Lander to MASCOT and beyond. *Acta Astronaut.* 93: 460–466.

56 Debus, A., Moura, D., Gave, G. et al. (2014). Two stage battery system for the Rosetta Lander. *Acta Astronaut.* 53: 623–632.

57 Lauretta, D.S., Balram-Knutson, S.S., Beshore, E. et al. (2017). OSIRIS-Rex: Sample return form Asteroid (101955) Bennu. *Space Sci. Rev.* 212: 925–984.

58 Bierhaus, E.B., Clark, B.C., Harris, J.W. et al. (2018). The OSIRIS-Rex spacecraft and the Touch-and-Go Sample Acquisition Mechanism (TAGSAM). *Space Sci. Rev.* 214: 107.

59 National Research Council of the National Academies (2011). *Vision and Voyages for Planetary Science in the Decade 2013–2022*. Washington, D. C: The National Academies Press.

60 Lorenz, R.D., Turtle, E.P., Barnes, J.W. et al. (2018). Dragonfly: A rotorcraft Lander concept for scientific exploration at Titan. *J. Hopkins APL Tech. Dig.* 324 (3): 374–387.

61 Howell, S.M. and Pappalardo, R.T. (2020). NASA's Europa Clipper – A mission to a potentially habitable ocean. *Nat. Commun.* 11 (1): 1311.

62 Ulloa-Severino, A., Carr, G.A., Clark, D.J. et al. (2017). Power subsystem approach for the Europa Mission. In: *E3S Web of Conferences*, vol. 16, 13004.

63 Smart, M.C., Krause, F.C., Ratnakumar, B.V. et al. (2020). The use of high specific energy 18650-size Li-ion cells with good radiation tolerance for Missions to the outer planets. In: *2020 Conference on Advanced Power Systems for Deep Space Exploration*. Pasadena, CA.

64 Heenan, T.M.M., Jnawali, A., Koki, M.D.R. et al. (2020). An advanced microstructural and electrochemical datasheet on 18650 Li-ion batteries with nickel-rich NMC811 and graphite-silicon anodes. *J. Electrochem. Soc.* 167: 140530.

65 Smart, M.C., Krause, F.C., Ratnakumar, B.V. et al. (2019). The impact of radiation exposure upon Lithium-ion batteries for future planned NASA missions to Europa. In: *235th Meeting of the Electrochemical Society (ECS)*, Dallas, TX.

66 Grey, M., Westlake, J., Liang, S. et al. (2018). Europa PIMS prototype Faraday Cup development. *IEEE Aerosp. Conf.* 2018: 1–15. https://doi.org/10.1109/AERO.2018.8396522.

67 Grasset, O., Dougherty, M.K., Coustenis, A. et al. (2013). Jupiter Icy Moons Explorer (JUICE): An ESA mission to orbit Ganymede and to characterize the Jupiter system. *Planet. Space Sci.* 78: 1–21.

7

Space Battery Safety and Reliability

Thomas P. Barrera and Eric C. Darcy

7.1 Introduction

A safe space-qualified lithium-ion battery (LIB) power system design will operate within its qualification requirements with an acceptable minimum risk for the possibility of injury, loss, or catastrophic outcomes. In high-reliability robotic and human-rated (crewed) spacecraft missions, safe LIB operation is required during both nominal and off-nominal operational scenarios. Safety requirements for LIB designs intended to support spacecraft electrical power systems (EPSs) generally rely on safety factors and safety margins established by relevant testing, relevant lessons learned, industry best practices, guidelines, and safety requirement standards to establish a known level of acceptable risk [1]. Industry experience has demonstrated that the high specific and volumetric energy density of LIBs can present severe safety hazards from an uncontrolled release of large quantities of energy in extremely short periods of time during abuse conditions. In some LIB designs, the rapid release of high energy may propagate with catastrophic consequences to personnel, flight hardware, and/or facilities [2–5].

Assessing safety margins and predicting reliability of a space LIB design is critical to assessing the likelihood that the spacecraft EPS will meet performance and safety requirements. This chapter discusses the key safety requirements, hazard types, causes, and controls associated with spacecraft LIB technologies. Mitigating the risk of cell-to-cell propagating thermal runaway (TR) in space LIBs by incorporating safe-by-design principles into the cell and battery designs is emphasized. In addition, practical approaches to understanding, analyzing, and predicting space LIB reliability are discussed to estimate the likelihood of the LIB meeting EPS performance requirements for the specified mission lifetime.

7.1.1 Space Battery Safety

Historically, a safe battery design for crewed space mission applications was defined as safe for ground personnel and astronaut crew to a) handle and use; b) use in the enclosed environment of a crewed spacecraft; and c) mount in adjacent unpressurized spaces. The intent of these battery safety

Spacecraft Lithium-Ion Battery Power Systems, First Edition. Edited by Thomas P. Barrera.

requirements has also been adopted into battery power systems used in unmanned spacecraft applications. As such, the approach to space battery safety requirements compliance has been governed by empirical determinations of safety margins by test and analysis. First published by NASA in 1985, the *Manned Space Vehicle Battery Safety Handbook* provided requirements and guidance for use by space battery manufacturers to implement safe battery designs used aboard the NASA Space Shuttle manned spacecraft system [6]. As described by NASA, safety characteristics of early NASA manned space applications of non-rechargeable and rechargeable battery-operated equipment were determined after the flight battery designs were complete or nearly complete. Cost and schedule impacts associated with late engineering assessments of battery safety characteristics were later mitigated by adopting systematic processes and procedures for identifying hazard sources and failure modes for the purposes of implementing battery-specific hazard controls as required.

To date, a wide variety of non-rechargeable and rechargeable battery cell chemistries have been used successfully with no reported on-orbit catastrophic safety incidents aboard NASA crewed, unmanned Earth-orbiting, or robotic planetary spacecraft missions. Today, nearly all unmanned Earth-orbiting satellites, robotic planetary spacecraft, and emerging launch vehicle (LV) mission applications use rechargeable LIBs to meet primary and secondary power demands. In support of these spacecraft missions, safety requirements for the design, test, operation, ground handling, transportation, and storage of LIBs are documented in the applicable domestic and international commercial industry standards, government agency specifications, and other industry-relevant sources.

7.1.2 Industry Lessons Learned

The effect of recent industry LIB safety incidents in the commercial passenger aviation and electrified transportation marketplace has had a significant impact on space LIB safety design and test requirements. The root causes of industry LIB safety incidents include undesirable electrical, mechanical, and/or thermal overstress conditions sufficient to defeat existing system, subsystem, LIB, and/or cell-level safety features [7]. Space industry experience has also shown that Li-ion cell-level hazard controls do not necessarily translate into multi-cell module or LIB-level hazard controls [8]. This field experience has provided an engineering basis for verifying safety hazard controls at the cell, module, battery, and EPS-level of integration [9]. For example, the global visibility of the 2013 Boeing Co. 787 Dreamliner™ commercial passenger aircraft rechargeable LIB and non-rechargeable Li battery safety incidents had a significant impact on how NASA viewed the risk of cell-to-cell TR propagation in space-qualified LIBs [10–13]. The severity of the Boeing 787 LIB safety incidents prompted the NASA Engineering Safety Center (NESC) to conduct independent studies into the susceptibility of NASA LIBs to cell-to-cell propagating TR in existing crewed spacecraft applications [14, 15]. Relevant LIB applications included the NASA International Space Station (ISS) LIB orbital replacement unit (ORU) and US astronaut extravehicular mobility unit (EMU) LIB applications [16–18]. The outcome of these studies resulted in a number of significant findings, which identified LIB design features that contribute to the severity of cell-to-cell propagating TR characteristics:

- Excessive thermal energy transfer from failed cell(s) to adjacent LIB cells in the same battery contributes to the likelihood of cell-to-cell propagating TR severity.
- Electrical connections between cells in parallel strings within a p-s LIB design architecture can promote in-rush and sustained currents that heat the failed cell(s) and if allowed to continue unabated, can heat adjacent cells to the point of TR.
- Vented TR products (ejecta) from failed cell(s) in TR can transfer significant amounts of heat and create electrical shorts.

Other contributing factors, such as high LIB specific energy, electrolyte flammability, and use of flammable materials in LIB construction, were also identified. Relevant lessons learned from safety incidents in LIB markets adjacent to space applications, such as commercial electronics, electrified transportation, and grid energy storage, are now commonly incorporated into applicable space battery standards, specifications, codes, and guidelines. Lessons learned from space LIB applications have been made public by NASA via a lessons learned database called the Lessons Learned Information System (LLIS) [19]. The LLIS documents applicable spacecraft battery and EPS-related lessons learned from a wide range of NASA programs.

7.2 Space LIB Safety Requirements

Compliance with the applicable battery safety practices, guidelines, and standards ensures that the relevant spacecraft LIB power systems will perform their intended operational function safely in the intended application and operational environment. Battery safety standards for space, automotive, industrial, stationary, and other commercial applications are available from multiple professional domestic and international organizations [20]. Domestic and global standards organizations, such as the American National Standards Institute (ANSI), International Electrotechnical Commission (IEC), International Organization for Standardization (ISO), Radio Technical Commission for Aeronautics (RTCA), and Underwriters Laboratories (UL), specialize in developing and publishing industry-consensus battery safety standards applicable to a diverse set of consumer and industrial applications. Professional organizations, such as the American Institute of Aeronautics and Astronautics (AIAA), the Institute of Electrical and Electronics Engineers (IEEE), and the Society of Automotive Engineers (SAE International), also specialize in developing battery safety standards relevant to their specific industry marketplace emphasis.

However, batteries designed to function safely in a space environment under a wide set of operating and non-operating conditions are subject to a unique set of safety hazard risks. Table 7.1 lists applicable and reference top-level parent battery safety standards common to space battery safety

Table 7.1 Safety standards, guides, and regulations for space Li-ion cells and battery design, test, evaluation, and transportation. 1- Crewed; 2-Unmanned or Robotic.

Document	Title	Organization	Mission type
JSC 20793	Crewed Space Vehicle Battery Safety Requirements	NASA	1
TM-2009-215751	Guidelines on Lithium-ion Battery Use in Space Applications	NASA	1,2
AFSPCMAN 91–710	Range Safety User Requirements Manual	DoD	1,2
RCC 319-19	Flight Termination Systems Commonality Standard	DoD	1,2
SSC-S-017	Lithium-Ion Battery Standard for Spacecraft Applications	US Space Force	2
ISO 17546	Space Systems-Lithium Ion Battery for Space Vehicles - Design and Verification Requirements	ISO	2
ECSS-E-HB-20-02A	Space Engineering – Lithium-Ion Battery Testing Handbook	ESA	2
DOT Title 49 CFR Part 173	U.S. Department of Transportation Hazardous Materials Regulations	US DOT	1,2
UN 38.3, Part III	Manual of Tests and Criteria, United Nations Recommendations on the Transport of Dangerous Goods	United Nations	1,2

certification [21–29]. These standards, guidelines, and regulations serve to provide the necessary guidance and requirements compliance approach to achieve safe performance, transportation, handling, and storage of cells and LIBs intended for unmanned, crewed, and robotic spacecraft applications. Finally, applying LIB safety standards to spacecraft programs often involves requirements tailoring in order to demonstrate compliance when considering unique LIB power system design complexities, design margins, life cycle cost, in-process controls, unique environments and other factors. Specifically, the tailoring process often involves modifying or replacing certain top-level parent safety standard requirements with alternate requirement solutions. The alternate requirement solutions must still demonstrate compliance and the same level of verification to at least meet the intent of the top-level parent safety standard requirements.

7.2.1 NASA JSC-20793

LIB designs intended to support crewed NASA spacecraft applications are required to comply with the NASA Johnson Space Center (JSC)-20 793 standard [21]. The JSC-20793 standard covers all battery chemistry types intended for use aboard crewed spacecraft, crew equipment, crewed (astronaut) spacesuits, and experiments. Examples of crewed spacecraft include LVs with crewed payloads, crew transport vehicles, and on-orbit crewed space stations. As discussed in Chapter 5, US astronaut spacesuit EMU non-rechargeable lithium batteries must also meet JSC-20793 requirements. Provisions for safe ground personnel handling and testing of batteries are also specified. Hazard sources, controls, and process requirements are described for each unique battery chemistry. In 2017, new requirements were added for multi-cell LIBs of sufficient energy (>80 Wh) would need to be evaluated and tested for the severity of a worst-case single-cell TR event and risk of cell-to-cell TR propagation in the relevant space application configuration and environment. Although the NASA JSC-20793 standard does not require all LIB designs to be non-propagating, utilizing test and analysis verification methods to quantify the severity of battery-level TR is required. These LIB TR evaluation requirements were partially derived from the National Transportation Safety Board investigative recommendations resulting from the Boeing 787 aviation LIB safety incidents and were later refined based on outcomes from multiple NESC-sponsored LIB projects [10].

7.2.2 Range Safety

The US Space Force (USSF) Space Systems Command (SSC) LIB requirements for satellite contractor performance in defense system acquisitions and technology developments are intended to ensure specified performance and safety of LIBs and to provide protection against degradation during integration, pre-launch, launch, and operational use on all types of spacecraft and electrical ground support systems [25, 30]. The SSC-S-017 standard further specifies that commercial off-the-shelf (COTS) and configuration-controlled Li-ion cell designs meet relevant Air Force Range Safety AFSPCMAN 91-710 and Department of Transportation (DOT) safety requirements [23, 28]. However, COTS Li-ion cell and cell lot safety testing from applicable DOT, UN, and UL safety testing may be used to augment SSC-S-017 COTS cell safety test requirements.

In support of US government, civilian, or foreign entity launches on USSF ranges, the AFSPCMAN 91-710 standard contains LIB safety requirements for ground and launch support personnel and equipment, systems, and material operations. Range Safety requirements for arrival, usage, test, packing, storage, transportation, or disposal of non-rechargeable lithium batteries and rechargeable LIBs on the launch ranges specify that:

- Safety devices (fuses, overpressure relief devices, overtemperature cut-off, reverse current blocking diode [non-rechargeable batteries only], current-limiting resistor, or other devices) shall be incorporated into the LIB design.

- Safety-venting shall be demonstrated by test to show that the venting operates as intended and that the vent is adequate to prevent cell/battery fragmentation.
- Range users shall have an operational plan for battery/cell handling that includes emergency and contingency operations for physical abuse incident and battery installation/removal.
- LIB designs should address requirements contained in RTCA DO-311A [31].

US-based Range locations include the Eastern Range at Cape Canaveral Space Force Station and the Western Range at Vandenberg Space Force Base. East and west coast Range Safety also dictates stringent LIB acceptance and qualification test requirements for reliable range processing of unmanned and crewed LVs utilizing flight termination systems (FTSs) [24]. The Range Commanders Council (RCC) 319-19 commonality standard establishes the design and test requirements for LV FTSs operating at major Department of Defense (DoD), Department of Energy (DOE), Federal Aviation Administration (FAA), and NASA launch and test facilities. Lot acceptance and qualification test requirements for FTS-dedicated battery power source types such as LIB, silver-zinc (Ag-Zn), nickel-cadmium (Ni-Cd), thermal, and lead-acid battery technologies are described.

7.2.3 Design for Minimum Risk

Battery safety requirements for crewed space applications generally differs from unmanned battery applications due to the presence of crew members, enclosed crewed environments, and mounting (or use) in unpressurized spaces adjacent to habitable volumes. Crewed spacecraft with habitable areas, such as the NASA ISS, Dragon 2, and Orion spacecraft, utilize fault tolerance requirements combined with a design for minimum risk (DFMR) approach to control catastrophic LIB safety hazard causes [32]. The DFMR process is used to address catastrophic battery hazards that cannot practically be controlled by a failure tolerance approach. If a DFMR approach is implemented, technical authorities accept that the process has demonstrated that the applied methods and measures have reduced the likelihood of occurrence and hazard severity to an acceptable level for the intended application. An example of applying DFMR principles to LIB safety certification for crewed applications is the use of thorough cell screening and qualification test methodologies to reduce the risk of latent cell internal defects that are known to cause energetic catastrophic TR conditions. Other examples of using a DFMR strategy in space LIB designs include the ability to prevent electrical shorts external to cells, but internal to LIB, rigorous qualification and acceptance testing on the Li-ion cell and LIB designs, and compliance with safety factors.

7.3 Safety Hazards, Controls, and Testing

Industry experience across various aerospace LIB applications has shown that LIB safety hazard severity is representative of specific Li-ion cell and battery design characteristics. In space applications, LIB hazards are also related to mission-specific operating, non-operating, storage, handling, and transportation conditions commonly documented in the LIB hazards analysis. In general, unmanned and crewed spacecraft applications of LIB-based EPS designs include a wide variety of safety requirements intended to mitigate identified LIB hazards risk.

Li-ion cell-level design characteristics, such as specific energy, electrolyte flammability, and internal safety device characteristics, are commonly used in identifying hazard types, causes, and controls. At the battery and EPS level of integration, hazard controls become increasingly critical to achieving compliant redundancy and fault protection design features in order to mitigate safety hazard risks. In general, Li-ion cell and LIB safety hazard types can be categorized as electrical, mechanical, thermal, and chemical by origin [33]. Examples of LIB safety hazard analysis include safe-by-design considerations to the structural integrity of the cell and battery housings; the possibility of gas generation, pressure increases, electrolyte leakage; prevention of short circuits (internal and external) and circulating currents; high battery temperature exposure; and verification of proper charge–discharge management.

7.3.1 Electrical

Electrical LIB hazards are characterized by sources that cause electrical stress to LIB components and, in some cases, the spacecraft EPS. The primary causes of LIB electrical stress hazards include overcharge, overdischarge, external short circuit (ESC), and internal short circuit (ISC) conditions. For example, continued charge–discharge cycling of overdischarged LIB cells can result in performance degradation followed by a catastrophic TR event. Uncontrolled electrical stress hazards can result in degraded LIB performance, harm to ground processing personnel, and damage to facilities or flight hardware. In crewed spacecraft applications, catastrophic outcomes may result in loss of crew or of a major crewed space system element [34].

7.3.1.1 Overcharge

Unlike heritage space rechargeable Ni-Cd and Ni-H_2 cell technologies, Li-ion cell chemistry involves overcharge reactions that are irreversible and energetic. As such, all Li-ion cell chemistry types are intolerable to overcharge conditions [35]. Relevant space Li-ion cell and LIB-level safety abuse testing clearly indicates that overcharge conditions result in venting, smoke, fire, and/or explosion. As such, the most critical hazard of electrical overstress for Li-ion cells is an overcharge condition. In terms of the EPS charge control of LIBs, overcharge occurs when the battery voltage exceeds the manufacturer's recommended upper charge voltage limit and/or when LIB charging exceeds 100% state-of-charge (SOC) as defined by the battery specification. In either case, cells can sustain overcharge failure that may result in current interruption devices (CIDs) activation (COTS 18650 Li-ion cells only) and/or cell venting. A catastrophic TR event will occur when cell charging exceeds the specified cell or battery upper charge voltage termination limit for an extended period of time [36].

As described in Chapter 4, LIB charge control by the spacecraft EPS is critical to maintaining the performance, safety, and reliability of the spacecraft system. Due to the extreme hazard risk associated with LIB overcharge conditions, unmanned and crewed spacecraft applications commonly require fault tolerance against overcharge. In crewed spacecraft applications, JSC-20793 specifies that LIBs containing less than 80 Wh of energy may be classified as having a medium battery risk classification (BRC). A high BRC only applies to high energy and/or power LIB designs that contain greater than 80 Wh of energy in a crewed application. High BRC LIBs require dual-fault tolerance protection against all catastrophic hazards, including overcharge. In practice, high-reliability unmanned satellites and robotic planetary missions utilizing LIB-based EPS charge control systems generally implement at least dual-fault tolerant protection against overcharge fault conditions.

7.3.1.2 Overdischarge

In Li-ion cells or batteries, overdischarge occurs when the cell is discharged below the manufacturer's recommended lower discharge voltage limit. The lower cell voltage discharge limit is typically at 0% SOC under normal LIB operating conditions. Although overdischarge creates electrical stress within a Li-ion cell, most ground experience with exceeding the lower voltage discharge limits have resulted in a benign, but degraded, cell condition. In general, repeated Li-ion cell overdischarge further promotes the dissolution of the copper anode current collector substrate. Upon repeated cell cycling to overdischarge conditions, copper may electrodeposit on to the cell cathode, anode, and separator materials [37]. Copper electrodeposition within the cell may eventually lead to cell-level ISC conditions. Similar to LIB overcharge protection, LIB charge–discharge management systems are designed to maintain failure tolerance against overdischarge conditions. As discussed further in Chapters 4 and 9, on-orbit spacecraft-level anomalies, such as loss of solar array power due to insufficient sun illumination, may cause catastrophic depletion of the available battery energy. Under these anomalous conditions, a LIB low-voltage condition may eventually occur in a manner that degrades into a battery overdischarge condition.

7.3.1.3 External Short Circuit

ESCs are caused by the direct contact between the positive and negative cell (or battery) terminals via a conducting material. External to the battery, ESCs can be caused by bridging between the power and signal connector pins or at the charging source. Causes of ESCs internal to the battery, but external to cells, may include faulty cell interconnections, wiring insulation, or the presence of conductive foreign objects debris (FOD). ESCs are more likely to occur during ground processing of flight LIBs due to the increased frequency of handling and transportation activities. The risk to ESCs also increases during LIB manufacturing steps involving handling of charged Li-ion cells during cell integration into strings, modules, and/or banks. Cell ESCs can result in cell heating, venting, or other catastrophic outcomes. As discussed in Chapter 2, ESC hazard controls include cell-level internal safety devices, such as positive temperature coefficient (PTC) current-limiting devices, and CIDs used in COTS 18650 cell designs. Fusible links incorporated at the cell and/or battery-level can also mitigate the severity of ESC events. By design, some space Li-ion cell designs may be case-negative or case-positive. To protect these types of cell designs from ESCs, it is a best practice to add at least two dissimilar layers of electrical insulating non-flammable material between all cells, modules, banks, battery chassis, and unfused power circuits. Double layers of electrical insulation are also incorporated into the spacecraft EPS harness and component unit designs to further protect against shorting hazards.

7.3.1.4 Internal Short Circuit

In general, an ISC is a cell-level condition caused by a direct electronic connection between the cathode and anode within a cell that results in an unintended path for current flow. Potential causes of ISCs include the presence FOD, native object debris (NOD), manufacturing defects, or other internal cell flaws [38, 39]. In contrast, graceful degradation of Li-ion cells is characterized by an increase in cell internal resistance, which commonly results in high-impedance ISCs. The outcome of cell-level graceful degradation is a non-energetic gradual loss of cell capacity. As such, most cell ISC conditions cause high-impedance ISCs that result in non-energetic failure modes. However, a sustained low-resistance ISC is more likely to result in an energetic catastrophic failure mode. Careful attention to the stability of cell open-circuit voltage (OCV) often enables ISC conditions to be detected early in the cell-level screening and acceptance test process. This is an important measurement for cell-level hazard mitigation since subtle high-impedance ISC conditions can degrade to low-impedance conditions during cell charge–discharge cycling, storage, or aging [40]. Nonetheless, industry experience indicates that cell screening should not be relied upon as the sole hazard mitigation for cell-level ISCs.

Space-qualified LIB cells are designed and manufactured with multiple in-process controls to mitigate the possibility of ISCs. Due to the high specific energy of custom large-format cell designs, extra rigor in design, manufacturing, and testing is required to mitigate the presence of ISCs which may degrade into an energetic ISC safety hazard. While most custom large-format Li-ion cell designs do not include the same types of safety features found in COTS Li-ion 18650 cell designs, custom large-format space cell designs have demonstrated long product histories without energetic ISC failures. Space industry manufacturers of custom large-format Li-ion cell designs routinely practice rigorous design and manufacturing configuration controls, cell production line audits, and extensive screening protocols to meet high product quality standards. The rigor of these space industry best practices provides a rationale for a very low likelihood and thus acceptable risk of occurrence for a cell or battery-level catastrophic TR event caused by a cell-level ISC.

7.3.2 Mechanical

As discussed in Chapter 3, LIB mechanical overstress conditions may result from exposure to structural-mechanical environments throughout the ground, launch, and on-orbit lifetime of a flight LIB. Mechanical environments such as LV vibration, shock, and acceleration, contribute to LIB mechanical stress factors. During ground processing, handling and transportation vibration, and shock environments also present mechanical stress to the LIB flight hardware. In some cases, mechanical overstress conditions may degrade LIB parts, seals, connections, and other interfaces which could result in physical degradation to the LIB.

For example, an abuse condition could result in mechanical stress to the battery cells, resulting in a terminal seal rupture or damage to internal cell weld joints and electrical connections. In addition to loss of electrical performance, internal mechanical damage to a battery or cell may result in an ISC or an ESC condition. Depending upon the resistance of the short circuit and cell SOC, this could result in an energetic vent or more severe hazard condition. Battery safety is maintained with regard to mechanical overstress by designing LIBs with mechanical load margins above cell-demonstrated load capabilities and by performing relevant battery qualification testing to verify requirements compliance. Compliance to the battery manufacturer's approved storage, handling, and transportation process documentation also mitigates the risk of mechanical overstress conditions. The documentation is specific to the LIB design and identifies any special tooling, fixtures, or test equipment needed to support safe storage, handling, and transportation during ground processing.

7.3.3 Thermal

As discussed in Chapters 2 and 3, space-qualified Li-ion cell designs exhibit optimal operating electrical performance when controlled at temperatures between 15 and 20 °C for most spacecraft missions. However, as discussed in Chapter 6, certain planetary mission spacecraft may specify LIB cells to operate at extremely low or high temperatures. As such, LIBs intended for planetary spacecraft missions may utilize unique electrode materials and electrolytes with low temperature or high temperature performance capability. Exposure to cold or hot temperatures may also occur during LIB ground handling processing or transportation activities. Thermal stress on cells and batteries generally occurs at low and high temperature extremes, as defined by the qualification limits unique to the LIB cell design [2]. However, to mitigate cell performance degradation or safety hazards, on-orbit LIB temperatures are controlled over narrow operating temperature ranges to maximize performance and mission service life. Figure 7.1 shows a notional safe voltage and temperature (V-T) operating window for a representative COTS 18650 Li-ion cell design. The safe V-T operating window is bounded by the cell temperature and voltage consistent with the effect of SOC on stored energy. The onset temperature of Li-ion cell TR has been shown to be dependent

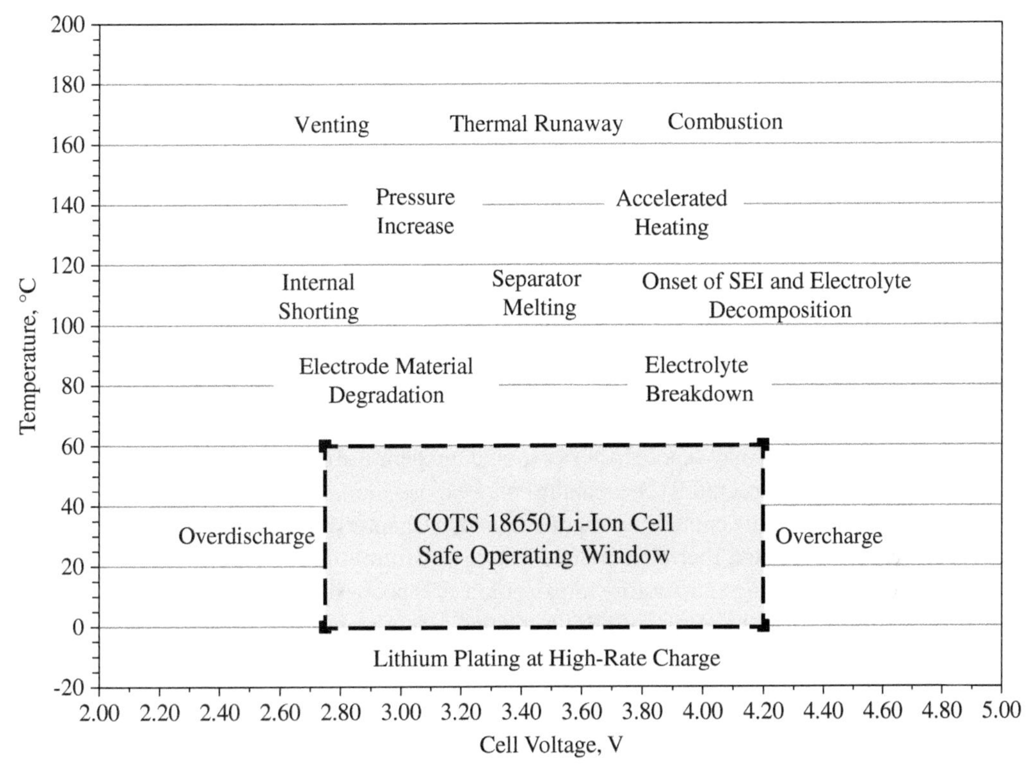

Figure 7.1 Notional safe voltage and temperature (V-T) operating window for a representative COTS 18650 Li-ion cell design.

upon cell chemistry, capacity, SOC, and aging. In general, cell surface temperatures in a closely packed multi-cell LIB should not exceed approximately 70 °C during operations due to cell internal electrodes being at much higher temperatures. Similarly, Li-ion cells exposed to environmental temperatures above approximately 90 °C may be at risk of a TR condition. Secondary factors such as cell age may also affect the severity of TR [41, 42]. Operating or exposing LIB cells outside the safe operating V-T window limits will degrade cell performance and increase the risk of catastrophic safety hazards.

7.3.3.1 Overtemperature

The hazard severity of exposing LIB cells to high (overtemperature) temperature conditions is dependent on the exposure duration, cell SOC, and surrounding environmental conditions. Short duration exposure of Li-ion cells to overtemperature conditions results in non-energetic degradation of cell electrode materials, which may cause loss of usable cell capacity that degrades on-orbit service life. In space applications, LIBs are required to meet all performance requirements when operated within the normal operating temperature range. However, when a space-qualified LIB is operated at moderately elevated temperatures outside the normal operating temperature range, permanent performance degradation may occur. High-reliability space LIB designs are also specified to demonstrate that the LIB can safely operate without permanent degradation or component failure after exposure to the given hot (or cold) survival temperature limit. Representative cold (minimum) and hot (maximum) survival temperature limits for a space LIB are −10 to +60 °C.

In general, exposure to high overtemperature conditions for an extended amount of time causes rapid cell materials breakdown and, under extreme high-temperature conditions, catastrophic TR conditions. LIBs may be exposed to out-of-specification, high-temperature environments during uncontrolled storage, transportation, or ground testing. Loss or degraded ground test or on-orbit spacecraft thermal control subsystem (TCS) capability will also expose the flight LIB hardware to overtemperature conditions. Various abuse-driven conditions such as overcharge, ISCs, ESCs, or high charge–discharge rates can result in high overtemperature conditions in LIBs.

7.3.3.2 Low Temperature

As described in Chapter 3, spacecraft EPS LIBs are thermally controlled to operate within the mission-specified operating temperature range. Although certain space LIBs are mounted within the spacecraft structure, thus minimizing exposure to extreme thermal environments, on-orbit anomalies or contingencies may expose batteries to cold temperature extremes. A degraded ground test or loss of on-orbit spacecraft TCS capability may expose the flight LIB hardware to out-of-specification low-temperature conditions. As mentioned in Chapter 3, primary and redundant LIB heaters with carefully located thermistors are effective in controlling cell and LIB temperatures within their normal operating temperature range. Li-ion cells not designed to tolerate extreme low-temperature operation may be at risk of lithium plating when operated at low temperatures and high charge rates [43–45]. To mitigate these hazards, space LIB EPS charge–discharge and/or battery management system (BMS) electronics are typically designed with conservative low-temperature limits. LIB heater set-points are set to avoid low-temperatures under all on-orbit operating conditions.

7.3.4 Chemical

The most flammable component of a Li-ion cell is the liquid organic electrolyte. As discussed in Chapter 2, space-qualified Li-ion cells generally use liquid organic electrolytes composed of ionic salts (such as $LiPF_6$) in mixtures of various cyclic and linear organic carbonate-based solvents with high flammability. Other combustible components in Li-ion cells include polymeric separators, cathode, and anode electrode materials. Unlike heritage Ni-Cd and Ni-H_2 space battery cells, Li-ion cells do not exhibit significant internal pressure growth over their service lifetime. During the lifetime of a Li-ion cell, small amounts of gas may be generated as a result of chemical and physical changes that occur during normal cell charge–discharge cycling. For example, irreversible pressure growth in space-qualified Li-ion cells has been shown to be less than 10 psia over extended periods of stressful low Earth orbit (LEO) cycling [46]. However, energetic Li-ion cell failures resulting from electrical, mechanical, or thermal abuse conditions may result in excessive byproduct internal gas generation.

Cell overpressure will result in cell swelling, venting, and/or case rupture, which commonly precedes the onset of catastrophic TR. The kinetics and thermodynamics of an energetic cell failure are dependent on the pressure, volume, and temperature (P-V-T) at which the failure occurs. Furthermore, the effect of cell chemical composition on the gaseous decomposition products may vary as a function of P-V-T. The oxygen concentration in the surrounding environment also has a significant effect on decomposition product components. In crewed spacecraft applications, release of toxic chemicals and gases into the enclosed environment of a spacecraft may contaminate the breathing air supply and cause operational problems for surrounding flight hardware. Release of these chemicals and gases external to a spacecraft may contaminate an astronaut space suit and other contamination-sensitive flight hardware (such as the solar arrays). As such, toxic gases are prohibited from venting into the habitable compartments of a crewed spacecraft. In addition, toxic gases are prevented from venting into non-habitable spacecraft environments if hazardous to flight components, wiring, or other sensitive hardware in the vicinity of the flight LIB installation. Furthermore, JSC-20793 requires that assessments be made by a toxicologist on the toxicity hazard level associated with the vented products of the mission-relevant Li-ion battery cell design [47]. These requirements also include crewed applications of non-rechargeable lithium cells capable of producing toxic sulfur dioxide, thionyl chloride, or hydrogen chloride gases. Although Li-ion cell gas composition studies may vary widely due to the cell chemistry, the method used to initiate TR, and the gas detection methodology, the majority of gas phase products detected are CO_2, CO, CH_4, and H_2 [48–51]. Trace amounts of other C2-C5 hydrocarbons are also commonly measured. The most toxic Li-ion cell gas phase decomposition product is HF, which may be detected in trace amounts as a function of cell chemistry, temperature, and other test conditions.

Leakage of electrolyte from a battery cell is a hazard which is included in a space LIB safety hazard analysis. As discussed in Chapter 2, Li-ion cells are hermetically sealed to meet safety, performance, and reliability requirements. Causes of electrolyte leakage may be faulty cell seals or leakage due to a cell overheating under electrical or mechanical abuse conditions. Electrolyte leakage hazards include formation of conductive ground paths between cells and metallic battery components, corrosion, and/or toxicity of electrolyte fluid and/or vapors. The primary hazard control for electrolyte leakage includes cell design features that meet mechanical hermeticity requirements. LIB-level design features such as protecting battery interior surfaces with non-electrically conductive coatings to prevent a short-circuit hazard, can mitigate the effects of cell electrolyte leakage on battery components.

7.3.5 Safety Testing

In order to comply with space industry Li-ion cell safety requirements, safety testing is conducted to establish thresholds where the flight cell design becomes unsafe over a range of charge–discharge voltages, currents, temperatures, and SOC [25]. The Li-ion cell bank, module, and/or LIB-level testing are also required to establish safety thresholds in the relevant LIB cell flight configuration. In general, cell-level safety requirements are verified during the cell-level qualification program, which generally precedes LIB-level qualification testing. However, cell-level internal safety devices (such as PTC, CID, or fuses) may be required and are verified on a lot sampling basis during cell acceptance testing. In Chapter 2, the importance of employing relevant Li-ion cell acceptance testing to evaluate the manufacturing workmanship of flight Li-ion cells prior to cell integration into flight batteries was discussed. Quantifying cell performance margins during acceptance level testing is a common first step in understanding cell-level safety margins. The effect of thermal, mechanical, electrical, and chemical interactions on performance margins is a primary consideration for safe Li-ion cell and battery design. Li-ion cell-level safety requirements are verified by test under known cell failure modes that include overcharge, overdischarge, overtemperature, ISC, and ESC conditions.

Lessons learned from industry LIB TR incidents indicate that worst-case safety testing in a relevant configuration and environment applicable to the intended spacecraft application is necessary to properly identify worst-case hazard safety risks. This lesson learned includes using flight-like test cells and batteries constructed from the same materials, components, and parts intended for use in the flight-qualified LIB design.

7.4 Thermal Runaway

Li-ion cell TR will occur when the rate of cell heat generation exceeds the rate of cell heat rejection, causing a rise in the cell temperature and pressure. As the cell temperature increases, the rates of reaction for self-sustaining exothermic chemical processes in the liquid and gas phases increases at a rapid rate. The cell internal pressure will simultaneously increase with temperature and other cell material decomposition processes. The next most common events include a decrease in cell voltage along with cell venting, electrolyte leakage, ignition of gas-phase flammable gases, combustion, smoke, sparks, fire, and rapid ejection of materials. Cell-level TR can result in a plasma-torch phenomena consisting of various combustion products, including molten Al, Cu, and superheated gases ejected at very high rates. These TR high-temperature combustion products can rapidly melt through adjacent cell metal enclosures and other battery components. Depending on cell chemistry and conditions for combustion, cell ejecta may be electrically conductive, leading to circulating currents inside batteries that creates additional hazards to adjacent personnel, flight hardware, and facilities.

7.4.1 Likelihood and Severity

In crewed NASA spacecraft applications, LIB (>80 Wh) qualification requirements include assessing the severity of a single-cell TR event. Assessing the risk of propagating TR in LIBs includes an analysis of the likelihood and consequences (severity) of various TR safety hazards. In general, technical risk is measured as the potential inability to achieve overall program objectives within defined constraints. The two risk components are (a) the likelihood of failing to achieve a particular outcome and (b) the severity of failing to achieve that outcome. Analytically, risk scoring is accomplished by calculating the numerical value given by the product of likelihood and severity, where the likelihood is the probability of a risk occurring within a stated timeframe. The hazard severity includes the negative impacts to personnel, flight hardware, and/or program success.

Traditionally, the likelihood of occurrence of a catastrophic cell TR event has been mitigated by implementing cell-level TR prevention protocols and LIB-level BMS controls [52]. Cell-level TR prevention protocols include relevant cell screening acceptance test methods such as charge retention and OCV testing to screen for the presence of ISCs. Other cell screening methods such as visual inspection, mass properties, DC resistance, AC impedance, and capacity measurements, are also effective as cell screening methods. LIB cell manufacturing processes and quality control measures are also effective in reducing the likelihood of TR hazard risk during normal or unintentional abusive LIB operations.

However, field experience has demonstrated that sources of Li-ion cell ISCs, such as latent FOD, NOD, or manufacturing defects, may be undetected during cell screening. Due to the consequences of a catastrophic LIB failure in a spacecraft application, LIB designs must consider opportunities to mitigate the severity of TR hazards. Redundant overcharge and overdischarge protection design features are examples of LIB-level BMS controls of TR safety hazard conditions.

The severity of a TR event is highest when a single Li-ion cell TR propagates to adjacent Li-ion cells packaged into a cell bank or module within a LIB. A cell-to-cell propagating TR event is a chain reaction that commonly results in a catastrophic LIB failure with the possibility of collateral damage to interconnecting systems. Causes of a single Li-ion cell TR include abuse conditions that result in excessive cell overtemperature conditions resulting from overcharge, overdischarge, ISC, or ESC hazards. Cell-level TR severity is dependent on Li-ion cell design

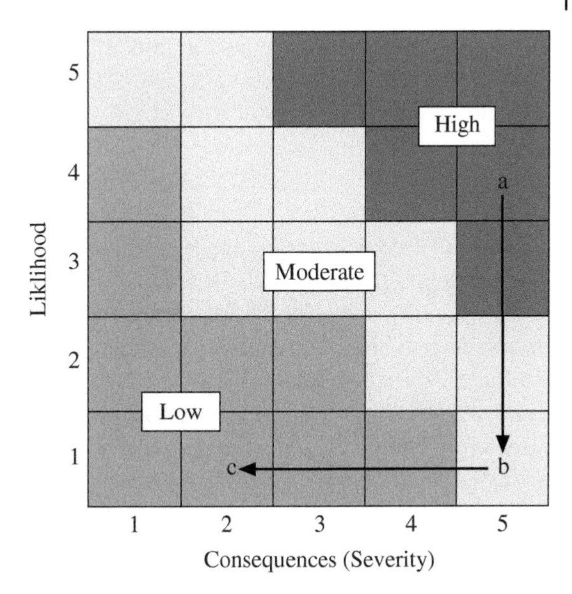

Figure 7.2 Five-by-five (5 x 5) risk cube for reducing the likelihood and consequences of thermal runaway. (a–b) Reduce likelihood of causes for thermal runaway through relevant cell screening protocols, and (b–c) Reduce consequences of thermal runaway via implementation of verifiable hazard controls.

features such as chemistry type, specific energy, safety devices, and geometry (form factor) [53, 54]. Li-ion cell SOC and aging also impacts the severity of TR hazards [41, 42, 55]. TR hazard severity is also highly dependent on the specific failure mode that caused the initial cell overtemperature condition. The severity of propagating TR can be mitigated by incorporating verifiable LIB design safety hazard controls. At the battery level, including safety design features that mitigate the severity of cell-to-cell TR is required for NASA crewed spacecraft applications of LIBs. Figure 7.2 illustrates a representative risk mitigation strategy for reducing the likelihood and severity of LIB TR in spacecraft applications [56]. Relevant Li-ion cell screening processes and procedures are used to reduce the likelihood for causes of TR, while implementation of verifiable safety hazard controls is used to reduce the consequences (severity) of TR. Understanding the severity of a possible LIB TR event facilitates an informed risk assessment for identifying risk mitigation opportunities.

7.4.2 Characterization

NASA crewed spacecraft applications utilizing high specific energy LIBs with catastrophic failure modes are required to be evaluated to determine the severity of a worst-case single-cell TR event. In addition, the susceptibility of the LIB design to cell-to-cell TR propagation in the intended application and on-orbit environment must also be evaluated. Evaluation includes test and analysis of the flight LIB design characterized in the relevant spacecraft configuration and on-orbit environment. However, the risk of injury to astronaut crew members due to an on-orbit catastrophic LIB TR event is not applicable for unmanned Earth-orbiting or robotic planetary spacecraft missions. As a result, certain unmanned spacecraft missions employing high specific energy LIBs or non-rechargeable lithium batteries may not be required to characterize the severity and/or consequences of a catastrophic LIB TR event for the intended on-orbit application.

A wide variety of Li-ion cell electrical, mechanical, or thermal abuse conditions have been shown to result in energetic TR events. For example, the energy released in a single-cell TR event may vary with specific cell design features, SOC, and failure mechanism (initiation) type [57]. Thus, characterizing the susceptibility of a multi-cell space LIB to propagating cell-to-cell TR requires an understanding of single-cell, module (if any), and battery-level TR behavior. As such, industry lessons learned indicate that cell or module-level TR test results facilitate the safe design of multi-cell LIBs.

7.4.3 Testing

Cell-to-cell TR propagation testing is a demonstration of the LIBs design tolerance to a single-cell TR event propagating to other cells within the battery. In addition to high standards for cell manufacturing quality, cell-level TR risk mitigation strategies include consideration of key cell design features such as chemistry, geometry, and internal safety devices. Cell-level TR severity may also be mitigated by battery design features such as materials of construction, cell packaging, protective fusing, venting capability, and other approaches. Operationally, BMS controls, such as cell voltage-temperature limits and battery-level charge–discharge inhibits for ground and on-orbit operations, can further mitigate the likelihood and severity of a TR event. In practice, cell-level safety requirements are verified by test under known cell failure modes identified in the LIB-level failure mode, effects, and criticality analysis (FMECA). Cell safety testing establishes the thresholds where the flight battery becomes unsafe over a range of mission-relevant charge–discharge voltages, currents, temperatures, and SOCs.

In practice, Li-ion cell and battery TR test outcomes can be highly variable due to the characteristic uncontrolled high energy release of cell contents in short periods of time. Wide variations in TR safety test procedures used in today's electric vehicle, aerospace, and commercial electronics LIB standards may also contribute to variable TR test outcomes. Other factors that contribute to variable TR test outcomes include differences in Li-ion cell designs (such as chemistry, specific energy, and internal safety features), variations in cell-level TR trigger methods, and types of TR failure modes [58].

7.4.3.1 Single Cell

Traditionally, non-destructive and destructive experimental techniques have been used to determine the thermal behavior of all types of cells and batteries. As described in Chapter 2, thermal analysis of space LIBs is required to ensure that battery temperature limits will not be exceeded during ground, launch, or on-orbit phases of the spacecraft mission. For the purposes of understanding the thermal behavior of Li-ion cells under abuse conditions, worst-case destructive testing is employed to best simulate safety hazard severity levels. Adiabatic and isothermal calorimetric techniques are commonly used to determine Li-ion cell heat generation rates during normal charge–discharge cycling or under various abuse conditions.

Accelerating rate calorimetry (ARC) results for a representative COTS 21700 format lithium manganese nickel cobalt oxide (NMC) Li-ion cell at 100% SOC are shown in Figure 7.3. The cell response to increasing ARC temperature reveals the characteristic stages associated with TR development under controlled adiabatic test conditions [50, 59–61]. Stage 1 (ambient temperature to 120 °C) is characterized by initial heating leading to the onset of cell decomposition and an increase in cell self-heating. During Stage 2 (120–170 °C), the cell is approaching TR, indicated by accelerating heat release, which causes internal cell pressure and temperature increases. Cell venting and release of internal cell material (ejecta) and smoke also occurs during Stage 2. Additional cell self-heating leads to Stage 3 (temperatures >170 °C), which is characterized by a rapid increase in temperature, cell internal material decomposition, and high temperature combustion of the cell electrolyte and electrodes. Figure 7.4 shows ARC profiles for a 36 Ah prismatic lithium colbalt oxide (LCO) Li-ion cell at increasing SOC. As SOC increases, the onset of cell decomposition and

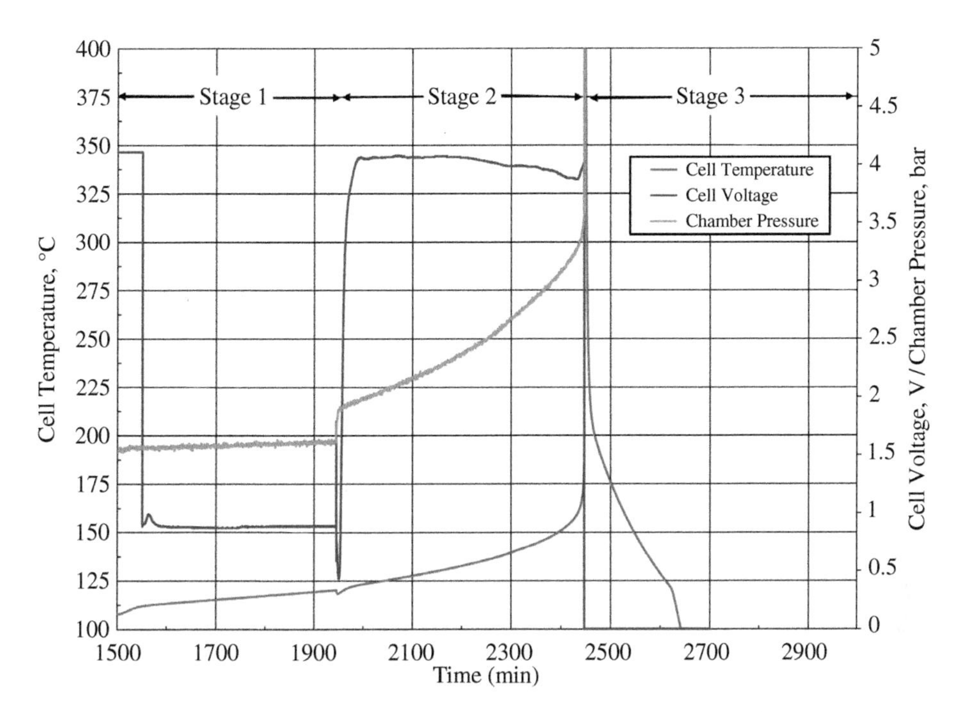

Figure 7.3 Thermal runaway characterization of a COTS 21700 format NMC Li-ion cell at 100% SOC. *Source:* Data courtesy of Thermal Hazard Technology.

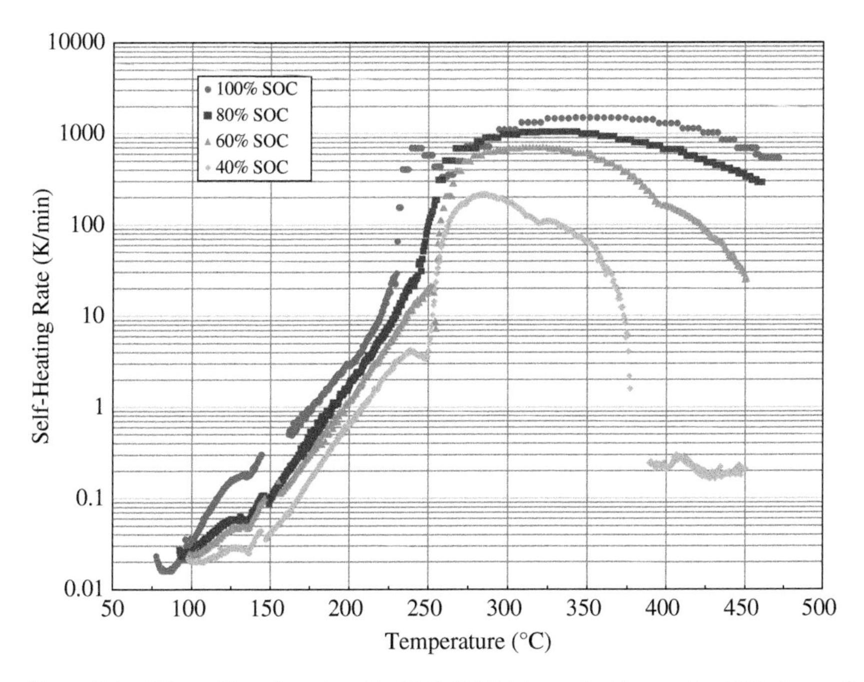

Figure 7.4 ARC profiles of a prismatic 36 Ah LCO Li-ion cell at increasing SOC. *Source:* Data courtesy of Thermal Hazard Technology.

TR temperature decreases along with an increase in the peak thermal response of the cell. The inverse relationship between Li-ion cell TR severity and SOC is directly applicable to establishing safe storage, handling and transportation requirements for LIBs [62].

7.4.3.2 Module and Battery

Whether cells are configured into series-parallel (s-p) or parallel-series (p-s) LIB architectures, industry experience has shown that single-cell TR severity is significantly different from module or battery-level TR behavior. More specifically, characterization of cell-level TR characteristics alone may be insufficient for determining module or battery-level TR characteristics. Therefore, high specific energy multi-cell LIB TR severity may be required to be characterized at the module and/or battery level for crewed mission applications. This approach is consistent with a test-like-you-fly (TLYF) philosophy, which dictates that worst-case safety testing in a relevant configuration and environment is necessary to properly determine safety hazard severity levels. For example, the NASA ISS 30s1p LIB ORU design was tested in a flight-like configuration and relevant space environment for susceptibility to TR hazards at the battery level of integration [16]. As discussed in Chapter 5, the ISS LIB ORU qualification program was successful in demonstrating compliance to non-propagating TR LIB design requirements [63].

7.5 Principles of Safe-by-Design

In order to reduce the risk of cell-to-cell propagating TR events, multi-cell LIB designs qualified for human space flight require unique design safety features. These safe design features include electrical, mechanical, and thermal design features that mitigate the severity of cell-to-cell TR propagation. Reducing the severity of TR safety hazards is necessary since cell-level ISC screening measures intended to prevent single-cell TR may not be 100% effective. This is accomplished by eliminating TR propagation to cells adjacent to the failed cell(s) within a multi-cell LIB. Chapter 2 further discusses various cell-level safety features commonly used in custom and COTS 18650 cell designs intended for spacecraft LIB applications. Aerospace industry experience has shown that relevant safety testing is required to establish thresholds where the flight cell and/or battery design becomes unsafe. However, meeting space industry safety requirement standards includes adopting a safe-by-design approach to cell, module, and battery-level design.

7.5.1 Field Failures Due to ISCs

Despite over 30 years of high-volume Li-ion cell commercial production with improving design and manufacturing quality controls, latent defect-induced cell ISCs still escape detection at the point of manufacturing and can lead to spontaneous catastrophic LIB field failures. All other root causes (overcharging, external short, overtemperature, or mechanical overstress) can be mitigated with redundant design features and use of qualified operational procedures. The defect-induced ISC failure is a rare event that cannot reliably be screened out at the cell's beginning of life due to its latency. Most field failures have occurred at high SOCs during charging, but many have occurred at rest or in storage with SOCs as low as 30%. The latency hazard was first demonstrated by implanting precisely sized and located Ni particles in 18650 cells, showing that some will pass acceptance screening and later, with further cycling, yield a TR response [40]. There are four general types of Li-ion cell ISCs: anode material-to-cathode material, anode current collector-to-cathode current collector, and cathode material-to-anode current collector, and anode material-to-cathode current collector ISCs. Research has found that the low-impedance pathway created by anode material-to-cathode current collector ISCs result in the most energetic and hazardous TR failures [64]. Risk mitigation measures for decreasing the risk of cell ISC-induced TR include cell design, manufacturing, test, and operational considerations.

7.5.2 Cell Design

Li-ion cell design selection is important for preventing defect-induced ISC field failures. Design features such as separator type and thickness, electrode winding geometry, current-collector tab routing, and active electrical part insulation can impact cell susceptibility to defects. Mature small format COTS cell designs in mass production with an absence of LIB recall history provide confidence in their design manufacturing quality. By contrast, custom large-format cell designs produced in small production lot quantities for space LIB applications may present certain risks for achieving consistent cell production quality. However, these risks can be mitigated when custom space Li-ion cell manufacturers provide configuration control and visibility into the cell specifications, design, manufacturing process, and qualification testing outcomes. In practice, custom Li-ion cell manufacturers routinely permit on-site cell production line audits, which enables verification of quality control requirements.

7.5.3 Cell Manufacturing and Quality Audits

To ensure uniformity of cell quality and performance, space LIBs should be manufactured with Li-ion cells sourced from one production lot. This is particularly critical with mass-produced COTS cell designs when it is contractually prohibitive for the spacecraft customer to impose cell design and manufacturing configuration controls on the cell manufacturer. Same model COTS cells from different date codes could have significant design changes and should be qualified separately. As discussed in Chapter 3, a single cell production lot is defined as all cells manufactured in a single or multiple production run manufactured from the same anode, cathode, electrolyte, and separator material sublots with no change in processes, drawings, or tooling, and within a defined production run or time period. For mass-produced COTS 18650 Li-ion cells such insight is rarely available, and hence a cell production lot is more commonly defined as a batch of cells made on a common production line and factory all in one or two consecutive days.

High-quality Li-ion cell production lines apply numerous preventative measures intended to reduce the occurrence of defects and prevent releasing defective cells. Measures include controlling impurities and contaminants in raw cell materials throughout the cell production process. For example, mechanical, vacuum, and magnetic filtering are commonly used in electrode active material mixing, coating, calendaring, slitting, and taping processes. These measures are also applied during cell winding, jellyroll or stack insertion into cases, electrolyte filling and doping, welding, and other process steps, and conclude with cell build completion. Verification into the effectiveness of these quality control measures is best gained when the cell manufacturer permits customer access for an on-site cell production line audit. Since high metallic particle counts, and poor contamination and humidity controls are known to contribute to high internal cell defect counts, continuous monitoring and inspection for manufacturing FOD and NOD is indicative of a high-quality Li-ion cell manufacturer.

A thorough Li-ion cell production line audit requires examination of all processes associated with cell manufacturing and acceptance testing. A typical cell production line audit includes, but is not limited to, examination of raw materials specifications, cell-level bill-of-materials, review of previous build data and non-conformances (if any), facility cleanliness measures, and personnel training records. Inspection and review of cell-level statistical process control data also facilitates a space Li-ion cell quality control audit. Space industry experience has shown that cell customer access is easier to obtain with manufacturers of custom space Li-ion cell designs versus access to COTS Li-ion cell manufacturers. The proprietary quality audit end-product includes observations and findings that are prioritized for closure before flight cell manufacturing commences. Access provided by the cell manufacturer including timely disposition of the quality audit findings is evidence of a commitment toward producing high-quality, safe, and reliable Li-ion space cells.

7.5.4 Cell Testing and Operation

Cell lot assessments to determine performance uniformity is dependent on the cell manufacturer's control of cell production quality. Per the JSC-20793 standard, initial COTS 18650 Li-ion cell lot assessments include cylindrical cell side wall rupture (SWR) characterization, computed tomography (CT), and destructive physical analysis (DPA) examinations. The SWR characterization test for cylindrical cell designs includes determination of a propensity to experience SWRs, spin groove ruptures, bottom and top ruptures, header assembly releases, jellyroll ejections, and nominal vents. In practice, several spacecraft LIB designs have failed passively propagation resistance (PPR) battery test campaigns due to the unexpected propensity of COTS 18650 Li-ion cell designs to experience cell can SWR during TR when mechanically supported in a battery configuration. The occurrence of a cell SWR was unexpected because mechanically unsupported single-cell TR testing indicated that cell designs rarely experienced SWR. Representing the cell's structural supporting features of a LIB design during safety testing is critical for relevant characterizations. NASA has developed best practices for initial flight cell lot assessments as a prerequisite for the more expensive cell screening and lot acceptance test activities [21]. These best practices include random sampling cells from the production lot to perform non-destructive performance and physical measurements to assess uniformity of the lot before lot acceptance testing (LAT) begins. Mass-produced COTS Li-ion cells are expected to show high uniformity, as specified in JSC-20793.

Initial cell assessments also include destructive testing of uniform performing cells selected from the initial sample. Destructive testing includes CT and DPA examinations and the characterization of SWR risk with relevant battery design features in-place. Given that the randomly selected samples conform to manufacturer specifications, as further detailed in JSC-20793, flight cell screening is performed on 100% of cells selected for custom flight batteries in crewed spacecraft applications. Flight cells are screened for OCV, self-discharge, visual inspection, mass, dimensional, capacity, and resistance measurements. Statistical data analysis of the test results determines lot acceptability. A failure to comply with production uniformity requirements indicates that the cells were built with poor manufacturing control processes.

After initial cell lot assessments, LAT is performed to qualify a space cell production lot for flight battery manufacturing. This includes demonstrating compliance to performance, quality, and safety test requirements to certify the cell production lot for spaceflight. Performance tests include cycle and calendar life tests, capacity performance, and OCV vs. SOC trending before and after environmental testing. The cell quality tests include seal integrity, vent and burst pressure, and CT examination followed by cell DPA as required. Safety tests include short circuit, overcharge, and calorimetry. Designing and operating a battery to function within failure thresholds identified during cell and module testing mitigates risk of cell ISC fault conditions.

7.6 Passive Propagation Resistant LIB Design

In light of preventative measures to mitigate risk of energetic cell ISC conditions, the latency potential of defect-induced cell hazards suggests that prevention cannot be assumed to be 100% effective. As such, NASA JSC-20793 requires LIBs over 80 Wh to be evaluated by test and analyses to assess the severity of hazards to any single-cell TR event. For LIBs that result in failure modes that are catastrophic to ground personnel, crew members, or mission, a best practice is to design the battery to be PPR. PPR LIB designs are able to limit the severity of any single cell TR event at worst-case credible mission operating conditions without active cooling and without causing TR propagation to adjacent cells [65].

Requirements for conducting a valid PPR test campaign on a space LIB design includes tests that are to be conducted on ground battery test articles that contain all the relevant design features of the flight battery. To be successful in certifying a PPR LIB design, no cell can degrade in performance by more than 5% of pre-test performance after a minimum of three separate trigger cell TR events at worst-case mission SOC and temperature. More specifically, cells adjacent to the trigger cell may not exhibit any soft shorting, current interrupt device (CID) activations, seal failures, or venting. In addition, no flames may exit the LIB enclosure. Non-compliance to these conditions may require design modifications and a repeat of the PPR test campaign.

7.6.1 PPR Design Guidelines

To increase PPR performance, a set of design guidelines are described. These PPR design guidelines are based on NASA and industry experience with test and thermal analysis results from various compliant and non-compliant PPR LIB designs [66, 67]. Additional findings from previous NASA TR testing across a number of space-qualified LIB designs include [14]:

- Direct cell-to-cell contact leads to propagation of TR via direct heat conduction.
- Effluents from cell venting carry sufficient energy to promote TR propagation.
- Prevention of catastrophic cell-to-cell TR propagation in a LIB design can be achieved for relatively low mass and volume impacts.

7.6.1.1 Control of Side Wall Rupture

The first guideline mitigates cell can SWR events during TR. Cylindrical COTS 18650 Li-ion cells are designed to vent gas and TR ejecta through the header vent or a bottom vent (if present). Breaches of the cell can wall or spin groove region can occur as cell internal pressure is relieved via metallic can melt-through, resulting in the high-energy release of gas, liquids, and solids (Figure 7.5). TR breaches that eject high momentum bursts of combusting cell material can rapidly transfer heat to adjacent cells, creating a cell-to-cell propagation condition. Furthermore, testing intended to characterize the vulnerability of a cell design to SWR while structurally supported in a non-flight configuration suggests that the likelihood of a cell SWR event can be underestimated [68]. Hence, it is critical to conduct cell SWR characterization studies with flight LIB

Figure 7.5 Catastrophic side wall rupture during TR of a representative Li-ion 18650 cell. *Source:* NASA.

relevant designs. Control and mitigation strategies for mitigating COTS 18650 Li-ion cell SWR risks include selecting cell designs with lower energy densities, thicker cell can walls, lower cell header assembly burst pressures, and cell designs with bottom vents. Additionally, structurally supporting the cell with exterior sleeves has been demonstrated to be effective at controlling SWR. Space industry experience has demonstrated that space-qualified custom large format Li-ion cells are not as susceptible to SWR due to their larger cell can wall thicknesses when compared to COTS 18650 Li-ion cell can wall thicknesses.

7.6.1.2 Cell Spacing and Heat Dissipation

LIB pack designs with adjacent cells in direct contact have a high likelihood of being susceptible to cell-to-cell TR propagating safety hazards. To mitigate this type of TR hazard, LIB designers have considered thermal isolation of cells within the LIB and/or maximizing cell heat conduction within the LIB. Cell spacing with thermal isolation attempts to store failure energy in the failed cell and may result in increased battery volume. Increasing cell-to-cell spacing may also increase the LIB-level voltage drop between the integrated cells and battery power output terminals. A high rate of heat transfer may cause a slow cascade of cell-to-cell TR propagation. Alternatively, maximizing heat transfer away from the failed cell to the other cells in the battery presents a lower volume impact to the battery design when a sufficient number of cells are available. This approach results in rapid high-temperature exposure to many adjacent cells rather than concentrating thermal energy on fewer cells for longer time durations.

7.6.1.3 Current-Limiting Cells

In a p-s LIB design, once a Li-ion cell fails and enters into TR, it will induce an ESC to all cells electrically connected in the same parallel string. Parallel-connected adjacent cells provide power to the ESC and generate internal heat as a result of the rapid electrical discharge, while also receiving heat from the failed cell. By reducing the internal heating due to rapid discharge, the risk of TR can be reduced. This behavior can be controlled with overcurrent protection at the cell level. Overcurrent protection can be implemented with cell fusible links or commercial fuses in the electrical connection between parallel cells or series cell strings connected in parallel. Fuse blow currents are selected to allow performance yet minimize the range of smart shorts that can occur in a TR cell. Smart shorts are types of short circuits that draw currents just below fuse blow current ratings in a LIB design. If sustained, smart shorts can generate hazardous amounts of heat within cells and propagate TR within a battery.

7.6.1.4 Ejecta Path

High temperature cell ejecta created during cell TR consists of flammable gases, liquids, and/or solids. Solid-phase cell ejecta composition can include melted aluminum, copper, and various types of carbonaceous materials. These ejecta materials are thermally and electrically conductive. Electrically conductive cell ejecta can lead to hazardous circulating currents between polarized cell surfaces and conductive materials inside the battery enclosure. Controlling the hazardous distribution of TR ejecta within a battery design can reduce heat transfer to cell, modules, and other battery components [69]. Use of non-conducting conformal coatings within the battery may also mitigate ejecta safety hazards.

7.6.1.5 Flame Suppression

For confined, crewed space battery applications, the battery enclosure must have vent ports with flame arresting properties against the TR event at any cell location within the battery. Flame arresting design features include adequate void volume, seals, a tortuous path to the pressure relief vents, and a flame arresting vent port. This is most challenging in LIBs with low cell quantities due to their low void volume, which results in high-pressure stresses on the vent port and enclosure seals. A combination of baffling blast plates can be orientated to produce a tortuous path for the ejecta. Vent ports with metal foams have been found successful in cooling the ejecta flames and quenching sparks prior to exiting. Use of non-flammable flame suppressant materials within the battery design may also mitigate flame safety hazards.

7.6.2 PPR Verification

LIB PPR to a single-cell TR event is verified by a combination of thermal analyses and test compliance methods. The PPR verification test plan includes the number of PPR LIB test articles to build for TR testing and how many trigger cell locations to designate. Smaller batteries with fewer cells enclosed with small void volumes are risky to populate with multiple trigger cell locations and thus are appropriate to have a dedicated PPR battery test article for each trigger test. Larger batteries with many cells and significant internal void volumes are less risky to populate with multiple trigger cell locations, as long as they are distributed and no cell is adjacent to multiple trigger cells. Table 7.2 lists the advantages and disadvantages of various TR trigger methods. The main objective of each trigger method is to initiate the TR response of defect-induced ISCs that lead to spontaneous and catastrophic field failures, and to do so with minimal relevant alterations to the flight battery and its predicted mission conditions in order to avoid over-test (false positive) or under-test (false negative) outcomes [70].

7.6.2.1 Trigger Cell Selection

An important step in testing for LIB TR characteristics is selecting candidate trigger cell(s) locations. To aid in trigger cell location selection, LIB-level thermal analysis is commonly required to predict the maximum (hot) and minimum (cold) cell temperature locations in the LIB. The worst-case (hottest) single cell(s) in the LIB are candidate locations for placing a trigger cell [71].

Table 7.2 Advantages and disadvantages of various test methods used to trigger single-cell TR in a LIB.

Trigger method type	Description	Advantages	Disadvantages
Mechanical	Crush; penetration	Easy to implement on batteries without an enclosure; no cell modification required.	Poor consistency; unable to control type of ISC; cell is vented prior to TR; electrode jellyroll or stack can be pinned down.
Electrical	Overcharge; external short circuit; high rate cycling at high SOC	Easy to implement; no cell modification required.	Requires electrical isolation of trigger cell from rest of cells in battery, obviating cell fusing verifications; TR triggered at much higher SOC than relevant application.
Thermal	Localized heat flux on trigger cell	Only minor external modifications to cell required.	Risk of thermal bias of adjacent cells at onset of trigger cell TR; must remove some structural and thermal protection to make room for cell heater.
Seeding	Device implantation into trigger cell	Able to select type and location of ISC without altering cell enclosure; replicates internal defect-induced ISC.	Requires modifying cell during cell manufacturing; only certain cell models are available.

Calorimetric measurements of the Li-ion cell design can aid in the LIB thermal analysis by providing the total heat generated under TR conditions. Cells located in corners of the LIB with few adjacent cells can be more likely to lead to propagation because neighboring cells can be subject to a greater heat load than the interior cells, which have more neighboring cells. In practice, internal cells that are triggered into TR via heating may also be more difficult to trigger into TR because the heat from the heater can be dispersed more readily to a greater collection of neighboring cells. Due to LIB structural support and heat sink variations in certain LIB designs, cells in edge locations also serve as candidates for trigger cells. In general, JSC-20793 requires that all types of cell locations (edge, corner, interior) be tested in the PPR LIB test campaign.

7.6.2.2 PPR LIB Unit Design and Manufacturing

In order to conduct a valid PPR test campaign on a LIB intended for a crewed spacecraft application, certain Li-ion cell and battery configuration criteria must be satisfied. For example, Li-cell mechanical spacing, interfaces with LIB components, and thermal interface heat sinking must mimic the intended LIB flight design features. In addition, the LIB cell's electrical topology, including bussing and current limiting devices, the cell ejecta path to the battery vent ports, and the battery void volume, must be representative of the flight LIB design. Other form, fit, and function aspects of all internal battery electronic circuitry must also be represented in a flight-like configuration.

Ideally, a dedicated PPR unit is allocated to each individual cell triggering test to facilitate conducting each PPR test run with a representative LIB unit. However, to reduce test resource allocations, multiple trigger cells can be included in one or multiple battery units if the battery void volume is sufficient. Multiple battery assemblies are normally manufactured if PPR testing of different trigger cell locations is required. When using heaters to trigger cell TR or to activate implantable ISC devices, careful heater placement is needed to minimize damage to battery components, or from exposing excessive temperature biases on cells adjacent to the trigger cell. Under worst-case PPR test conditions, the internal clocking of the trigger cell ISC device is orientated toward the most vulnerable direction within the test battery since the trigger cell is more likely to experience spin groove or bottom ruptures in that selected orientation.

Key PPR LIB unit manufacturing steps include preparing cells with or without non-conductive wrap and with tabs, as the flight battery design dictates. Thermal analysis may be performed to aid in placement of the thermal sensors. Placement of thermal sensors in the path of cell TR ejecta could result in very high transient measurements and/or failure of the sensor. Using flight-like parts, components, materials of construction, and manufacturing procedures to build the test PPR LIB unit enables verification of PPR compliance with relevant requirements.

7.6.2.3 PPR LIB Test Execution

Test planning includes battery charge, discharge, and TCS equipment for pre-conditioning the PPR LIB test article. Worst-case TR testing conditions are evaluated based on program-relevant ground processing environments and mission-unique on-orbit operational conditions. For example, NASA astronaut EMU spacesuit LIBs are commonly stowed in pressurized volumes, yet are operated in space vacuum environments [15]. As such, limited NASA PPR LIB test experience with conducting PPR testing both in space vacuum and ambient pressures indicates that testing in ambient pressure conditions is a credible worst-case test condition [72]. The PPR battery test measurements include battery voltage, current, cell temperatures, video, and infrared camera photography. In addition, trigger cell heater (or nail) current and voltage are recorded. If using a battery test article with multiple trigger cells, battery servicing (such as vent port cleaning) may be required in between cell triggering test runs. After each test run, a rest and temperature conditioning period is needed to bring the battery back to initial conditions for subsequent testing (if required). Once all testing is completed without propagation, the PPR units are discharged to measure remaining capacity. These data are used to evaluate the PPR LIB design performance in mitigating the severity of cell-to-cell TR propagation.

7.6.2.4 Post-Test Analysis and Reporting

Post-test forensics include physical analysis of the test battery to perform visual inspections of the trigger cell, adjacent cells, ejecta pathways, and blast plates, as well as examining which cell current limiters have activated. Next, the battery PPR unit can be disassembled for charge–discharge cycling to verify cell, module, and/or bank capacity. As a figure of merit, adjacent cells that degrade in capacity by >5% (of nameplate) are considered to be damaged. Damage can also be from thermal stress, seal gasket damage, CID activation, venting, or other observations.

Test reporting of a PPR test campaign includes PPR unit design descriptions, along with rationale for any deviations from a flight-like design. A record of the test battery manufacturing history is also recorded to maintain configuration control over future design changes or testing. Pre-test battery conditioning to initial conditions is also documented. For each test run, battery voltage, cell temperatures, heater current, and voltage (or nail power supply) are analyzed. Video and infrared camera data are also analyzed and compared to relevant test observations. Post-test analyses of physical, electrical, and thermal data are examined to determine whether test conditions were adequate and if adjustments to test procedures are warranted. If not, then thermal simulations validated with test data can guide changes to the battery test article design to achieve PPR characteristics.

7.6.3 Case Study – NASA US Astronaut Spacesuit LIB Redesign

In 2015, the NESC sponsored an assessment of the likelihood for catastrophic cell-to-cell TR propagation in the NASA US astronaut Long Life Battery (LLB). The assessment objective was to further develop TR severity reduction measures across the existing LIB designs supporting NASA government-furnished LIBs used to power the US astronaut EMU. The first-generation LLB design consisted of 80 COTS 18650 Li-ion cells secured at their ends by insulating capture plates and connected into five 16p banks with electrochemically-etched nickel bus plates to form a single rectangular 16p5s 38 Ah battery brick (Figure 7.6). A printed circuit board (PCB) assembly was placed on the top edge of the brick, electrically connecting the five 16p banks in series, and provided termination points for the charge–discharge connectors (Figure 7.7). Anodized aluminum side plates were fastened over the bus plates and provided structural fastening points for the battery brick to the bottom of the housing.

Figure 7.6 Original first-generation US spacesuit EMU 16p5s LLB design. *Source:* EnerSys/ABSL.

Although from 2010 to 2015, the original LLB design had successfully supported the ISS US astronaut spacesuit EVA power requirements, design deficiencies relative to the new NASA PPR design guidelines were identified [14]. The original LLB cell was a 2.4 Ah 18650 Li-ion

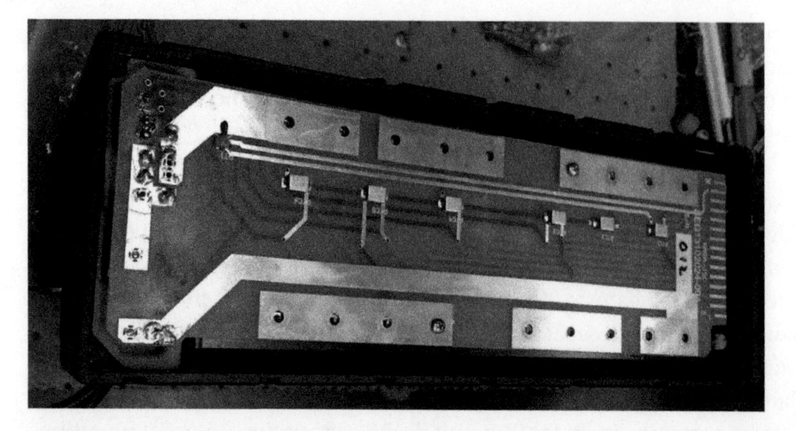

Figure 7.7 The printed circuit board that attaches to the top of the original LLB battery brick design, electrically connects the five 16p banks in series, and inserts into the metal housing. *Source:* NASA.

Figure 7.8 Full propagation of TR after a single cell is triggered in the first-generation spacesuit LLB battery. Large amounts of gas generation and particulates were released for over 15 minutes with housing surfaces reaching 350 °C. *Source:* NASA.

cell (180 Wh/kg and 650 Wh/l) design, which was space-qualified to the NASA JSC-20793 standard. The first LLB design deficiency identified was the presence of solid side plates that provided no spacing for a cell TR ejecta vent path. Next, although none of the cell cans were in direct contact with one another, the main heat dissipation path for a failed cell in TR was through radiative heat transfer to adjacent cells and through the thin nickel bus plates, which was deemed to be inadequate for protecting adjacent cells [73]. In addition, no cell overcurrent protection was included to isolate parallel cells from an energetic TR condition.

Nevertheless, TR verification testing was performed on a full-scale flight-like EMU spacesuit LLB with a patch heater placed on a peripheral cell in the cell brick, while enclosed inside the battery housing. Trigger cell TR was initiated in the selected trigger cell with a low power profile to the heater (15 W for 15 minutes; 20 W for 10 minutes; and 32 W for 3 minutes) to simulate a worst-case smart ISC that draws a sustained amount of energy from the 15 parallel cells (in the same cell bank as the trigger cell) over a long period of time. Heater power was sufficient to initiate TR in the trigger cell within 28 minutes followed by catastrophic cell-to-cell TR propagation during the next 15 minutes (Figure 7.8). No flames exited the housing and lid seals, but the external surface temperatures of the LLB measured in the range of 250 to 350 °C for more than 30 minutes (Figure 7.9). The post-test

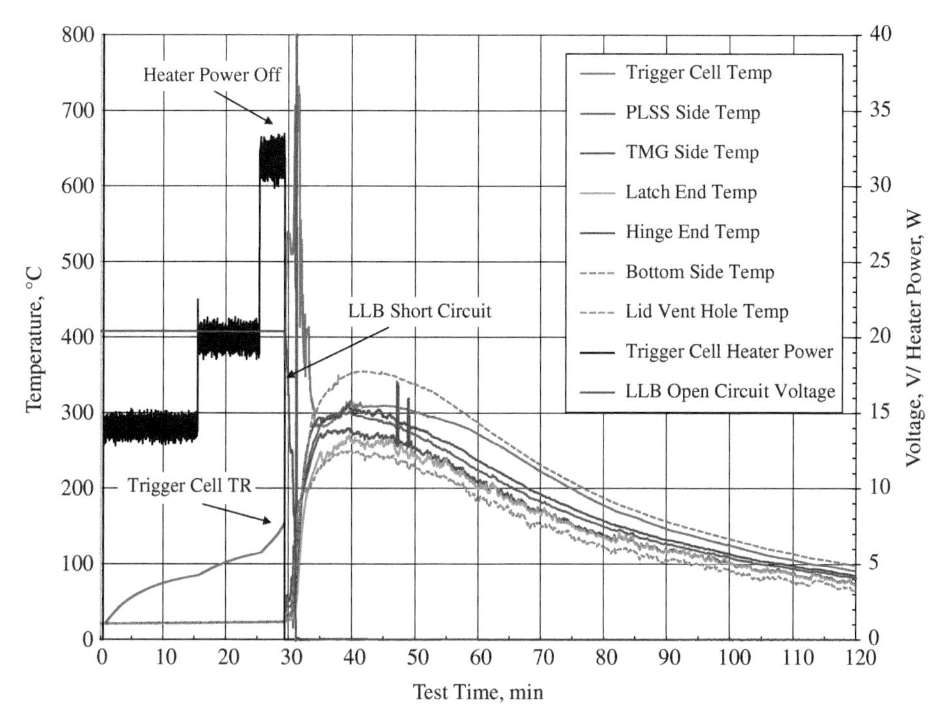

Figure 7.9 LLB test battery TR propagation test showing external LLB surface temperatures during catastrophic cell-to-cell TR propagation. Trigger cell temperature exceeded 900 °C. Measured LLB external surface temperatures varied between 250 and 350 °C [12]. *Source:* Courtesy of NASA.

analysis determined that the quantity of gas generation and particulates would present a catastrophic hazard risk to ISS crew members. In addition, the extreme heat generation would present a high safety risk to high-pressure components nearby in the astronaut spacesuit EMU backpack. Post-test forensics of the LLB test battery indicated that enough hot TR ejecta had accumulated next to the two cells adjacent to the failed trigger cell to initiate catastrophic cell-to-cell TR propagation.

The next iteration of the LLB redesign was with new side plates with vents holes intended to disperse hot TR ejecta away from cells adjacent to the trigger cell (Figure 7.10). In subsequent LLB-level TR testing, the new spacing gap between the side plate and the inside wall of the battery housing was found to be too narrow to sufficiently disperse the TR ejecta. As a result, interim testing resulted in LLB-level catastrophic cell-to-cell TR propagation.

Figure 7.10 Redesign iteration of LLB cell brick. Side plate cell vent holes were insufficient for preventing catastrophic cell-to-cell TR propagation. *Source:* NASA.

The new PPR LLB design, named LLB2, consists of five series banks of 14 COTS 18650 Li-ion cells in parallel (14p5s) in a cell brick assembly where the cell bank series connections are achieved on a circuit board

Figure 7.11 LLB2 PPR battery design with gas permeable vent ports on test stand. *Source:* NASA.

fastened to its top side. The LLB2 design utilizes 2.6 Ah 18650 LCO cells exhibiting a specific energy of 200 Wh/kg and energy density of 660 Wh/L at 100% SOC. Under these cell-level performance characteristics, the cell design was determined to present an extremely low risk of failing due to a SWR. A thin mica sleeve was placed on the cylindrical walls of each cell, which is then overwrapped with Kapton tape. These assemblies are placed in precisely-machined bores of an aluminum heat sink with 1.5 mm as the thinnest cell-to-cell spacing distance. This provided a reliable heat transfer path for rapidly distributing the conducted heat of any TR cell into numerous layers of adjacent cells. Cell ends were bonded to fiberboard plates that insulate the thin nickel bus plates connecting the 14p banks. Electrical isolation was achieved with fusible links located in the negative bus plate to the cell terminal weld pad. Due to the precision of chemical etching, thin fusible links provided consistent no-blow and always-blow currents.

By design, ejected TR material is dissipated through a channel gap between the cell positive terminals and the inside wall of the battery housing. Cells are axially oriented in the same direction in this cell brick assembly. Once dispersed in the positive channel, the hot ejecta temperature decreases as it contacts the aluminum housing wall. The hot ejecta further disperses around the gaps between the cell brick assembly and the housing, following a tortuous path to the flame-arresting semi-impermeable membrane vent port on the negative side. Compliant to all PPR safe-by-design guidelines, the LLB2 PPR test campaign demonstrated that cells adjacent to the trigger cell were well-protected from TR. The TR events from the trigger cells did not degrade adjacent cells and resulted in trace amounts of gas venting through the new vent ports (Figure 7.11). On May 21, 2018 the first set of space-qualified LLB2 flight batteries were launched by an Antares 230 LV to the NASA-ISS aboard the ninth flight (OA-9E) of the Orbital ATK (later acquired by Northrop Grumman Space Systems) uncrewed resupply Cygnus spacecraft.

7.7 Battery Reliability

Predicting the reliability of a space-qualified LIB is critical to estimating the probability that the spacecraft EPS will meet performance requirements during the specified mission design lifetime. As described in Chapter 4, the EPS is a mission-critical subsystem that enables overall spacecraft system function and operation. In general, the reliability of the power management and distribution (PMAD) electronics, solar array, and battery energy storage hardware is used to determine spacecraft EPS reliability estimates. Depending on program architecture and requirements, solar array mechanisms used to deploy the solar arrays post-launch may be included in the EPS-level reliability analysis. On-orbit spacecraft operating conditions such as orbit type, environments, power duty cycle, and mission duration, have a significant effect on EPS component reliability estimates. The spacecraft EPS reliability is therefore based on the probability (or likelihood) of the EPS components functioning together successfully during the specified mission duration.

7.7.1 Requirements

In general, battery reliability requirements are documented in the applicable battery or EPS specification. The reliability allocation from the EPS is distributed between the PMAD, solar array, and battery in a manner needed to meet spacecraft-level reliability requirements. The on-orbit service life battery reliability allocation may vary with the number of cell strings (parallel–series or series–parallel) allowed to fail during the mission, as well as other battery component wear-out rate estimates. Battery mission life reliability requirements are expressed as a quantitative statement of the minimum acceptable reliability value at spacecraft end-of-life (EOL) [74]. Alternatively, the spacecraft battery reliability requirement is commonly specified by demonstrating that the reliability prediction meets or exceeds the probability of mission success requirements. Battery reliability requirements may be expressed in terms of a probability of success or mean-time-between-failure (MTBF) values. Table 7.3 lists a common set of battery specification reliability requirements for a representative Earth-orbiting satellite LIB power system.

Table 7.3 LIB reliability requirements for a representative Earth-orbiting satellite EPS.

LIB requirement	General description	Compliance approach
On-orbit service life	The minimum acceptable reliability shall be 0.9990 for 5 years on-orbit, calculated for 1 failed cell string, a space flight environment, and a baseplate temperature of 20 °C.	Battery cell wear out and failure rate estimates are obtained from relevant cell life cycle testing; other battery component failure rates are calculated per applicable reliability design handbook(s). Failure rates from single event upsets and micrometeoroid impact are considered.
Single point failure	The battery design shall minimize the occurrence of credible single point failures.	Credible single point failures are identified and analyzed for likelihood and consequences to battery, EPS, and spacecraft operations. Conduct FMECA, as required.
Failure propagation	The battery design shall ensure that a failure within the battery shall not propagate in a manner that permanently degrades performance of the battery or causes performance degradation in another unit (beyond the allowable cell string failures).	Internal battery design redundancy features. Conduct FMECA, as required.
Parts derating	The battery shall meet performance over the design life accounting for component electrical and thermal stresses.	Comply with industry EEE parts derating standards for worst-case hot and cold stressing environments, radiation exposure, and aging. Conduct parts stress analysis, as required.

7.7.1.1 FMECA, WCCA, and Parts Stress Analysis

Predicting the probability that the spacecraft LIB will meet performance requirements during the specified mission design lifetime is accomplished by conducting reliability analyses and testing. There are a number of different types of battery reliability analyses typically required to support high-reliability Earth-orbiting, planetary, and LV spacecraft missions [75]. The most common reliability analysis conducted on LIBs supporting unmanned and crewed spacecraft applications is a FMECA. The battery FMECA is a qualitative analysis that employs a bottoms-up approach to systematically identify the potential failure modes of every LIB subassembly, component, and part to include failure mode causes, effects, and criticality [76, 77]. Common LIB hazard causes include, but are not limited to, overcharge, overdischarge, ISCs, ESCs, and overtemperature conditions [78]. Depending on battery location within the spacecraft, risk to impact or collision with micrometeoroid and orbital debris (MMOD) or ground-tracked objects may also be a hazard cause to LIB failure. The effect of a given failure mode on battery-level operation is also analyzed, along with collateral impacts to higher levels of battery power system integration. The failure criticality analysis is the process by which each potential battery failure mode is ranked according to the combined influence of severity and probability of occurrence. The FMECA is also used to identify battery-level single point failures and to establish technical recommendations for increased battery safety design features that permit a reduction to the consequences of a critical single-point failure. As such, the battery FMECA analysis is completed early in the battery and EPS design process.

LIB designs utilizing analog or digital electronic parts may require additional reliability analyses to verify circuit function under worst-case operating extremes rather than the nominal operating values. Examples include LIB designs with integrated BMSs, bypass switches, thermistors, fuses, or heaters. In a worst-case circuit analysis (WCCA), a circuit analysis is performed on identified battery electronic parts under worst-case temperatures (hot and cold), voltages, and radiation extremes. Battery WCCA outputs are used to verify that the LIB meets performance requirements without overstressing the battery electronic components [79]. Finally, a parts stress analysis that evaluates battery parts overstress characteristics may be required. The objective of a battery parts stress analysis is to decrease parts failure rates by reducing parts stress levels [80]. Reducing parts stress levels (or derating) is the reduction of a parts applied stress, with respect to its rated stress, for the purpose of providing a performance margin between the applied stress and the demonstrated limit of the part's capabilities. Maintaining the parts derating margin reduces the occurrence of stress-related failures and aids in increasing the part's reliability [81, 82].

7.7.1.2 Hazard Analysis

Alternatively, a battery hazards analysis utilizes a top-down approach to identify LIB safety hazards that pose a potential risk situation capable of causing injury to personnel, damage to the environment, or damage to critical hardware. Risk is then assessed by identifying controls needed to eliminate or reduce the associated risk to acceptable levels. A typical LIB hazard analysis report will include a detailed list of hazard causes and controls used to mitigate the safety risk associated with each hazard. The report also identifies the verification methods used to ensure the effectiveness of the hazard controls, as well as the methods used to ensure that the battery meets the safety requirements for flight. This includes the types of tests, analyses, inspections, or procedures used to verify each hazard control. The JSC-20793 standard requires a unique hazard analysis report for non-rechargeable and rechargeable lithium battery applications supporting crewed spacecraft missions [83].

7.7.2 Battery Failure Rates

Historically, there has been limited availability of open-source data for EPS-related launch or on-orbit spacecraft failure types and rates [84]. Although selected on-orbit (and ground) failure, anomaly, and discrepancy data for some government and civil satellite programs are reported, root causes of on-orbit commercial satellite failures are generally not disclosed. However, estimates from reported on-orbit failure rate data indicate that approximately 27 to 49% of all satellite subsystem failures are attributable to EPS (PMAD, solar array, and batteries) components [85, 86]. The wide range of failure rates is attributable to varying satellite orbits, power levels, operating conditions, payload types, and mission durations. Figure 7.12 shows the distribution of EPS component failure rates for Earth-orbiting and planetary mission spacecraft since 2000. Battery failures include anomalies that caused complete loss of satellite mission capability, limited loss of redundancy to satellite operations, or minor failures that did not have a significant impact on satellite operations [87]. Battery failure rates are distributed across LIB, Ni-H_2, and Ni-Cd battery technologies.

7.7.2.1 Failure Rate in Time

LIB cell failure rate data inputs to reliability models are derived from relevant on-orbit data and/or cell ground life cycle testing (LCT), which provides wear-out rates as a function of battery charge–discharge cycling and environmental exposure. As discussed in Chapter 8, cell-level real-time and accelerated LCT data can be used as inputs to reliability models and analysis. Reliability predictions for non-cell battery components, such as connectors (power, signal, and test), fuses, temperature sensors, bypass switches, relays, heaters, BMS electronics, cell terminal connectors, or other electronic piece-parts not found from industry failure rate handbooks, may be estimated from relevant on-orbit or ground data representative of the intended battery application [88, 89].

Battery component reliability estimates are based on its failure rate in time (FIT), which is expressed in the number of failures per million hours of operation [90]. LIB cell FIT rates vary with cell manufacturing quality of construction, cell chemistry, charge–discharge cycling characteristics, operating temperature, storage conditions, and depth-of-discharge (DOD). In general, space-qualified LIB cells operating within their design-specific requirements range are commonly

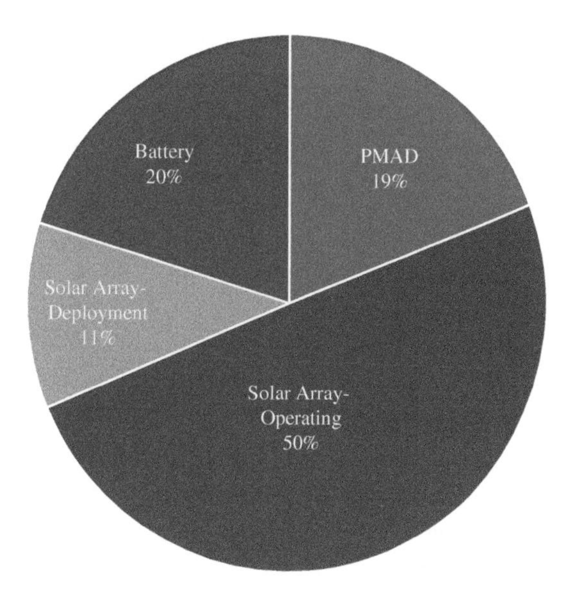

Figure 7.12 Distribution of spacecraft EPS component failure rates in LEO and GEO orbits. *Source:* Courtesy of Seradata, Ltd.

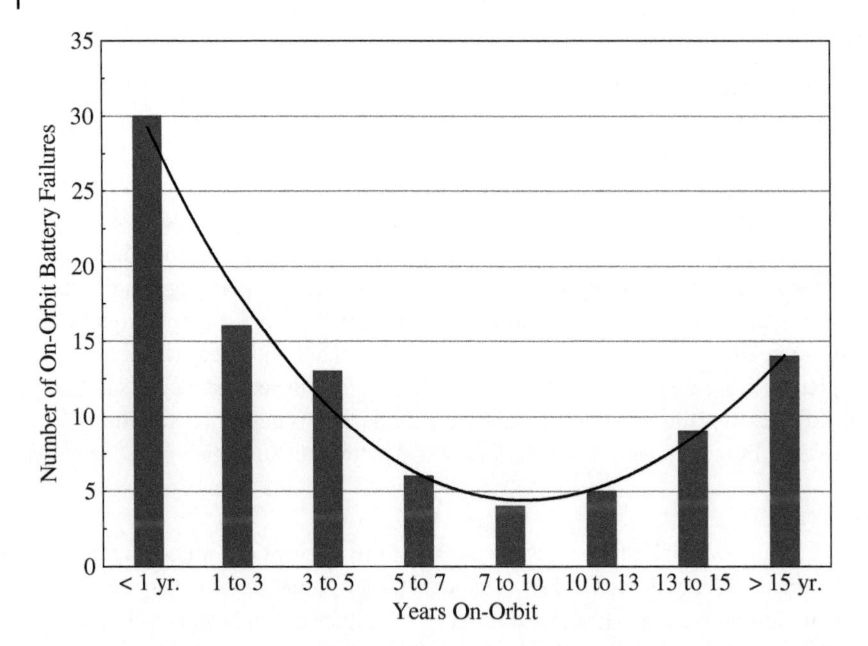

Figure 7.13 Effect of on-orbit mission service time on spacecraft battery failure rates. *Source:* Courtesy of Seradata, Ltd.

expected to exhibit non-energetic graceful degradation (non-catastrophic) failure modes. Graceful cell performance degradation is characterized by the gradual loss of useful cell capacity resulting from internal resistance growth [91–93]. On-orbit satellite battery experience has also demonstrated that capacity fading is the most common degradation mode of cycle life performance.

7.7.2.2 Failure Rate Characteristics

Figure 7.13 shows the effect of on-orbit mission service time on spacecraft battery failure rates collected since 2000 [87]. These data represent Earth-orbiting and planetary spacecraft missions operating with various rechargeable battery types. The on-orbit battery failure rate follows a parabolic- or "bathtub"-shaped trend over the battery mission design life of the spacecraft. The initial high infant mortality failure rates are typically caused by battery workmanship-related latent defects not detected during ground acceptance testing, or other EPS component failures causing early mission loss of battery recharge capability. Infant mortality rates decrease with battery cycle life reaching a lower failure rate for longer on-orbit service times. In the lower failure rate region, the battery useful life is consumed as a function of on-orbit cycle life. Battery useful life failures are extremely rare, and as such, are random in nature. Finally, the increasing battery failure rate near EOL is due to long-term wear-out mechanisms associated with the battery-unique design and operating conditions. As discussed in Chapter 8, estimating LIB service life performance degradation characteristics under mission-specific operating conditions is performed via cell (or battery) LCT.

7.8 Summary

Safety and reliability requirements are common key elements found in space LIB requirement specifications. Requirements for safe LIB design can vary between Earth-orbiting satellites and robotic planetary mission spacecraft. Spacecraft LIB hazards caused by electrical, mechanical, and thermal abuse conditions can be mitigated and controlled by complying with aerospace industry relevant best practices, guidelines, and standards. Industry lessons learned indicate that the objectives of LIB safety testing include testing in relevant environments with flight-like hardware configurations representative of the intended mission-specific application. In space applications, reducing the severity and consequences of catastrophic cell-to-cell propagating TR remains a significant focus for space cell and LIB designers. Recent developments in PPR LIB design guidelines have been successful in effectively mitigating the risk of a catastrophic cell-to-cell TR propagation event. In addition, spacecraft EPS requirements such as bus voltage, charge management, fault tolerance, operating temperature, and mission duty power loading, also have a significant impact on battery safety and reliability. On-orbit spacecraft experience has demonstrated that in addition to safe LIB operations, on-orbit conditions (such as orbit type), environments, power duty cycle, and mission durations have a significant effect on LIB-based EPS reliability.

References

1 Jeevarajan, J.A. (2009). Battery safety. In: *Safety Design for Space Systems, Chapter 16* (ed. G.E. Musgrave, A.M. Larsen and T. Sgobba), 507–548. Elsevier.

2 Bandhauer, T.M., Garimella, S., and Fuller, T.F. (2011). A critical review of thermal issues in lithium-ion batteries. *J. Electrochem. Soc.* 158: R1.

3 Jindal, P. and Bhattacharya, J. (2019). Review – understanding the thermal runaway behavior of Li-ion batteries through experimental techniques. *J. Electrochem. Soc.* 166: A2165.

4 Mauger, A. and Julien, C.M. (2017). Critical review on lithium-ion batteries: are they safe? Sustainable? *Ionics* 23: 1933–1947.

5 Chen, Y., Kang, Y., Zhao, Y. et al. (2021). A review of lithium-ion battery safety concerns: the issues, strategies, and testing standards. *J. Energy Chem.* 59: 83–99.

6 Trout, J.B. (1985). *Manned Space Vehicle Safety Handbook*. NASA JSC-20793.

7 Feng, X., Ouyang, M., and Lu, L. (2019). Overview about accidents: selected lessons learned from prior safety-related failures of Li-ion batteries, chapter 12D. In: *Electrochemical Power Sources: Fundamentals, Systems, and Applications: Li-Battery Safety* (ed. J. Garche and K. Brandt), 571–602. Elsevier.

8 Jeevarajan, J.A. and Winchester, C. (2012). Battery safety qualifications for human ratings. *Electrochem. Soc. Interface* 21: 51.

9 Gitzendanner, G., Puglia, F., Ravdel, B. et al. (2018). Safety enhancements for large format, high power, lithium-ion cells and batteries. *ECS Meet. Abstr.* MA2018–MA2002 480.

10 National Transportation Safety Board (2014). *Aircraft Incident Report: Auxiliary Power Unit Battery Fire Japan Airlines Boeing 787–8, JA829J Boston, Massachusetts January 7, 2013*. NTSB/AIR-14/01, PB2014–108867. Notation 8604.

11 Japan Transport Safety Board (2014). *Aircraft Serious Incident Investigation Report*. All Nippon Airways Co., Ltd, JA804A, JTSB AI2014–4.

12 Air Accidents Investigation Branch (2015). *Report on the Serious Incident to Boeing B787–8, ET-AOP London Heathrow Airport 12 July 2013*, Aircraft Accident Report 2/2015. Department for Transport Air Accidents Investigation Branch.

13 Swain, B., Bauer, M., and Chapin, T. (2014). Boeing 787 battery investigation. In: *NASA Aerospace Battery Workshop*. Huntsville, AL.

14 Iannello, C.J., Button, R.M., Darcy, E.C. et al. (2017). *Assessment of International Space Station (ISS)/Extravehicular Activity (EVA) Lithium ion Battery Thermal Runaway (TR) Severity Reduction Measures*. NASA/TM-2017-219649/Volumes I and II, NESC-RP-14-00942.

15 Iannello, C.J., Barrera, T.P., Doughty, D. et al. (2018). *Simplified Aid for Extra-Vehicular Activity Rescue (SAFER) Battery Assessment*. NASA/TM 2018–219818, NESC-RP-14-00963.

16 Dalton, P.J., Schwanbeck, E., North, T., and Balcer, S. (2016). International space station lithium-ion battery. In: *The NASA Aerospace Battery Workshop*. Huntsville, AL.

17 Darcy, E. (2014). Thermal runaway severity reduction assessment & implementation: on LREBA. In: *NASA Aerospace Battery Workshop*. Huntsville, AL.

18 Darcy, E. (2015). Safe, high performing Li-ion battery designs: summary of 2015 findings. In: *NASA Aerospace Battery Workshop*. Huntsville, AL.

19 Lessons Learned Information System. http://www.llis.nasa.gov (accessed 6 July 2022).

20 Ruiz, V., Pfrang, A., Kriston, A. et al. (2018). A review of international abuse testing standards and regulations for lithium ion batteries in electric and hybrid electric vehicles. *Renewable Sustainable Energy Rev.* 81: 1427–1452.

21 Russell, S. (2017). *Crewed Space Vehicle Battery Safety Requirements*. NASA JSC-20793.

22 McKissock, B., Loyselle, P., and Vogel, E. (2009) *Guidelines on Lithium-ion Battery Use in Space Applications*. NASA/TM-2009-215751.

23 United States Space Force Command Manual (2020). Range Safety User Requirements. AFSPCMAN 91–710.

24 Range Safety Group (2019). *Flight Termination Systems Commonality Standard*. Document, 319–19

25 United States Space Force Space Systems Command Standard (2022). *Lithium-ion Battery Standard for Spacecraft Applications*. SSC Standard SSC-S-017.

26 ISO (2015). *Space Systems-Lithium Ion Battery for Space Vehicles - Design and Verification Requirements*. ISO 17546.

27 European Cooperation for Space Standardization (2015). *Space Engineering – Lithium-Ion Battery Testing Handbook*. ESA-ESTEC, ECSS-E-HB-20-02A.

28 U.S. Department of Transportation Hazardous Materials Regulations. Title 49, Code of Federal Regulations, Part 173, Section 173.185.

29 Manual of Tests and Criteria, United Nations Recommendations on the Transport of Dangerous Goods, Part III, Section 38.3, 7th edn., 2019.

30 Ang, V. (2013). Architectures for Lithium Ion Based Power Subsystems. The Aerospace Corporation, TOR-2013-00295.

31 RTCA (2017). Minimum Operational Performance Standards for Rechargeable Lithium Batteries and Battery Systems, DO-311A.

32 NASA procedural requirements (2022). *Human-rating requirements for space systems*. NPR 8705.2C.

33 Levy, S.C. and Bro, P. (1994). *Battery Hazards and Accident Prevention*. New York: Plenum Press.

34 Lyness, C. (2019). *Sources of risk*, chapter 7A. In: *Electrochemical Power Sources: Fundamentals, Systems, and Applications: Li-Battery Safety* (ed. J. Garche and K. Brandt), 145–166. Elsevier.

35 Juarez-Robles, D., Vyas, A.A., Fear, C. et al. (2015). Overcharge and aging analytics of Li-ion cells. *J. Electrochem. Soc.* 167: 090547.

36 Leising, R.A., Palazzo, M.J., Takeuchi, E.S., and Takeuchi, K.J. (2001). A study of the overcharge reaction of lithium-ion batteries. *J. Power Sources* 97–98: 681–683.

37 Guo, R., Lu, L., Ouyang, M., and Feng, X. (2016). Mechanism of the entire overdischarge process and overdischarge-induced internal short circuit in lithium-ion batteries. *Sci. Rep.* 6: 30248.

38 Mikolajczak, C., Kahn, M., White, K., and Long, R.T. (2011). *Lithium-Ion Batteries Hazard and Use Assessment*. New York: Springer.

39 Huang, L., Liu, L., Lu, L. et al. (2021). A review of the internal short circuit mechanism in lithium-ion batteries: inducement, detection and prevention. *Int. J. Energy Res.* 45: 15797–15831.

40 McCoy, C., Offer, D., Barnett, B., and Sriramulu, S. (2012). Safety Technologies for Lithium Ion Cells. In: *Proceedings of the 45th Power Sources Conference*, 1–4, ISBN: 978–1–63266-165-4.

41 Waldmann, T., Wilka, M., Kasper, M. et al. (2014). Temperature dependent ageing mechanisms in lithium-ion batteries – a post-mortem study. *J. Power Sources* 262: 129–135.

42 Borner, M., Friesen, A., Grützke, M. et al. (2017). Correlation of aging and thermal stability of commercial 18650-type lithium ion batteries. *J. Power Sources* 342: 382–392.

43 Lin, H.P., Chua, D., Salomon et al. (2001). Low-temperature behavior of Li-ion cells. *Electrochem. Solid-State Lett.* 4: A71.

44 Smart, M.C. and Ratnakumar, B.V. (2011). Lithium-ion cells effects of electrolyte composition on lithium plating in lithium-ion cells. *J. Electrochem. Soc.* 158.

45 Smart, M.C., Ratnakumar, B.V., Chin, K. et al. (2003). Performance characteristics of lithium-ion technology under extreme environmental conditions. *1st International Energy Conversion Engineering Conference.* Portsmouth, Virginia.

46 Bang, H.J., Thillaiyan, R., and Quee, E. (2013). Mechanical force and pressure growth in Li-Ion cells during LEO cycling. In: *NASA Aerospace Battery Workshop.* Huntsville, AL.

47 Garcia, H., Lam, C.W., Langford, S., and Ramanathan, R. (2014). *Guidelines for assessing the toxic Hazard of spacecraft chemicals and test materials.* National Aeronautics and Space Administration JSC 26895.

48 Kumai, K., Miyashiro, H., Kobayashi, Y. et al. (1990). Gas generation mechanism due to electrolyte decomposition in commercial lithium-ion cell. *J. Power Sources* 81–82: 715–719.

49 Golubkov, A.W., Fuchs, D., Wagner et al. (2014). Thermal-runaway experiments on consumer Li-ion batteries with metal-oxide and olivin-type cathodes. *RSC Adv.* 4: 3633–3642.

50 Roth, P.E., Crafts, C.C., Doughty, D.H., McBreen, J. (2004). *Advanced Technology Program for Lithium-Ion Batteries: Thermal Abuse Performance of 18650 Li-Ion Cells.* Sandia National Laboratories Report, SAND2004–0584.

51 Ribiere, P., Grugeon, S., Morcrette et al. (2012). Investigation on the fire-induced hazards of Li-ion battery cells by fire calorimetry. *Energy Environ. Sci.* 5: 5271–5280 .

52 Shack, P., Iannello, C., Rickman, S., and Button, R. (2014). NASA perspective and modeling of thermal runaway propagation mitigation in aerospace batteries. In: *NASA Aerospace Battery Workshop*. Huntsville, AL.

53 Walker, W.Q., Cooper, K., Hughes, P. et al. (2022). The effect of cell geometry and trigger method on the risks associated with thermal runaway of lithium-ion batteries. *J. Power Sources* 524: 230645.

54 Walker, W.Q., Bayles, G.A., Johnson, K.L. et al. (2022). Evaluation of large-format lithium-ion cell thermal runaway response triggered by nail penetration using novel fractional thermal runaway calorimetry and gas collection methodology. *J. Electrochem. Soc.* 169: 060535.

55 Ren, D., Hsu, H., Li, R. et al. (2019). A comparative investigation of aging effects on thermal runaway behavior of lithium-ion batteries. *eTransportation* 2: 100034.

56 Lamb, J. and Jeevarajan, J.A. (2021). *New developments in battery safety for large-scale systems. MRS Bull.* 46.

57 Walker, W.Q., Rickman, S., Darst, J. et al. (2017). Statistical characterization of 18650 format lithium ion cell thermal runaway energy distributions. In: *NASA Aerospace Battery Workshop.* Huntsville, AL.

58 Walker, W.Q., Darst, J.J., Finegan, D.P. et al. (2019). Decoupling of heat generated from ejected and non-ejected contents of 18650-format lithium-ion cells using statistical methods. *J. Power Sources* 415: 207–218.

59 Roth, P.E. (2008). Abuse response of 18650 Li-ion cells with different cathodes using EC:EMC/LiPF6 and EC:PC:DMC/LiPF6 electrolytes. *ECS Trans.* 11: 19.

60 Doughty, D.H. and Roth, P.E. (2012). A general discussion of Li ion battery safety. *Electrochem. Soc. Interface* 21: 37.

61 Feng, X., Zheng, S., Ren, D. et al. (2017). Key characteristics of thermal runaway for Li-ion batteries. *Energy Procedia* 158: 4684–4689.

62 Yayathi, S., Walker, W., Doughty, D., and Ardebili, H. (2016). Energy distributions exhibited during thermal runaway of commercial lithium ion batteries used for human spaceflight applications. *J. Power Sources* 329: 197–206.

63 ISS Safety Requirements Document (2019). The National Aeronautics and Space Administration, International Space Station Program, SSP 51721.

64 Santhanagopalan, S., Ramadass, P., and Zhang, J. (2009). Analysis of internal short-circuit in a lithium-ion cell. *J. Power Sources* 194: 550–557.

65 Russell, S. and Darcy, E. (2017). Lessons learned maturing thermal runaway tolerant lithium ion battery designs. In: *NASA Aerospace Battery Workshop*. Huntsville, AL.

66 Darcy, E., Darst, J., Walker, W. et al. (2018). Driving Design Factors for Safe, High-Power Batteries for Space Applications. In: *Advanced Automotive Battery Conference*. San Diego, CA.

67 Barrera, T.P., Bond, J.R., Bradley, M. et al. (2022). Next-generation aviation Li-ion battery technologies—enabling electrified aircraft. *Electrochem. Soc. Interface* 31: 69.

68 Darcy, E. and Petrushenko, D. (2021). Battery Relevant Cell Side Wall Rupture Characterizations. *Advanced Automotive Battery Conference*, San Diego, CA.

69 Darcy, E., Darst, J., Walker, W. et al. (2019). *Modeling Anchored by Key Tests to Established Driving Design Factors for Safe, High Power/Voltage Batteries*. Hamburg, Germany: Energy Storage For Aerospace.

70 Barrera, T.P. (2017). *Battery Design Impacts on Thermal Runaway Severity under Worst-Case Test Conditions*. Battery Safety 2017. Arlington, VA.

71 Wilke, S., Schweitzer, B., Khateeb, S., and Al-Hallaj, S. (2017). Preventing thermal runaway propagation in lithium ion battery packs using a phase change composite material: an experimental study. *J. Power Sources* 340: 51–59.

72 Dalton, P. and North, T. (2017). ISS Li-Ion main battery TR propagation test. In: *NASA Aerospace Battery Workshop*. Huntsville, AL.

73 Torres-Castro, L., Kurzawski, A., Hewson, J., and Lamb, J. (2020). Passive mitigation of cascading propagation in multi-cell lithium ion batteries. *J. Electrochem. Soc.* 167: 090515.

74 Zahran, M., Tawfik, S., and Dyakov, G. (2006). LEO satellite power subsystem reliability analysis. *J. Power Electron.* 6 (2): 104–113.

75 ECCS-Q-ST-30C (2017). *Space Product Assurance – Dependability*. ESA-ESTEC Requirements & Standards Division.

76 MIL-STD-1629A (1980). *Procedures for Performing a Failure Mode, Effects, and Criticality Analysis*.

77 ECSS-Q-ST-30-02C (2009). *Space Product Assurance – Failure Modes, Effects (and Criticality) Analysis (FMEA/FMECA)*, ESA-ESTEC Requirements & Standards Division.

78 Hendricks, C., Williard, N., Mathew, S., and Pecht, M. (2015). A failure modes, mechanisms, and effects analysis (FMMEA) of lithium-ion batteries. *J. Power Sources* 297: 113–120.

79 Sugathan, R. et al. (2015). Worst Case Circuit Analysis of a New Balancing Circuit for Spacecraft Application. *International Conference on Power and Advanced Control Engineering*.

80 MIL-STD-975M, "*NASA Standard Electrical, Electronic, and Electromechanical (EEE) Parts List*", 5 August 1994.

81 MIL-STD-1547B, "*Electronic Parts, Materials, and Processes for Space and Launch Vehicles*", 01 December 1992.

82 ECCS-Q-ST-30-11C, "*Space Product Assurance: Derating – EEE Components*", ESA-ESTEC Requirements & Standards Division, 23 June 2021.

83 NASA JSC 26943, "*Guidelines for the Preparation of Payload Flight Safety Data Packages and Hazard Reports*", February, 1995.

84 Castet, J.F. and Saleh, J.H. (2009). Satellite and satellite subsystems reliability: statistical data analysis and modeling. *Reliab. Eng. Syst. Saf.* 94: 1718–1728.

85 Tafazoli, S. (2009). A study of on-orbit spacecraft failures. *Acta Astronaut.* 64: 195–205.

86 Frost and Sullivan (2004). *Commercial Communications Satellite Bus Reliability Analysis.*

87 Seradata. www.seradata.com (accessed 3 March 2022).

88 MIL-HDBK-217F (1991). *Military Handbook: Reliability Prediction of Electronic Equipment.*

89 ECSS-Q-HB-30-08A (2011). *Space Product Assurance – Components Reliability Data Sources and Their Use.* ESA-ESTEC Requirements & Standards Division.

90 MIL-HDBK-338B (1998). *Electronic Reliability Design Handbook.*

91 Isaacson, M.J. and Gibbs, C.G. (2012). *Analysis of Life Degradation Mechanisms for Li-ion Batteries in GEO and LEO Life Tests.* The Aerospace Corporation, Space Power Workshop.

92 Xiong, R., Pan, Y., Shen, W. et al. (2020). *Lithium-ion battery aging mechanisms and diagnosis method for automotive applications: recent advances and perspectives. Renewable Sustainable Energy Rev.* 131: 110048.

93 Woodya, M., Arbabzadeha, Lewis, G. et al. (2020). Strategies to limit degradation and maximize Li-ion battery service lifetime – critical review and guidance for stakeholders. *J. Energy Storage* 28: 101231.

8

Life-Cycle Testing and Analysis

Samuel Stuart, Shriram Santhanagopalan, and Lloyd Zilch

8.1 Introduction

The commercial electronics, electrified transportation, and space industries widely recognize life-cycle testing (LCT) as the preferred compliance method to verify rechargeable lithium-ion battery (LIB) life expectancy (service life) requirements. Also known as endurance testing in the commercial aviation industry, space LIB LCT demonstrates that the LIB has the capability to perform within battery specification limits for the service life of the spacecraft mission. The goal of spacecraft LIB LCT is to baseline and execute test protocols that envelope the intended application non-operating and operating conditions in a flight-like manner. Flight-like test conditions include mission-specific battery load profiles, depth-of-discharge (DOD), charge–discharge control methodologies, and operating temperatures at different phases of the spacecraft mission. The objectives of satellite LIB LCT also include flight-like test configurations that mimic the electrical, mechanical, and thermal flight battery-to-spacecraft interfaces.

To achieve these LCT objectives, government and commercial satellite stakeholders have committed significant investments toward Li-ion cell and battery LCT program initiatives. These programs have employed a wide variety of space-qualified Li-ion cell and battery test articles subjected to mission-specific real-time and accelerated LCT. This testing has been successful in supporting all types of Earth-orbiting and planetary mission applications across the space industry. This chapter discusses the objectives, requirements, planning, and procedures representative of battery LCT and analysis. Special topics such as test design, equipment, planning, data acquisition, analysis, and modeling are discussed in terms of satellite LIB LCT requirements.

8.1.1 Test-Like-You-Fly

Implementing test-like-you-fly (TLYF) principles in Li-ion cell and battery LCT is a best-practice used to ensure that relevant on-orbit battery operating conditions are properly incorporated into the LCT plan [1, 2]. TLYF requirements are derived from a thorough systems engineering evaluation of the expected mission-specific satellite operating environments and configuration. Electrical power subsystem (EPS) battery, solar array, and power management and distribution (PMAD) component TLYF conditions are typically derived from predicted mission-specific launch conditions, orbit type, and mission life requirements. However, on-orbit flight experience has shown that actual mission environments and operational conditions may deviate from ground

Spacecraft Lithium-Ion Battery Power Systems, First Edition. Edited by Thomas P. Barrera.

predictions. As such, the effect of certain on-orbit operational and environmental (such as combined radiation dosage) conditions on satellite systems may be difficult to replicate during ground test activities.

Due to program cost, schedule, and/or technical constraints, demonstrating compliance to EPS battery service lifetime requirements may be prohibitive to complete prior to launch. For example, it is cost and schedule prohibitive to conduct a 10-year real-time low Earth orbit (LEO) battery LCT to demonstrate compliance to a 10-year battery mission service life requirement. This inherent constraint in meeting a TLYF approach for satellite battery qualification is especially prevalent in long-life missions. To gain confidence in the battery design, additional complementary tests (such as accelerated LCT) are implemented to help analyze the real-time TLYF data. For example, a real-time TLYF LCT plan may be combined with accelerated and calendar (storage) life testing to demonstrate compliance to ground and on-orbit service life requirements in a cost and schedule efficient manner.

8.1.2 Design of Test

There are multiple approaches to design-of-test (DoT) strategies. These approaches include trial-and-error, use of previously used LCT DoT designs, and DoTs based upon detailed EPS-level analysis [3–6]. The DoT ensures that all needed data parameters are identified and properly documented for performance verification to build the desired confidence in flight battery performance. Formal configuration management is important to detail and document requirements that define the DoT, test plan, and procedures. Since space battery LCTs may be executed for many years, configuration-controlled documentation is necessary to understand early test decisions, which may be impacted at a later time by obsolescent test equipment parts, materials, or personnel changes. Additionally, it may be challenging to achieve a proper DoT for space batteries when only one or two battery test articles are available for testing. In practice, since test cells are typically more available than batteries for LCT activities, a cell-level LCT DoT is commonly implemented. Cell-level LCT results are then used to predict battery-level performance characteristics. Processes and procedures for combining similar test results to achieve higher confidence levels, such as Bayesian inference techniques, LCT history, prior on-orbit performance history, and other on-orbit relevant experience are available.

8.1.3 Test Article Selection

The test plan establishes the requirements used to define the population of representative cells and/or battery test articles to support the LCT DoT. By contrast, Chapter 3 describes cell screening and selection criteria used to choose flight cells for battery manufacturing. Test results may be suspect if key parameters of LCT testing are not representative of the planned flight battery on-orbit operation. An approved set of LCT requirements for selecting Li-ion cell test articles is typically established before test article selection is finalized. Test processes and procedures for selecting which cells are chosen for LCT and which cells will be used for flight battery manufacturing are also key steps in the DoT. Pass–fail requirements for test article defects are documented in the test plan and executed via the test procedure. Cells found to fail inspection may be removed as candidates for LCT activities.

For example, visual inspection of flight cells may reveal defects that may predispose selected flight cells to become candidates for LCT articles. In general, test cell selection may involve a number of program-specific requirements applicable to the population of cells available for flight battery manufacturing, LCT, or other purposes. As discussed in Chapter 2, space-qualified Li-ion cells are available as custom-designed cells manufactured specifically for satellite applications, or as commercial off-the-shelf (COTS) cells mass-produced for commercial electronics applications. To adhere to TYLF principles, cell test articles are chosen from the same manufacturing build lot as the flight cells.

8.1.4 Personnel, Equipment, and Facilities

High-reliability satellite systems may require many years of real-time cell (or battery) LCT to demonstrate compliance to flight battery service time requirements. As such, conducting a LCT program creates various personnel, equipment, and facilities challenges to the performing test organization. Specific aspects of satellite battery LCT program capability include, but are not limited to:

- Test facilities – Conducting long-term space battery testing requires dedicating a test laboratory that meets test plan requirements. For example, test laboratories require heating, cooling, humidity control, chilled water supply, pressurized air and dry nitrogen sources, electrical power interfaces, and physical and data security resources. Facility utility power quality, stability, and uninterruptible power supplies (UPSs) are chosen for compatibility with LCT article interfaces. Reliable Internet connectivity and data acquisition infrastructure is also integral to test facility readiness. Facility safety considerations include allowances for multiple egress points, emergency ventilation, zoned smoke and fire detection, and fire suppression capabilities.
- Test equipment – Space battery LCT typically requires some custom-designed specialty equipment. Although commercial test sets are available for LCT, additional custom fixtures may be required to simulate the battery test article mechanical and thermal environments and physical interfaces. Previously, custom-built test equipment was commonplace for space battery LCT. However, employing commercially available LIB test hardware can aid in reducing LCT complexity and cost.
- Test operators – Labor for tests running constantly may require daily or multiple work shifts (24 hours) including on-call test operator availability. Personnel are trained and qualified as test technicians to operate test facilities under normal and off-normal conditions.
- Scheduled maintenance – Equipment maintenance and periodic calibration is a requirement for long-duration LCT activities. A modular approach to the test design with readily available equipment spares minimizes down-time associated with procuring and qualifying replacement equipment. Equipment calibration drift is recorded appropriately so that proper data interpretation can be performed. Modifications to the test system are verified to work properly before LCT resumes after test interruptions. Pre-test calibrations of measurement devices aids in avoiding unscheduled test interruption. For example, current measuring shunts may need recalibration every six months, while digital multimeters may only require annual recalibration.
- Data acquisition – Data acquisition during LCT produces significant amounts of data in short periods of time. Concurrent review and analysis of LCT-generated data aids in verifying that test procedure execution, test equipment function, and test article performance are in compliance with test plan requirements. After data acquisition, LCT requires many resources for accessing, analyzing and reporting results. Raw data are transferred in customer-requested formats for subsequent analysis activities.
- Data analysis – Real-time test article state-of-health (SOH) assessments includes reviewing data for trends in battery voltage, capacity, DC resistance, and temperature. Equally important, data analysis includes monitoring data trends for operating conditions that impact a safe test environment for personnel, hardware, and facilities.

- Safety – As discussed in Chapter 7, industry Li-ion cell and battery safety incidents in test and deployed environments may have catastrophic outcomes. As such, test equipment and facility safeguards are a critical component of LCT test plans and procedures. Safety incidents involving high specific energy Li-ion cells or batteries can harm test personnel, damage test articles, and shut down test facilities.

8.2 LCT Planning

Spacecraft programs routinely conduct cell- and/or battery-level LCT to verify on-orbit battery service life requirements. These data assist in mitigating program identified risks associated with satellite performance and reliability. Before LCT begins, a DoT is created and baselined in support of test planning activities. Early test planning is essential to a successful LCT program. Several DoT protocols may be used to develop specific test plans and procedures for a cell- or battery-level LCT evaluation. In addition to characterizing battery life-cycle performance, other information about the flight battery may be characterized such as:

- Cell, module, and battery design risks and opportunities,
- Battery life-cycle performance requirements after storage and handling, and
- Battery capability to meet life-cycle performance requirements under mission specific charge–discharge voltage limits, operating temperatures, DOD, and loads.

In addition, there are other unknown and unexpected operating conditions that testing may address, such as safety certification, on-orbit battery anomaly investigations, mission fault recovery scenarios, and mission life extension evaluations.

8.2.1 Test Plan

A battery LCT plan documents the requirements from which test procedures are developed. The LCT objectives and success criteria are critical elements to the LCT plan. The test plan also identifies test article quantities, allocations, sequence of testing, and the DoT baseline. Critical test resources such as special test equipment, facilities, safety equipment, tooling, data acquisition systems, materials, and supplies are identified in the test plan. Cell and battery test conditions, such as end-of-charge voltage (EOCV), end-of-discharge voltage (EODV), charge and discharge rates, depth-of-discharge (DOD), operating temperatures, dynamic environments (shock, vibration, acceleration), and test tolerances are derived from the planned use of the battery in the intended spacecraft mission application. The test plan also specifies test intervals for conducting capacity and DC resistance measurements to monitor test article SOH. The American Institute of Aeronautics and Astronautics, as well as the US Department of Defense, publishes requirements documents relevant to satellite battery testing that may be tailored to the program-unique requirements [7, 8]. For more complex testing, multiple test plans covering only specific aspects of the life-cycle evaluation may be considered.

8.2.2 Test Procedures

Once test plans are created, detailed test procedures specific to the LCT evaluation are developed. Test procedures are written with the intent that properly trained personnel can execute the procedure in a successful and safe manner. Procedures are written for performing a specific test on a selected test article and to perform the given test (such as capacity and DC resistance measurements) with parameters described in the applicable test plan. Test procedures are reviewed and validated to ensure safety by identifying risks to safety hazards and risk mitigations. Test operators are trained and certified to execute specific tests or types of tests in the documented procedures. Traceable revision change logs to test plans and procedures are documented and configuration managed.

8.2.3 Test Readiness Review

After test plans and procedures have been approved and test operators certified to perform the various tests, a test readiness review (TRR) is conducted. The TRR is a gated event that provides an opportunity to verify that the test plan objectives can be achieved with the identified resource requirements and procedures. The TRR also ensures that the test team is properly trained and staffed to execute the LCT. This includes securing the proper test resources such as test equipment, test articles, and facilities. TRRs have entrance (and exit) criteria that define specific actions and information needed to ensure that testing can begin. TRRs are attended by all test team members including safety and quality personnel, test management representatives, test technicians, engineers, and the test program customer. A team consensus agreement that the TRR exit criteria have been met is documented before the test team is authorized to begin LCT activities. The TRR outcome may be a pass or conditional pass with actions to disposition prior to beginning the LCT.

8.2.4 Sample Size Statistics

Custom space Li-ion cells are manufactured in low-volume production lots and quantities. By contrast, COTS cell-based LIBs are produced in large volumes where many flight cell assets may be procured. Consistent with the quantity of test articles available, a cell sampling plan based on available engineering data and analyses is documented as part of the test plan and procedures. The purpose of the cell sampling plan is to determine the quantity of test articles required to create a statistically significant LCT program result. There are multiple commercial guides and standards available for determining statistical sample sizes [9–12].

8.3 Charge and Discharge Test Conditions

Defining LCT charge and discharge load profiles is a critical LCT requirement. It may be challenging to baseline a pre-launch on-orbit battery load profile, since actual flight conditions may encompass a wide range of load variations. Hence, although nominal load variations may not change the battery LCT performance in a significant manner, ground LCT load profile requirements are baselined with a margin to compensate for actual on-orbit flight battery load variations. Chapter 3 further describes the orbit types, expected DOD, cycle life, and operating temperature ranges for satellite batteries.

8.3.1 Charge and Discharge Rates

Test cell charge–discharge rates have significant DoT implications for LCT planning. The intended on-orbit battery charge–discharge rates are typically bounded by cell manufacturer recommendations and/or cell qualification requirements. Since test battery charge and discharge currents can vary widely, test equipment must be specified to deliver peak power levels with a positive thermal and electrical design margin. The test design also includes a battery thermal management capability via thermal control test equipment. This capability is designed to maintain test battery thermal control at elevated charge–discharge rates, peak power demand, or during lower ambient temperature operation.

8.3.2 Capacity and DOD

Quantifying test cell and battery capacity is critical to specifying the ground test article DOD levels. During cell and battery acceptance and qualification testing, beginning of life (BOL) capacity data are measured under mission-relevant EOCV, EODV, temperature, and charge–discharge rate conditions. Usable cell capacity measurements may also be sensitive to non-operational conditions such as storage environment, storage time, manufacturing lot-to-lot variation, and test measurement accuracy.

To mitigate technical risks caused by cell capacity variation, satellite battery and cell capacities are commonly de-rated to support LCT requirements. A simple approach is to test a Li-ion cell as if its available capacity was a fraction of the nameplate or actual measured capacity. Test designs may consider differentiating between actual and de-rated capacities to gain an understanding of the expected test outcomes. During LCT, the frequency of performing test battery capacity measurements is chosen to minimize uncharacteristic capacity degradation that may result from the required higher DOD conditions required to perform real-time capacity measurements.

During Li-ion cell and battery charge–discharge cycling, DOD and operating temperature can cause chemical and physical changes in the cells that directly impact cycle and calendar life. LCT designs accommodate the sensitivity of Li-ion cells tested at high DOD when simulating geosynchronous orbit (GEO) satellite load profiles. Battery DOD is also important for LEO satellite profiles that may have a low DOD yet have a large number of charge–discharge cycles. Testing at high DOD stresses test cells and batteries such that LCT design carefully incorporates controls for EOCV, EODV, and capacity limits. Battery manufacturers commonly recommend specific guidelines for cell DOD based on their qualification and acceptance specifications. However, satellite-specific charge–discharge rates and thermal environments are usually implemented when specifying cell or battery LCT DOD limits.

8.3.3 Voltage Limits

Li-ion cell manufacturers recommend EOCV, EODV, and charge–discharge rate limits to meet cell performance and safety requirements. However, cell performance can vary within EOCV and EODV limits as a function of the charge–discharge rate, DOD, and temperature. Since test cells and batteries are expected to gracefully degrade during LCT, ground-test equipment adjustable EOCV and EODV limits are utilized to maintain a safe range of charge–discharge rates. As such, incorporating test equipment with adjustable voltage limits increases real-time test flexibility. To avoid overcharge and overdischarge conditions, test equipment and software voltage limits are designed to protect the cell and battery test articles. To meet the intent of TLYF objectives, test article voltage limits are commonly chosen based on relevant on-orbit battery operating conditions.

8.3.4 Charge and Discharge Control

As discussed in Chapter 4, most satellite bus and payload components discharge under constant power operating conditions. However, it is challenging to compare constant power discharge test results across different test facilities, power levels, and battery manufacturers. To overcome this, ground-test battery charging is usually performed at a specified constant-current level followed by a taper charge performed at a constant voltage to the specified cell or battery EOCV limit. In order to meet the intent of TLYF principles, certain battery LCT plans employ constant-power discharge profiles. Although some satellite components operate as constant-resistant loads, constant-resistance load profiles are generally not used in cell or battery LCT programs. Maintaining cell-to-cell voltage balancing is another test variable that can significantly affect the LIB cycle life. There are several methods used to achieve cell balancing in a space battery. Chapters 3 and 4 further describe charge control protocols, cell bypass approaches, and cell-to-cell balancing techniques.

8.3.5 Parameter Margin

Margin stack-up is a key LCT design parameter that requires an analysis when implementing LCT load profiles. This is a common practice when developing life-cycle profiles, where each successive element in a complex system adds additional voltage, current, power, and thermal margin to each test parameter. The addition of margin on subsystems is a method to ensure that, as a voltage source, a battery will be able to operate with load uncertainty. However, excessive added margin can lead to unrealistic LCT results. The caution for margin stacking is that, when combined with all other test margins, the stated test load profile can be much more stressful for the battery than necessary. Although such a test profile can be considered conservative, ground test results may not represent the planned flight battery conditions. Proper battery LCT design includes an analysis of these margins to assess risk of unexpected false-positive or false-negative test outcomes.

8.4 Test Configuration and Environments

The effect of on-orbit satellite thermal environments on LIB performance has a significant impact on the battery performance characteristics and service life. An objective of LCT design is therefore to simulate the on-orbit thermal interface characteristics between the flight LIB, satellite structure, and thermal control subsystem. There are three main approaches to transfer heat actively or passively from the LCT battery test article that simulates on-orbit flight battery thermal interface and control subsystem conditions. An industry best practice is to utilize temperature-controlled chambers, thermal vacuum (TVAC) chambers, or cold plates to simulate the flight battery thermal interface to the satellite bus structure.

8.4.1 Test Article Configuration

TLYF principles also apply to packaging and mounting of cell LCT articles. Replicating the mechanical packaging of cell test articles in the same manner as integrated into the flight battery mechanical design is an important test design consideration. Previous ground and on-orbit nickel-cadmium (Ni-Cd) and nickel-hydrogen ($Ni\text{-}H_2$) satellite flight battery experience demonstrated that cell packaging compression or mounting is needed to meet mission service life requirements. As discussed in Chapter 3, mitigation of prismatic or elliptical-cylindrical Li-ion cell swelling is achieved by applying cell manufacturer recommendations for compression force as a function of SOC. However, Li-ion cells with cylindrical form factors benefit from the radial hoop strength of their cases to provide the internal mechanical support they require to meet structural requirements. Nonetheless, cylindrical cells have mounting structure or attachment devices used in the flight battery design that may be replicated in the LCT mechanical interface.

LCT ground experience has demonstrated that the Li-ion cell axes mounting orientation does not affect cell performance or safety. As such, LCT test article orientation is not typically required to mimic the planned pre-launch flight battery orientation. Hence, cell and battery test orientations are chosen for the working convenience of personnel supporting the testing. Test cell packaging is commonly chosen to mimic the predicted heat transfer characteristics represented by the flight battery design. If applicable, cell swelling and contraction during charge–discharge cycling is also an important consideration in LCT mechanical packaging design.

8.4.2 Test Environments

The launch phase of a satellite mission exposes the flight battery to unique mechanical, dynamic, and thermal environments prior to on-orbit operations. As described in Chapter 3, flight battery qualification (or proto-qualification) testing is used to verify flight battery requirements for surviving launch and on-orbit temperatures, shock, vibration, acceleration, acoustic, and space radiation environments. In accordance with TLYF principles, conducting a subset of pre-launch and launch environment conditions can be performed on the battery life test articles prior to commencing ground LCT activities. These pre-LCT environmental tests may include testing to bound battery acceptance-level vibration, acceleration, and launch vehicle stage-separation shock conditions. This testing is particularly important if evaluating cell or battery designs that have no relevant on-orbit heritage. In addition, simulating pre-launch flight battery charge–discharge cycling (burn-in) that occurs during spacecraft ground processing activities may be considered during pre-LCT planning.

As with launch environments, other on-orbit environments can be evaluated for applicability to a specific space battery design. In a ground test environment simulating the actual range of all operating conditions a space battery will experience may be prohibitive, so it may be feasible to subject life test articles to a suitably representative subset of these conditions. For example, on-orbit temperature cycling, electrical loads, and capacity measurements can be reasonably simulated in ground test evaluations.

However, certain mission-unique environmental conditions (such as on-orbit radiation dosage) are difficult to accurately simulate. Periodic exposures of various radiation types and energy levels over the mission lifetime, fluences, and total doses are considered and then a suitable simulated exposure is applied to the ground test batteries if deemed necessary. Usually this is performed for convenience before the ground LCT begins to avoid interrupting the battery charge–discharge cycling that has commenced.

8.4.2.1 Temperature Controlled Chambers

A test battery set-point temperature can be controlled using a temperature chamber operating at constant temperature and ambient pressure conditions. This test approach has the benefit of simplicity with control-point capability at the outer case of the cell or battery test article. The ability of the constant-temperature chamber to respond by heating or cooling the battery to maintain a set-point temperature is achieved by attaching a control thermocouple to the test article in a location representative of the flight battery design. Figure 8.1 shows an example LCT battery in a temperature-controlled chamber. As shown, LCT test articles are positioned inside a temperature-controlled chamber with attention made to air flow patterns, wire routing, personnel access to test articles, and chamber door clearances.

8.4.2.2 Thermal Vacuum Chambers

Space environmental thermal conditions are simulated in TVAC chambers. TVAC chambers are configured with conductive and radiative heat transfer surfaces to simulate the on-orbit flight battery operating conditions. Using TVAC chambers for simulating the on-orbit thermal environment is consistent with TLYF principles. Test design for this approach considers what absolute chamber pressure is maintained to approximate on-orbit heat transfer and space vacuum conditions. Periodic TVAC chamber and vacuum pump maintenance are accommodated, as well as test

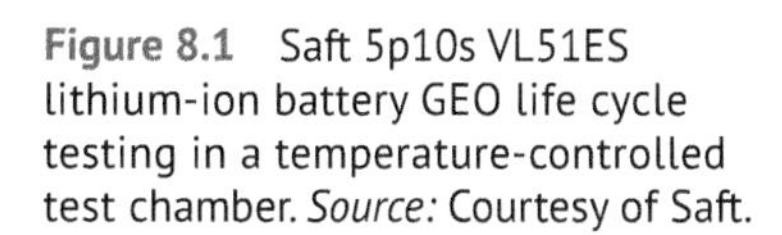

Figure 8.1 Saft 5p10s VL51ES lithium-ion battery GEO life cycle testing in a temperature-controlled test chamber. *Source:* Courtesy of Saft.

interruptions when physical access to the test battery is necessary. If supporting test equipment is to be co-located with the test battery, it must be compatible for use under space vacuum environmental conditions. Other benefits of TVAC chamber testing include cell-level leak detection opportunities and improved containment in the event of a catastrophic battery safety mishap.

8.4.2.3 Cold Plates

Another approach that contains most of the benefits of temperature and TVAC chamber environments is to mount the battery test article on to an active cold plate inside an electrically non-conductive and thermally insulative container. Cold plates are regulated using a circulating working fluid (usually a water–ethylene glycol mix) that is heated or chilled as needed to maintain the test battery at a desired temperature set-point. Electrically non-conductive insulation is placed around the outer battery surfaces to minimize convective heat transfer, thus providing a degree of vacuum-like thermal isolation.

Use of cold plates for battery LCT is a TLYF approach to representing flight batteries interfaced to spacecraft thermal control subsystem cold plates. This requires the ground-test battery to be outfitted with cooling plates and/or fins that allow attachment to the temperature-controlled, thermally conductive plate that transfers heat from the battery to the test environment. The insulating container around the battery isolates the battery from circulating air currents and can be designed to provide a degree of safety hazard protection to nearby test personnel and equipment. Figure 8.2 is an example of an LCT battery mounted on a cold plate.

8.5 Test Equipment and Safety Hazards

A long-term LCT program may present a risk to test equipment and software to exhibit degraded performance due to wear-out or other unexpected operating conditions. Industry experience has shown that commonly experienced test anomalies with test equipment hardware and software can

Figure 8.2 LCT LIB mounted on a temperature-controlled cold plate. *Source:* NASA.

be effectively mitigated by following certain test best practices. Similarly, safety hazard risks during test can be mitigated by following organizational safety, quality, and reliability requirements.

8.5.1 Test Equipment Configuration

A best practice to reduce test equipment compatibility issues before starting an LCT is to first perform test setup simulations. Test setup simulations are designed to mimic the planned LCT using a surrogate non-LCT type battery. The surrogate battery is specified with similar LCT battery asset capacity, charge–discharge, and operating temperature characteristics. In addition to performing pre-test simulations, high-quality electrical connections are necessary when testing satellite batteries. Using a TLYF philosophy, it is preferable to replicate the electrical interface connections representative of the flight battery-to-spacecraft interface. This approach has the added benefit of verifying flight-like electrical connections prior to executing the actual LCT. Since the test battery terminals may be exposed to thousands of thermal cycles during battery charge–discharge cycling, electrical and mechanical cell-level terminal connections are commonly in compliance with battery manufacturer specifications [13].

8.5.1.1 Hardware

Designing a test setup involves selecting, procuring, and implementing equipment and hardware resources that meet test plan requirements. Test controller equipment requirements include real-time testing capability, proper handling of software interrupts and hardware latency, long-term operating reliability, availability of replacement parts and components, commonality of use, and technical support availability. Certain battery testing laboratories prioritize commonality between multiple sets of testing hardware to allow for reduced spare parts inventory and minimize operator training costs for achieving test system proficiency and consistency. Custom equipment design and fabrication also requires expertise for specialized equipment and understanding unique requirements of a spacecraft battery LCT. Though custom test hardware is capable of meeting LCT requirements, it is more expensive and time consuming to design, fabricate, qualify, and maintain this type of equipment. Figure 8.3 shows an example of a custom space battery LCT facility.

Figure 8.3 An example custom satellite battery LCT facility. *Source:* Naval Surface Warfare Center – Crane.

Procuring a commercial test set involves similar requirements for definition and validation to ensure that the LCT system will meet test plan requirements. Fault scenarios are best described during the requirements development phase of the test set procurement process to allow manufacturers to properly accommodate a specific test. For example, if an LCT goal is to simulate a satellite electrical EPS grounding configuration, the test system must be capable of performing with those specific grounding and common mode connections. Another example is the unexpected loss of main power to the test system. Even with an UPS, unexpected loss of power may result in either enabling the test to continue uninterrupted or allow the test set to safely disconnect from the test article. Since some satellite batteries are designed for high charge–discharge rates, sizing a smaller UPS only to allow a safe shutdown is sometimes preferred to reduce test costs but may result in occasional test interruptions.

Special test equipment hardware, such as cell compression and mounting fixtures, signal and power connections, cooling plates and heat spreaders, fittings for chilled water supplies, temperature chambers, humidity measurement devices, battery enclosures, cut-out switches, and relays, may require special requirements definition. Hardware items known to exhibit time-dependent wear-out or loss of functional characteristics (such as bolts, nuts, screws, and thermocouples) can be positioned for ease of timely diagnosis and replacement. Implementation of failsafe-type monitors are used to verify that the test-set is operating as intended. Hardware-based protections (such as individual cell or battery overcharge detection via voltage comparators) for off-nominal battery conditions and operator errors are commonly included in LCT designs.

Typical LCT equipment requirements include the capability to collect and store large quantities of data over a wide range of data acquisition rates and short-term high sampling frequencies. For example, a single eight-cell LIB test pack can require over 30 signals to be measured at slow data rates of between 0.05 and 1 Hz for very long periods and fast data acquisition during special tests measured at rates of 100–1000 Hz. Adaptive sample rates can reduce the data burden such that not every data point needs to be recorded. Signals for detecting the onset of possible faults and failures can be sampled continuously at fast sample rates to ensure faults are identified and communicated to test personnel in a timely manner.

8.5.1.2 Software

A battery LCT usually requires a suite of software applications and an operating system environment that safely and reliably performs testing. Since multiple software applications will be required, attention to identifying software operational risks is a best practice. Anomalies in any of the major software components such as test system control for applying loads and charging, communications data transfer capability, and test operator user interfaces can negatively impact LCT activities. LCT test software is typically designed for real-time test environments and operation. Unique data bus interfaces may also include architectural trades between CAN, Ethernet, PXI, or SCXI standard devices. The combined hardware and software subsystems can also be designed to avoid data loss due to measurement latency, buffer overrun, and accommodate the large volume of data generated at various sampling rates.

Test set design and validation includes operating system requirements for periodic updates accommodating patches for errors and security vulnerabilities. If test systems are enabled to have remote access so that test operators do not have to always be physically present, proper security protocols and updates will likely be necessary to manage external access to the test system. Since software anomalies are most likely to occur after changes have been made, a best practice is to first apply and verify any software patches, versions, and revisions in a simulated pre-test environment prior to LCT execution. Software anomalies can have a wide range of possible outcomes, so it is important for LCT operators to develop criteria for normal operations and document a course of action when off-nominal conditions occur.

Likewise, software designed to conduct the life test sequence of control, collecting, storing, and analyzing data all interact in a controlled architecture to prevent unintended consequences. Test software is typically designed to be fail-safe in unintended or unexpected situations. For example, after a power outage, the test hardware can be designed to automatically reset quickly. The test software may also be required to restart nominally in a safe condition. Software requirements also include verification of proper component operation prior to initiating test restart after interruption by an unexpected power transient or failure.

LCT data usually consist of a substantially large collection of records that document measurements that would be prohibitive to replace if damaged or lost. Information technology risks such as hardware and software failures, malware, and inadvertent test system commands can cause loss of LCT data. A threat analysis for the LCT program can be conducted to identify risks according to test facility requirements. Common mitigations include limiting or preventing physical and network access to live test systems, selective data access for test personnel, data access logging, malware and virus protection, hardware and software intrusion detection and protection, and redundant offsite backups of the data storage servers.

The ability to edit recorded data can be tightly restricted or prevented altogether. If revisions or corrections are needed, a new copy can be used and labeled as such. Annotation fields to add explanatory comments to a dataset can be edited in a separate but linked file that avoids write access to the dataset itself. Another protective approach that prevents unauthorized, inadvertent modification or disclosure is to use protected software that builds a subset of data based on query. This derivative dataset allows authorized test personnel the ability to edit and analyze data without fear of corrupting the original test data set.

8.5.2 Test Safety Hazards

Test readiness includes special attention to safety awareness and certification for test personnel, hardware, and facilities. Readiness includes training, shift scheduling, certification, safety, and constraints awareness across all aspects of the planned LCT activity. For example, incorporating lessons learned from previous tests on similar battery test articles reduces safety hazards risk. This also includes evaluations of facilities emergency procedures and first responder rules of engagement. Performing any type of LIB testing requires proper planning for worst-case safety conditions and scenarios.

Test operator errors are one of the more common of all safety risks. Test personnel are involved in almost every procedural step of LCT such as battery test pack manufacturing, intercell connections, electrical connections from the cell/pack to the test system, software programming, test monitoring, and system maintenance. As such, safety hazard controls that reduce the likelihood of operator error can be implemented in LCT test set-up and procedures. Additional risk mitigation steps include proper personnel training to increase the operator's understanding, knowledge, and experience with the LCT plan and procedures. Specific personnel training in LIB technologies with an emphasis on safety is a common best practice for test organizations.

In addition to proper test personnel training, there are best-practice strategies that can be followed to reduce the risk of operator errors. One of these best practices is for test operators to adhere to the two-person rule. The two-person rule is implemented when the operators need to perform a procedure that can negatively impact the test or has a higher level of hazard exposure. Having two trained and certified test technicians present to perform a given operation not only reduces the possibility of an error but also provides a level of safety if an error or fault does occur. It is also a best practice for test operators to record any procedural adjustment to the test system in a test log with a date, time, and description of the operation performed. The test log serves as a time-stamped historical record of test events for future reference as needed. A test log can aid in verifying completion of test procedure steps, serves as a record of test interruptions, and documents test system troubleshooting activities.

Additional industry best practices include operator presence and monitoring while real-time testing (such as capacity measurements) is executed. On-site test personnel monitoring may not be practical for long-duration LCT, thus designing an LCT test program with the capability to monitor real-time testing from remote locations may be advantageous. This approach provides test operators with the ability to monitor testing when not physically present at the test facility.

8.5.2.1 Test Articles

Test cell or battery performance anomalies can be caused by a number of factors, including latent test article manufacturing defects or unintended off-normal operating conditions. Although test cells are typically screened for defects, even with the most effective cell screening protocols, a defective cell may be inadvertently chosen for LCT. However, an underperforming test cell can still provide engineering data that may be useful to the LCT stakeholders.

In practice, there have been LCT mishaps where test cells have exceeded their upper temperature control limits, which led to undesirable thermal operating conditions. Under these test conditions, an increased temperature was detected above the pre-determined temperature limit set to suspend testing. LCT that is suspended manually, or automatically by the test system, typically places the test article into an open-circuit voltage (OCV) condition and prevents further testing from occurring until the problem that triggered the suspension has been resolved. Once testing is suspended, test articles may be prevented from experiencing irreversible damage if the test anomaly is detected in advance of exposure to excessive temperatures. As such, it is a best practice to put at least one temperature sensor on each individual cell test article while continuously monitoring current and voltage. With battery packs manufactured from custom large-format Li-ion cells, individual cell-level electrical connections are often feasible. However, the complexity of measuring individual cell voltages increases greatly with a larger number of cells integrated into the battery test article. For example, it may be prohibitive to measure individual cell voltages in multi-cell LIBs designed with COTS 18650 Li-ion cells or similar. Finally, a best-case scenario for cell failure due to wear-out is in a graceful performance degradation manner near the end of its design life. Under these LCT conditions, the cell (or battery) test article is expected to exhibit increased internal resistance, capacity loss, decreased EODV, temperature increases, or other degraded performance trends.

8.5.2.2 Equipment Induced

Test equipment safety and reliability is an important requirement to conducting a successful LCT activity. Test anomalies can be caused by the charge–discharge test equipment, a test controller, or other secondary test system component. For example, industry experience has shown that several laboratory Li-ion test battery safety mishaps have occurred due to improperly connected or mislabeled voltage sense wires. In these cases, the test battery was being charged and the feedback system reported the wrong voltage and current such that the battery continued charging when it should have stopped. This caused the battery to reach an overcharge condition, causing a catastrophic failure. A timeout-limited failsafe monitor can stop a test if hardware or software fails. These timers work by starting a pre-determined countdown time period after a communications or control event occurs. If communication has not resumed by the end of the safety period countdown, the safety timer independently suspends the test. Normally, routine updates from a properly functioning test system restarts the failsafe monitor timeout period such that the countdown period does not expire and testing continues normally. It is likewise important to choose proper safety control limits that stop a test before test article failure occurs. Types of upper and lower safety thresholds that can be implemented include temperature, voltage, and current limits based on test plan requirements. Safety limits based on test time are effective in detecting when a planned charge or discharge is taking much longer than predicted. A timed safety limit is also useful to detect if communication between critical equipment components is no longer occurring, which may indicate a possible test equipment anomaly. When a test article condition is reached that is outside the range of the expected parameters, the test system triggers a test suspension that stops the battery test article from charging or discharging. Real-time operator notifications via electronic e-mail, text, or other means can facilitate a rapid operator response to test anomalies. These test protocols aid in maintaining test personnel safety, safeguards test article SOH, protects test facilities, and can minimize test downtime.

Another best practice is to incorporate test systems with separate limits set for different test phases and cycles. For example, during a typical LEO charge–discharge cycle the battery temperature may only reach 28 °C (for a cold plate controlled to 20 °C) so a reasonable upper temperature alarm limit could be set at 35 °C. During a capacity cycle, however, the battery temperature may reach >35 °C and unintentionally trigger test suspension. If different safety limits are implemented, then the upper temperature limit during a capacity cycle may be increased to approximately 40 °C, thereby not improperly suspending a normal cycle. Each limit can be examined sequentially to verify that unnecessary test interruptions are minimized to maintain LCT SOH and safety.

8.5.2.3 Laboratory Induced

Anticipating laboratory-induced safety risks is another key aspect of maintaining safe and reliable LCT activities. Industry experience has shown that even highly trained test operators and the most reliable test system may be at risk to anomalies or failure. Example test laboratory anomalies include alternating current (AC) mains power outages, loss of a phase for two- or three-phase AC powered equipment, voltage drop brown-outs, environmental temperature swings due to seasonal changes, environmental-control outages, accidental water damage, and problems caused by co-located laboratory activities.

Fire department approved smoke and carbon monoxide detection systems are mandatory when conducting long-term LIB LCT. To mitigate the adverse impacts of a laboratory safety incident, high-value tests can be conducted in separate laboratory locations with sufficient spacing between test batteries to minimize the loss of test assets or data. To protect recorded LCT data, data is backed up frequently and to off-site storage facilities. If test data is stored only to the co-located test system or not backed-up, irreversible data loss may occur if the test system is damaged due to an operator error, water damage, fire, or other laboratory mishap.

8.5.2.4 Test Control Mitigations

Test control mitigations are systems or features that either predict or detect circumstances that will have a negative impact on the test activity. These mitigations often incorporate protocols to suspend an LCT to prevent additional hardware damage from occurring. Test suspensions should be minimized to adhere to a TLYF philosophy, but the capability to prevent or stop a catastrophic safety incident has the highest priority when compared to the baseline test objectives. Control mitigations are implemented by hardware or software solutions such as upper and lower control limits placed on voltage, current, and temperature measurements. In addition, mitigations can include charge control system monitoring, redundant scanning computers, timeout circuits, back-up monitoring systems, and failsafe monitor timers. Each of these systems monitor the test and when off-nominal conditions are detected, testing will be suspended or interrupted to avoid an undesirable consequence. These mitigations are often automated, resulting in an automatic test system shutdown whenever the safety limits are exceeded. One of the best approaches to mitigate these risks is to employ trained operators who have the skills to examine the real-time LCT activities. This man-in-the-loop approach has the added benefit of not only being able to suspend the test in case of impending failure but can also identify other more subtle non-limit related issues. As such, the test operator can manually suspend testing in a manner that minimizes safety impacts to personnel, hardware, and facilities.

8.5.2.5 Physical Mitigations

Physical mitigations are those that do not necessarily prevent a safety mishap from occurring but minimize the severity and consequences of an undesirable event. These are test features such as fire alarms, first responder intervention, and active fire suppression systems. Fire suppression systems include physical barriers designed to protect the test article such as a test chamber enclosure or containment box. These physical barriers serve to physically prevent unintended physical contact with the test pack, which may cause an external short between a charged battery and another object.

Physical barriers may also act to absorb or deflect the energy release of initial deflagration during a catastrophic battery test mishap. They also limit atmospheric oxygen sources from being readily available to the pack, thus mitigating the severity of fire or combustion events. To proactively remove oxygen from the battery test environment, commercial inert gases can be used to backfill the container or continuously flow over the pack during testing. Other physical barriers and increased spacing between cells can mitigate the severity and consequences of a catastrophic cell-to-cell propagating thermal runaway (TR) condition. Temperature control systems, such as a test chamber or a cold plate, facilitate removing heat from the system and thus can mitigate the severity of a TR safety hazard. Fire sensors and alarms placed in close proximity to the test article serve to provide early warning in case of a TR safety incident. Finally, limiting or replacing flammable materials with non-flammable materials in the test area is critical to decrease the hazard severity and limit undesirable consequences resulting from a test laboratory fire.

8.6 Real-Time Life-Cycle Testing

Subsequent to baselining test plan requirements, test system completion, and a successful TRR, real-time LCT can begin. Beyond test plans and procedures, industry experience has demonstrated that test execution is a combination of real-time decision making supported by test technicians, test engineering management, and satellite program stakeholders. The objective of real-time LCT is to verify the life expectancy of the selected Li-ion cell (or battery) test article design. Data obtained from a cell-level LCT is commonly used to predict the life expectancy of LIBs made from those cells. In addition, real-time LCT demonstrates mission-relevant performance capability in terms of charge-discharge cycling and calendar life aging. Real-time LCT requirements are typically the longest time element of an LIB cell qualification test program.

8.6.1 Test Article Selection

Since LCT occurs over extended periods of time, proper test cell selection and matching is critical to meeting test plan requirements. Selecting test cells for LCT includes analyzing cell manufacturer acceptance, qualification, and other relevant engineering test data. Receiving and inspection data are also analyzed to facilitate cell selection for cell matching into test article modules or batteries. Cell matching commonly includes analysis of an as-received cell-level capacity, AC impedance, DC resistance, self-discharge rate, and other relevant acceptance test data. Other pre-test data commonly collected at the start of LCT includes test cell mass, dimensions, OCV, X-ray radiography, computed tomography (CT) scans, and detailed photographs of the test setup. These data enable before and after test comparisons in support of LCT data analysis and reporting.

8.6.2 Test Execution and Monitoring

As described in Chapter 3, satellite LIBs are sized using worst-case EPS performance requirements based on end-of-life (EOL) spacecraft performance. However, BOL LCT battery performance data provides a necessary initial baseline for subsequent comparative analysis. Real-time LCT data trending is based on BOL capacity, DC resistance, and other performance data obtained from execution of the LCT procedures. Analysis of these data is performed on a periodic basis to assess test article real-time SOH under the given LCT conditions.

Assessing test article SOH includes procedures for conducting real-time LCT capacity measurements at various discharge rates at a fixed temperature. Depending on test requirements, DC resistance is also measured during the real-time capacity measurements. Consideration of the frequency of capacity measurements versus how these measurements impact unintended useful capacity loss to the test articles is germane to the SOH assessment. Capacity and DC resistance measurements during periodic capacity testing are consistent with test procedure current, voltage, and temperature limits. Capacity and DC resistance SOH measurements on LCT test articles simulating real-time test conditions are conducted per program requirements with a periodicity aligned to mission-specific TLYF guidelines. SOH measurements for LCT test articles simulating real-time GEO test conditions are typically conducted at the beginning and end of each simulated Fall and Spring solstice period. Capacity and resistance SOH measurements for real-time LEO testing is based on spacecraft program-unique requirements.

8.6.3 LCT End-of-Life Management

Ground LCT data is commonly used to support future mission operations planning to extend the on-orbit mission life beyond the intended spacecraft design life. On-orbit operational and ground test experience has demonstrated that maintaining and extending LIB service life can be accomplished via various changes to battery operating conditions. For example, decreasing the LIB test article DOD may increase the predicted cycle life. A more common strategy to increase LIB cycle life is to decrease the battery EOCV. Although a decrease in EOCV results in decreased usable capacity, it also lowers cell degradation rates, which may also increase cycle life. The trade between changing LCT battery DOD or EOCV is dependent on LCT resource limitations and test plan requirements.

8.7 Calendar and Storage Life Testing

Real-time battery LCT programs simulate the flight battery's expected non-operational and operational characteristics under mission-relevant conditions. By comparison, calendar life testing measures the effect of time-dependent non-operational cell (or battery) performance loss on cycle life. There are two main categories of calendar life testing: (i) ground characterization of cell capacity degradation that occurs after battery manufacturing and up to spacecraft launch day and (ii) cell capacity degradation that occurs due to time-dependent mechanisms experienced during on-orbit battery non-operating periods. Calendar life testing measures these capacity degradation effects to predict on-orbit battery performance. In addition to Li-ion cell chemistry type, LIB calendar life primarily depends on battery voltage, SOC, and temperature during storage and non-operational time periods. As discussed in Chapter 4, calendar life capacity loss predictions are commonly used in spacecraft EPS power budget and energy balance analyses.

For satellite batteries with missions that do not require a high cycle life (such as GEO orbits), yet have long orbital solstice (non-operating) periods, calendar life capacity degradation can become a significant contributor to battery capacity degradation. As with other types of battery non-operating periods, it is important to understand the effect of extended ground storage periods on LIB capacity in order to meet battery and spacecraft EPS specification requirements.

8.7.1 Calendar Life

The effect of on-orbit calendar life on cell capacity degradation is challenging to assess due to the real-time interactions of on-orbit environments on LIBs in long-life satellite applications. For example, certain satellite Earth orbits, such as GEO and highly elliptical orbit (HEO), have long solstice periods during which the satellite battery experiences little or no discharge. Operationally, satellite LIBs exposed to long solstice periods are maintained at voltages and temperatures that minimize battery capacity loss due to the effects of calendar life. Combined with charge–discharge cycling, incorporating calendar life testing into the LCT plan enables a TLYF worst-case approach to demonstrating compliance to on-orbit service life requirements. As such, combining mission-relevant charge–discharge cycling and non-operational solstice periods in GEO LCT results in greater cell capacity loss and internal resistance growth than if only charge–discharge cycling was performed. For these reasons, cell calendar life effects are included when specifying LCT conditions. Solstice periods are defined in the LCT plan and executed in the procedures in real-time or using an accelerated life testing methodology.

8.7.2 Storage Life

Pre-launch flight cell and battery storage time is the primary contributor to LIB capacity loss during ground processing activities. Although there is a small number of charge–discharge cycles experienced by the flight battery during acceptance, qualification, and satellite ground testing, non-operational storage time dominates the pre-launch portion of the battery life cycle. Controlled storage conditions help to maintain flight battery SOH by minimizing capacity degradation mechanisms that occur at various temperatures and SOC. Cell manufacturers recommend storage conditions based upon their specific cell chemistry, design, and other characteristics. Storage life test data are therefore used to measure the cell capacity loss under controlled storage conditions. These data are then used to support pre-launch BOL battery capacity predictions and as inputs to the EPS power budget and energy balance analyses.

8.7.3 Test Methodology

Previously, heritage space Ni-Cd and Ni-H_2 batteries were extensively evaluated using ground test and on-orbit data to support mission objectives. When Li-ion cells were baselined for spacecraft applications, calendar life characteristics were re-evaluated to demonstrate compliance to mission-specific requirements. Multiple LCT studies demonstrated that, when Li-ion cells are cycled at lower charge–discharge rates or lower DODs, irreversible capacity degradation is minimized [14–18].

For LCT orbital calendar life evaluations, representative degradation factors are identified and a test sequence is constructed to simulate those factors. Storage life testing is performed by maintaining a representative sample of cells at a range of reduced SOC at constant temperatures. Each test cell is periodically measured for voltage decay resulting from self-discharge and inspected for evidence of electrolyte leakage. If a cell is found to have an out-of-family decrease in OCV, maintenance charging to compensate for capacity decay can be performed.

A test cell that exhibits an out-of-family self-discharge rate may have an internal short. These cells are typically removed from the storage life testing since they are not representative of cells that will be selected for integration into a flight battery. An example Li-ion cell storage test would include discharging cells to the manufacturer-recommended storage SOC, connecting voltage monitoring leads for OCV measurements, and placing the test cells into a controlled-temperature chamber. Data are reviewed and analyzed periodically for OCV loss trends. Test duration is based upon the cell (or battery) specification storage life requirements. Observed self-discharge rates are typically compared to self-discharge rates recorded during cell screening, acceptance, or qualification testing. Cells that exhibit out-of-family voltage decay during their storage life testing may require modifying the cell screening process to remove these cells before they are accepted for use in future flight battery builds or LCT. Although flight cells may be in controlled storage for years prior to flight battery integration, it is expected that these cells will exhibit different levels of self-discharge during storage.

8.8 Accelerated Life-Cycle Testing

Given the resources necessary to conduct real-time battery LCT, it is sometimes preferable to accelerate the LCT to reduce test time. Accelerated LCT is employed to screen cells for infant-mortality (workmanship) characteristics, gain early confidence in cell performance, and obtain relevant data for early verification of predictive cell (and battery) performance models. In order to implement an accelerated LCT plan, test tailoring may be considered as allowed by the space battery specification or relevant industry standards. Accelerating LCT reduces LCT time duration so that battery compliance to service life requirements is demonstrated in a substantially shorter period of time. As such, accelerated LCT is an effective risk mitigation strategy to support pre-launch battery service life preditions. To effectively accomplish accelerated life testing, risk mitigation plans, which ensures that the test battery will not be overtested, are documented in the LCT test plan.

8.8.1 Accelerated Life Test Methodologies

Various types of accelerated LCT has been performed on many space battery technologies, with results that can be compared to real-time LCT results. Conducting accelerated LCT requires that the battery test article performance being investigated is characterized in real-time prior to testing. Common examples of battery-accelerated LCT methods are to increase the battery DOD, increase the charge and discharge rates, or reduce the solstice (non-operational) time periods [19].

For LIBs, test parameters that may be considered to accelerate cell (or battery) LCT include, but are not limited to:

- Increase charge–discharge rate – Most commonly used in LCT programs supporting LEO spacecraft missions due to the large number of real-time cycles in LEO orbits.
- Increase DOD – Commonly used in support of LEO and GEO LCT testing. Increasing DOD increases the delivered capacity (per cycle) and accelerates performance degradation.
- Increase EOCV – Increasing EOCV above the mission-specific battery EOCV accelerates test article performance degradation.
- Decrease solstice period – Most commonly used in LCT programs supporting GEO, HEO, and planetary spacecraft missions due to the long battery non-operational periods during solstice orbital periods.
- Increase operating temperature – Not commonly used as a means to conduct accelerated LCT for space programs due to uncertainties in the effects of temperature on the cell-level Arrhenius kinetics and mass transfer rates.

One of the most common accelerated LCT methods used in satellite LIB LCT is accelerating the cycling charge–discharge rates to reduce the overall test time. A risk with this method is that it can also cause performance degradation that is not representative of real-time LCT. Another common accelerated LCT method is to reduce the simulated solstice periods test time for some missions. For example, GEO (or HEO) LCT can be semi-accelerated by shortening the time (solstice period) that the battery is maintained at relatively constant SOC and temperature. In a semi-accelerated GEO LCT, the solstice orbital period is shortened while the eclipse orbital period is conducted in real-time. Increasing battery DOD has also been used to simulate worst-case operating conditions to accelerate LCT outcomes. Though not comprehensive, data from accelerated LCT of space LIBs continue to be used to support all types of Earth-orbiting and planetary missions. Confidence in these empirical approaches improves as more real-time and accelerated test data become available from other LIB market applications [20, 21]. Quantitative models relating real-time to accelerated LCT data can be used to increase confidence in LIB service life estimates. Table 8.1 is a representative LCT real-time and accelerated test matrix for LEO and GEO orbit types.

8.8.2 Lessons Learned

Accelerated battery aging can introduce unexpected operating conditions that are not well-controlled in LCT. For example, large acceleration factors increase the risk of potential over-test conditions that may lead to false negative test results. When examining accelerated LCT data, EODV voltage trends, internal resistance growth, and capacity loss can be significantly different from that

Table 8.1 Test matrix for representative cell LEO and GEO real-time and accelerated LCTs.

Orbit type	LCT type	DOD, %	Eclipses per day	Solstice days	Temperature, °C
LEO	Real-time	10–30	12–16	None	20
LEO	Accelerated	20–40	12–16	None	20
LEO	Accelerated	10–30	24–32	None	20
GEO	Real-time	60–70	1	138	20
GEO	Semi-accelerated	60–70	1	1	20

obtained during real-time LCT. Due to the wide variations in available Li-ion cell chemistries, form factors, manufacturers, and design features, a conservative application of accelerated testing conditions is warranted. The primary objectives are to avoid accelerating LCT protocols, which may accelerate degradation mechanisms that do not represent real-time electrochemical processes. Non-Arrhenius behavior further complicates determining a suitable level of acceleration. LIB electrochemical processes can vary widely throughout service life, thus accelerating processes that normally take place in real-time may not be practical for a high-reliability spacecraft application.

8.9 Data Analysis

Long-duration battery LCT programs can generate large data file sizes in short periods of time. Data-set file size is dependent on the data sampling rates and number of voltage, temperature, and current measurements per test article. Business-suite software may not be capable of importing and reading large data files without additional data parsing applications. The most useful LCT data sets are those that are capable of not only supporting the intended verification of satellite LIB cell lifetime performance, but also of other useful analyses such as modeling and simulation that can aid battery lifetime predictions.

8.9.1 LCT Data Analysis

Single charge–discharge cycle data analysis is a common method used to analyze selected LCT charge–discharge cycles. A single time-series graph is commonly used to analyze multiple LCT cell data sets for performance variations between cells as a function of cycle number. Figure 8.4 is an example of a single cycle plot that shows how a battery cell voltage responds to load current in a

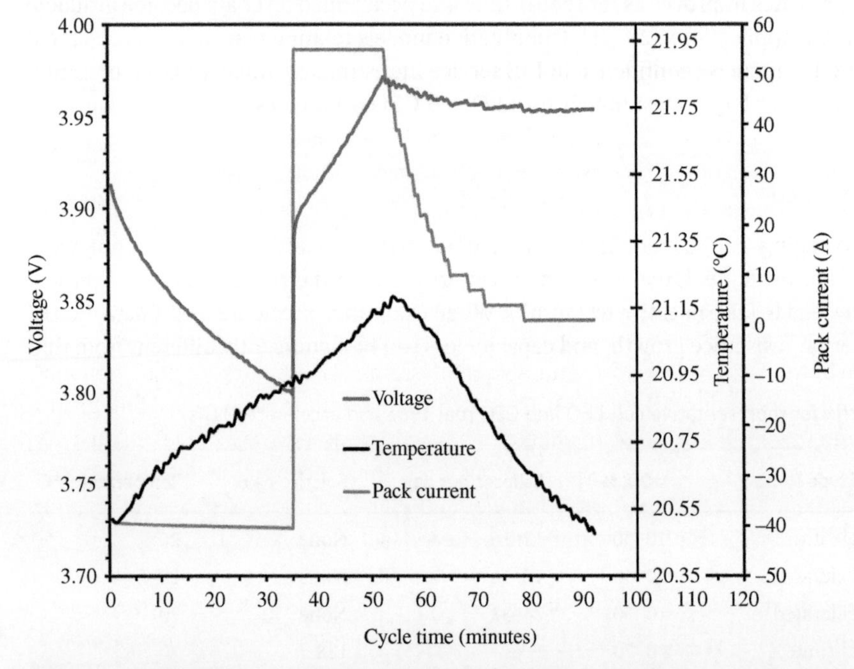

Figure 8.4 Representative satellite battery cell single charge–discharge cycle data. *Source:* Courtesy of NASA/Public Domain.

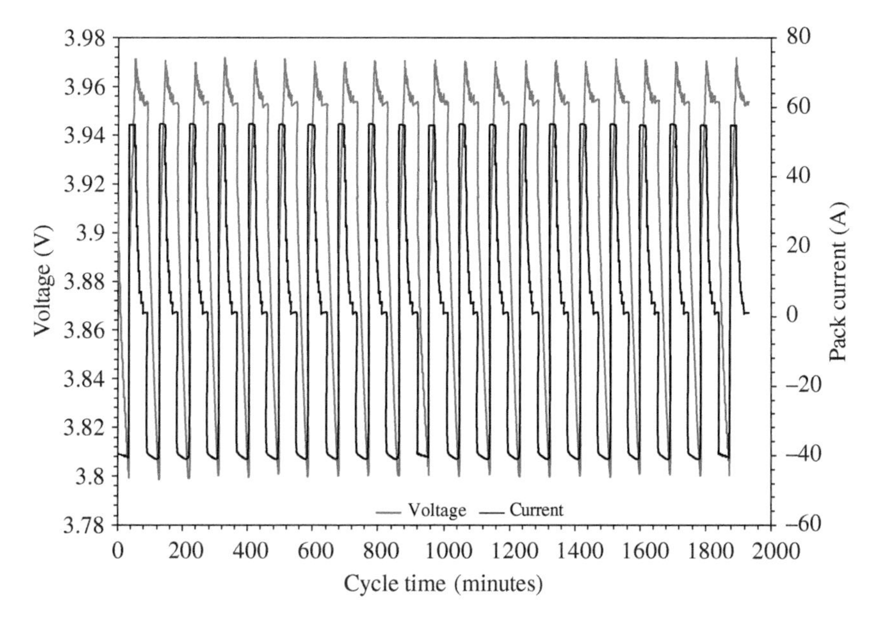

Figure 8.5 Representative satellite LIB cell charge–discharge cycling during real-time LCT. *Source:* Courtesy of NASA/Public Domain.

constant-temperature test chamber. These LCTs are useful to compare BOL to EOL single cycles in the same test battery to examine how the cells are performing.

Figure 8.5 is a representative example of a cell-level charge–discharge LCT at constant temperature. A common means to graphically represent LCT cell SOH characteristics is to trend EOCV, EODV, and temperature as a function of cycle number. SOH trending also includes analysis of cell LCT capacity and DC resistance data as a function of cycle number. This analysis is performed by selecting data points from a single cycle (such as the EOCV and EODV) and comparing them to cycle numbers. At constant environmental temperature and DOD, expected LCT SOH trend data includes cell EODV, DC resistance, and capacity as a function of cycle life. An increase in cell resistance corresponds to a decrease in measured cell capacity as a function of cycle life.

8.9.2 Trend Analysis and Reporting

Each type of orbit cycle has specific characteristics that depend on battery operating conditions that are recorded to monitor battery SOH. For LEO LCT, the EODV is expected to decrease with cycle life. The slope of the EODV versus cycle number analysis may be used to predict the remaining number of useful charge–discharge cycles (Figure 8.6). The duration of the constant-current portion of charge and the minimum value of charge current can be used to determine the capacity throughput for a given cycle number. As test cell internal resistance increases, the duration of the constant-current segment decreases with cycle number and EODV. With a continued increase in cell resistance with charge–discharge cycling, the battery reaches the cut-off EOCV limit for constant-current charging progressively earlier.

In Figure 8.6a, the discharge duration is constant during each cycle. With aging, the battery EODV decreases. In Figure 8.6b the ampere-hour throughput during each day is fixed and as a result the EODV during a given season exhibits a parabolic-shaped profile. With aging, the EODV for a given day of the season decreases. In GEO battery cycling, the EODV versus number of cycles

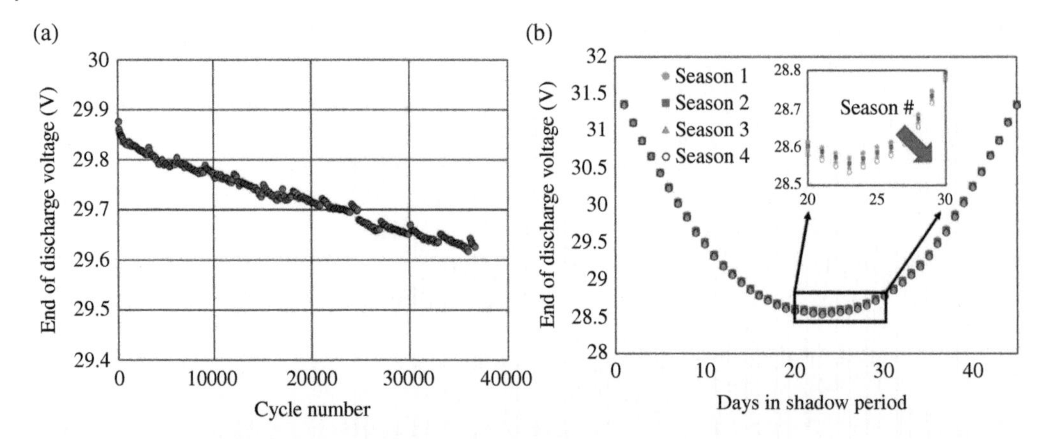

Figure 8.6 Sample profiles of battery EODV parameters: (a) LEO cycling under 10% maximum DOD at 20 °C with an EOCV of 32 V and (b) GEO cycling under 60% maximum DOD GEO cycles at 20 °C ambient temperature with an EOCV of 32 V. *Source:* Courtesy of NASA/Public Domain.

measured during the eclipse test period follows a parabolic-shaped profile corresponding to the duration of daily DOD. Variability in EODV across different cells connected in series within a test pack is monitored for cell-to-cell variability with charge–discharge cycling. Trends in heat generation or temperature data are also used to assess battery SOH.

Periodic data reviews are used to analyze LCT test conditions, test article SOH, test anomalies or interruptions (if any), and other LCT program relevant information. Satellite battery LCT is typically conducted within battery qualification limits for voltage, DOD, current, and temperature. These test conditions provide opportunities to conduct real-time data reviews and analysis for SOH trending and out-of-family performance. In practice, the following LCT cell (or battery) measurements are analyzed to trend SOH:

- Voltages – Cell voltages are measured and monitored to trend capacity, internal resistance, and cell-to-cell voltage divergence changes as a function of cycle life.
- Capacity – Cell capacity test data can be compared to previous capacity measurements to determine how individual cells are performing with calendar and cycle life. If there is a test plan requirement, adjustments to test cell EOCV and EODV limits may be employed to manage test article capacity degradation and SOH.
- DC resistance – Cell internal resistance is calculated from LCT voltage and current data that utilize real-time DC resistance measurement techniques.
- Temperature – Cell temperature is trended for variations as a function of load current and cycle life.

Life test reports are issued periodically as required in the test plan. The reports include a brief description of the cell design characteristics, LCT test profile, control temperature, and other relevant test information. Examples of data representations include LCT trends of EOCV, EODV, capacity, temperature, and DC resistance as a function of charge–discharge cycles.

8.10 Modeling and Simulation

In addition to graphical analysis of LCT data, modeling and simulation can be an integral part of the LCT data evaluation process. Models provide a methodology to describe battery performance characteristics based on LCT results that facilitate predictions of expected cell and battery behavior. Models can also be used to assess cell and battery designs to determine how well the test articles are performing compared to design expectations. Defining which types of models and simulations aids in identifying the needed data to construct battery models when designing, implementing, and conducting the LCT evaluations.

Some degree of battery modeling is common for satellites and is increasingly adopted throughout the design, test, and operational phases of the satellite battery engineering life cycle. These models can take many forms. For satellite batteries, empirical or semi-empirical models are derived from least squares fits of linearized data and non-linear regression fits of test data [22, 23]. Engineering data-fitting techniques have been defined for Li-ion cells as well, which has enabled satellite battery model capability and use [24–32]. Cell and battery design-focused models can incorporate various degrees of physics-based simulations covering thermal, mechanical and electrochemical first principles. Model-based systems engineering (MBSE) approaches can also implement various types of descriptive behaviors to assist with LIB LCT design and analysis.

Simulation of satellite battery performance is a useful tool during the early stages of LIB design. Sometimes it is computationally intensive to achieve a multi-scale model suitably inclusive of micro-scale electrode mechanisms through cell, pack, and battery-level operable mechanisms. This is particularly true when degradation and side reactions are considered. Simulations can be as simple as regressing key parameters that are monitored periodically to detailed phenomenological studies of degradation mechanisms. Reduced parameter approaches can simplify otherwise intractable computations with acceptable losses in accuracy.

Certainly, not all satellite mission types will warrant modeling as part of the LCT, especially shorter operational durations. Depending on the objectives and data available, suitable mathematical models have been deployed with varying degrees of success in space applications. For any of these modeling and simulation approaches, LCT data provide for their foundation and validation.

8.10.1 Modeling and Simulation in Battery-Life Testing

The capability to assess and predict on-orbit satellite LIB degradation is one of the more valuable uses of an LCT-derived model. Simulation of battery degradation for space applications has been a particularly well-suited proposition for several reasons:

- Satellite load profiles are well-defined, compared to automotive or consumer electronics applications. Orbit-to-orbit (LEO) or eclipse-to-eclipse (GEO) load profiles can be similar to one another.
- LCT cells are fully characterized with minimal cell-to-cell variability, operating temperature, and voltage ranges all within cell qualification limits.
- There are often years of application-relevant data available to parameterize and validate the models.
- Interpreting well-designed accelerated aging LCT data to compare various degradation modes can help reduce duration or number of test conditions by identifying operating parameters that control battery cycling performance.

Some key challenges in building life models specific to space batteries include:

- Extremely long cycle life requirements compared to other consumer electronics and the resulting need to predict battery performance over thousands of cycles.
- Multiple aging mechanisms, especially those associated with a long calendar life in addition to cycling-related degradation.
- Very limited number of cells are available for characterization and in several instances the cells are custom-made, which in some cases limits access to historic statistical performance data.
- The need to balance collecting life-cycle data under actual orbit conditions against exposing the cells to worst-case operating conditions.

With these requirements and trade-offs in place, several types of models have been developed to describe life-cycle performance of LIBs in space applications. These are classified into several groups based on the degree to which the degradation parameters can be traced back to fundamental changes in cell material components.

8.10.2 Empirical Approaches

Empirical models fit experimental data sets to pre-determined equations that relate measurable variables such as power, temperature, voltage, and current. To match experimentally observed trends among these variables, parameters in the empirical models are tuned using a regression scheme. Some general forms have already been established through flight experience and can be easily fitted with LCT data. For instance, capacity loss during calendar aging follows a square-root dependence on time, whereas during cycling, capacity loss is usually correlated against the amp-hour throughput. Correction factors for temperature and/or SOC dependence are incorporated in the model in order to fit test data. While fitting empirical models to cycle life data, it is useful to separate out capacity loss due to calendar aging for the same duration, to get better estimates of degradation from factors related to cycling. This can be especially useful for GEO missions where long solstice orbital periods (non-operational) are present.

A simple, but powerful example of an empirical model is the NTGK model [27]:

$$j'' = Y(T, \text{DOD}, n) \times \left(U_{\text{eq}}(T, \text{DOD}, n) - V_{\text{cell}}\right)$$

Here the cell voltage (V_{cell}) is related to the current (j'') using two sets of look-up tables: one for the open-circuit potential, $U_{\text{eq}}(T, \text{DOD}, n)$ and another for the cell internal resistance, given by the inverse of $Y(T, \text{DOD}, n)$. These functions in turn depend on the ambient temperature (T) and DOD defined either based on cell capacity or as a specified voltage limit, and the number of cycles (n). Some sample results comparing a model calibrated against Li-ion cell data collected at multiple C-rates are shown in Figure 8.7.

Cell degradation is monitored by tracking changes to the fitting parameters (in this case, functions U and Y), to match cycling data periodically. The main advantage of this approach is the

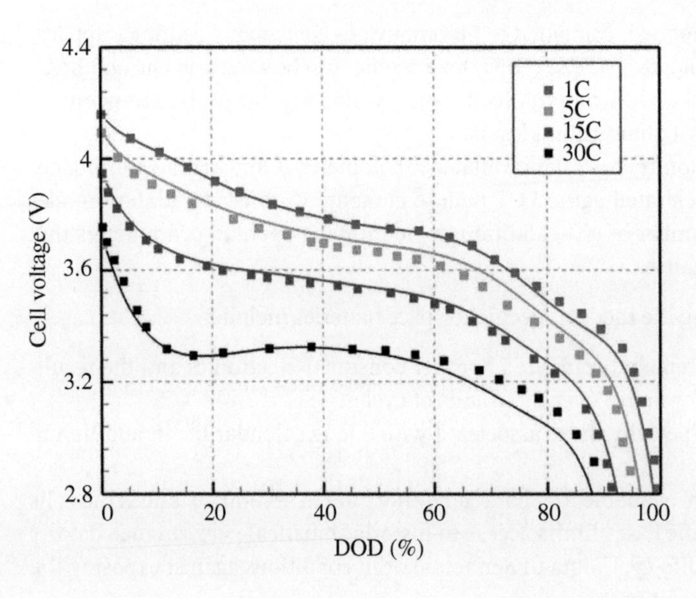

Figure 8.7 Sample fit for an empirical model. Symbols: experimental data; solid lines: NTGK model predictions.

limited need to characterize cell components. These empirical models can be used to readily parameterize data from multiple Li-ion cell chemistries and formats. The disadvantages are twofold:

- Models are good at interpolating cell-level performance but require actual test data on the relevant cells across wide voltage and temperature operating windows. Predictions outside the range of test data usually lack accuracy.
- Parameters used to tune the model response to experimental results are empirical and are hence difficult to trace back to specific degradation mechanisms, although there are some exceptions. This limits the ability of the model in determining remaining useful life, since the extent of degradation depends on cell history.

8.10.3 First Principles of Physics-Based Models

Physics-based models relate a component-level response to cell performance using well-established relationships popularly referred to as first principles. Some common examples include the use of Fick's laws to simulate mass transport limitations and Butler–Volmer or Marcus theory to explain reaction kinetics at each of the electrode–electrolyte interfaces [27, 33].

These models track internal states that retain some physical significance (such as lithium concentration at the surface of each electrode). The use of physics-based models allows for identification of degradation mechanisms that control cell aging and thus can assist in developing mitigation strategies if the observed degradation rates are higher than anticipated. For instance, by knowing the key degradation modes at different operating conditions, the trade-offs between reducing the voltage window and operating at a different temperature can be compared in order to achieve a target rate of capacity decay after a few months on-orbit. This is called a mission life extension strategy.

The disadvantage to developing detailed phenomenological models is the need to extensively characterize cell material and component-level properties such as diffusivities, resistances, and conductivities. Most often access to cell components and/or their properties is difficult to obtain from cell manufacturers. However, usually cell disassembly is sufficient since new cells have undergone some formation and aging processes. This establishes the need for disassembling fresh and aged cells in a clean and safe environment. It is still not trivial to develop a high-fidelity first principles model encompassing all the degradation processes coupled with a thermal model for each cell chemistry and battery design. There are several unexpected and irregular degradation mechanisms that may surface during a mission that often make modifying the physics-based models to reflect such anomalies very challenging.

8.10.4 Systems Engineering Models

LCT provides the best data for model development and validation. When LCT is designed, parameters important for model development are identified, and those values are recorded and reported during the actual battery testing. These data can then be used to compare predictions to actual ground test and flight battery performance. MBSE can use life-cycle data in various models to help design and verify the relevant satellite systems to meet mission-specific requirements.

Modern satellite development relies, at least in part, on MBSE to enable improved communications and information flow between the product teams working on the satellite's systems, subsystems, and components. Requirements for management software, like the Dynamic Object-Oriented Requirements System (DOORS), allows the product team to uniformly define and communicate component requirements. Additional model functionality can be achieved with a component-level model (such as System Markup Language, SysML) to provide a mechanism that helps ensure requirements of the battery are suited for the satellite. These models can also be extended or coupled with other models to provide more realistic simulations of the battery's capabilities in worst-case operating scenarios. For example, a model of the satellite battery could be implemented in MATLAB™ or similar modeling software and coupled to a SysML model. This allows the entire satellite model to simulate various operating conditions and provide results that reflect changes in the battery from BOL through EOL.

8.10.5 Models for Tracking Test Progress

Degradation of space LIBs with aging can be traced to several physical phenomena. Known processes at the anode contributing to loss of cyclable capacity include growth of an interfacial layer (commonly referred to as the solid-electrolyte interphase or SEI), plating of lithium, and/or electrical isolation of the active material from the current collectors due to poor mechanical adhesion. Similarly, at the cathode, irreversible phase changes, build-up of an oxidative film that adds to the resistance of the cell, and poor electronic conductivity of the active material resulting in overheating of the cells, have all been observed as factors contributing to degradation of cell capacity. Additional factors such as electrode dry-out, changes in electrolyte concentration, and issues with wetting or mismatch in porosities across cell electrodes are often categorized as cell-level engineering issues that affect cell performance [33].

The models outlined in previous sections are used together with test data to evaluate LIB aging characteristics. With empirical models, this process entails following changes to critical model parameters. Since these parameters are specified in the model as functions of cycling conditions, the task of tracking test results is reduced to regressing coefficients in the Y and U_{eq} functions from voltage versus DOD (or time) data obtained under multiple cycling conditions. Degradation in the cells is then attributed to different factors (excessive cell heating, chemical degradation of the electrodes, corrosion of the tabs, etc.) based on historically observed trends in the coefficients with cell aging.

The first principles-based models capture degradation of cell performance from various factors by tracking changes in physically observable quantities like porosity, loss of lithium, or mechanical isolation of electrode particles, by regressing these parameters from cycling data. One example is shown in Figure 8.8, where performance degradation due to most of these processes can be broadly grouped into loss of active (cyclable) charge carriers and loss of the active material [32]. The surface concentrations of cyclable lithium normalized by the thermodynamic limit for the Li-ion concentration for the respective electrodes are defined as the stoichiometric lithium content within each electrode (x_p and x_n) [33]. The thermodynamic limit is the maximum allowable stoichiometric lithium content that can be extracted without the host lattice undergoing an irreversible change. The value of x_n is close to 0 for a completely discharged cell, indicating that the anode is devoid of Li^+ at the beginning of charge and increases as the cell is charged.

There are two possible aging scenarios for the anode: (i) the loss of cell capacity can be attributed to depletion of cyclable Li-ions due to parasitic side reactions or (ii) loss of electrical contact due to mechanical isolation in regions of the active anode material coatings. In the former scenario (i), there are adequate sites to store lithium but there is a deficiency of charge carriers as the cells age. Alternatively, even if the side-reactions are controlled to a minimum, the active sites that hold the cyclable lithium are decreasing. Using a physics-based model, x_n is defined for a fully charged cell

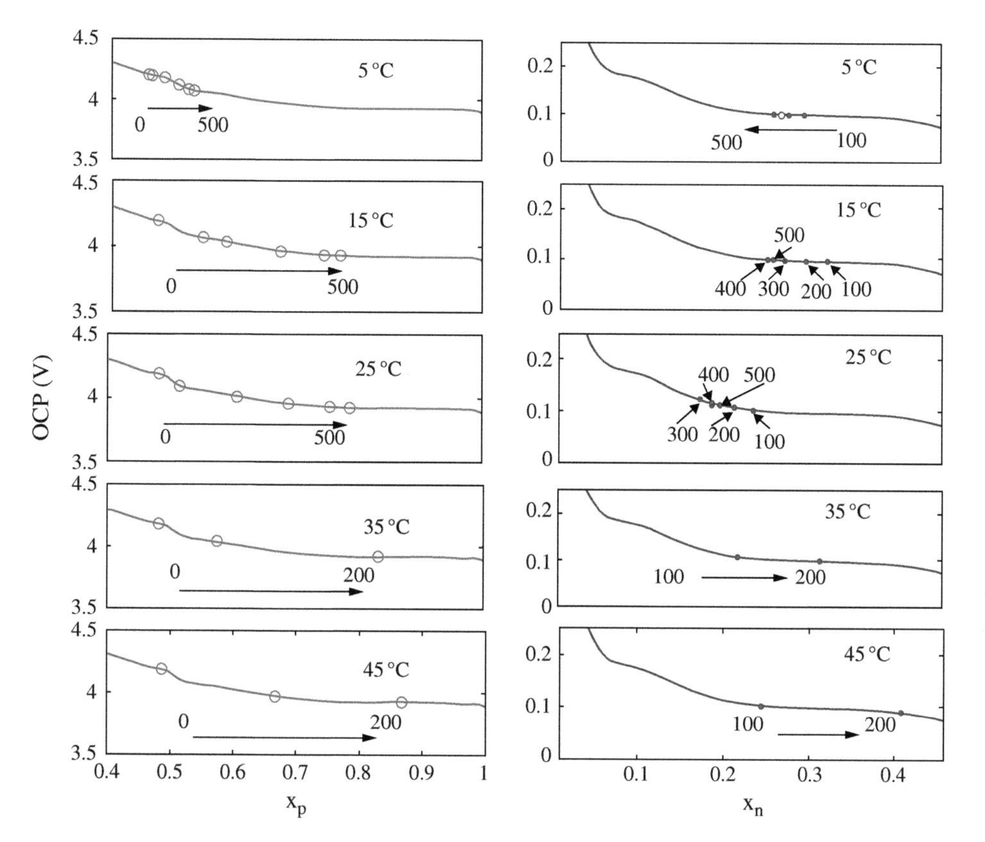

Figure 8.8 Example results tracking surface lithium stoichiometry. Here, positive and negative electrode stoichiometries as a function of cycle number at different temperatures calculated using a physics-based model are shown. The cathode stoichiometry increases with cycling at all temperatures indicating a loss of cyclable lithium whereas degradation of the anode is more complicated with lithium loss controlling degradation at the lower temperatures and mechanical isolation of active material from the electrodes being the predominant degradation mechanism at the higher temperature for these cells. *Source:* S. Santhanagopalan, NREL.

as a parameter ($x^0{}_n$) to track cell aging. The thermodynamic limit for Li-ion concentration within each electrode is dictated by the number of active sites. If the cell degradation follows (i), the value of $x^0{}_n$ decreases with cycle number, since the number of active sites does not change significantly in this instance and the surface concentration of cyclable Li^+ is depleting. With cycling, if the number of active sites becomes the bottleneck (ii), the value of $x^0{}_n$ increases since there are a greater number of cyclable Li^+ available per active site.

For the example shown in Figure 8.8, at 5 °C the lithium loss is the predominant mode of degradation. With an increase in temperature, site losses become increasingly prominent with cycling while the capacity loss during the initial cycles is still attributed to lithium loss. Beyond 35 °C for instance, site loss is the controlling mechanism for aging. The cathode shows an increase in the value of the SOC at the beginning of discharge ($x^0{}_p$) with cycling indicating progressively lower amounts of lithium being cycled. The spread in the value of $x^0{}_p$ increases with temperature, indicating a higher rate of degradation with temperature for this data set.

Other examples include monitoring cell resistance with cycling to regress transport and kinetic rate constants that signify a reduction in power capability with cell aging.

8.10.6 Parameterization Approaches

Every model has parameters that are dependent on the cell chemistry, build, and/or operating conditions. Determining parameter values that are accurate under practical constraints is required for any of the modeling approaches outlined in earlier sections. Different stakeholders rely upon models to answer different sets of queries: in the context of battery life-cycle analysis, cell developers use models to compare chemistries and determine cell and/or component dimensions while systems engineers try to interpret on-orbit data to estimate remaining useful battery life. Accordingly, there are different approaches to parameterizing the models.

For a given Li-ion cell chemistry the physics-based models typically employ model parameters measured from experiments that are independent of the cell design and validate the models using the same set of parameters across multiple cell formats. Empirical models are often employed when access to component-level information is limited. Hence, model parameters are obtained by regressing cell- or module-level data.

8.10.7 Data Requirements

Although cell teardowns and characterization of cell components are time consuming, most cell parameters can be reasonably measured with a few additional cell samples. Data required to parameterize the empirical models are usually power or voltage versus time at various discharge rates, operating temperatures, and voltage ranges obtained from relevant cell cycling. Periodic capacity measurements and pulse DC resistance measurements are used to validate and/or recalibrate model parameters. While the actual application may not impose stringent data requirements on the batteries, model development usually has some additional considerations:

- The voltage or power range across which model parameters are measured must be sufficient to perturb each parameter to perform a sensitivity analysis. For example, for most cell chemistries, if the cycling protocol is restricted to a small voltage and/or SOC range close to 50%, the solid-phase diffusivity of Li^+ ions is not very sensitive to constant current charge–discharge data within this regime since the cell performance is controlled by kinetics and electrolyte-phase transport under these operating conditions. Instead, the test plan considers a wider voltage range or includes pulse measurements within the original operating range.
- Parameters such as the ratio of negative-to-positive active material mass (n/p ratio) effectively participate in the reaction and/or definition of C-rates change as the cell capacity changes with extensive cycling and/or calendar aging. An effective way to account for these changes is to periodically recalibrate the models. The test plan accounts for these measurements, either in terms of additional downtime from cycling (or storage), additional cells required, or both.
- Both transient and frequency domain impedance measurements should have appropriate sampling rates: short-term measurements often last only a few seconds and lower sampling rates may result in loss of a pertinent current–voltage response. Too high a data acquisition rate often results in noisy data and/or loss of communication between the cells and the control unit. The range of frequencies used in electrochemical impedance measurements reflect time-constants of processes studied. Resistance and thermal measurements account for variations in contact resistances between the cells and probes.
- Variations in the data are due to changes in the ambient temperature, clamping pressure, or humidity, so downtime during long-term cycling must be accounted for when parameterizing the models.

8.10.8 Lifetime and Performance Prediction

Validated models can be used to help predict the effects of satellite operation on battery performance. Modern models have greatly refined the ability to forecast battery performance. A key advantage of physics-based models is the ability to conduct what-if analyses by varying operating constraints or estimating robustness for a chosen set of operating conditions. In essence, this task transforms analyzing battery performance to tracking changes in model parameters using phenomenological relationships such as Fick's Laws.

Following evolution of these parameters with cycling allows expected cell resistance growth or decreases in usable capacity to be predicted. To illustrate this, Figure 8.9 shows simulation

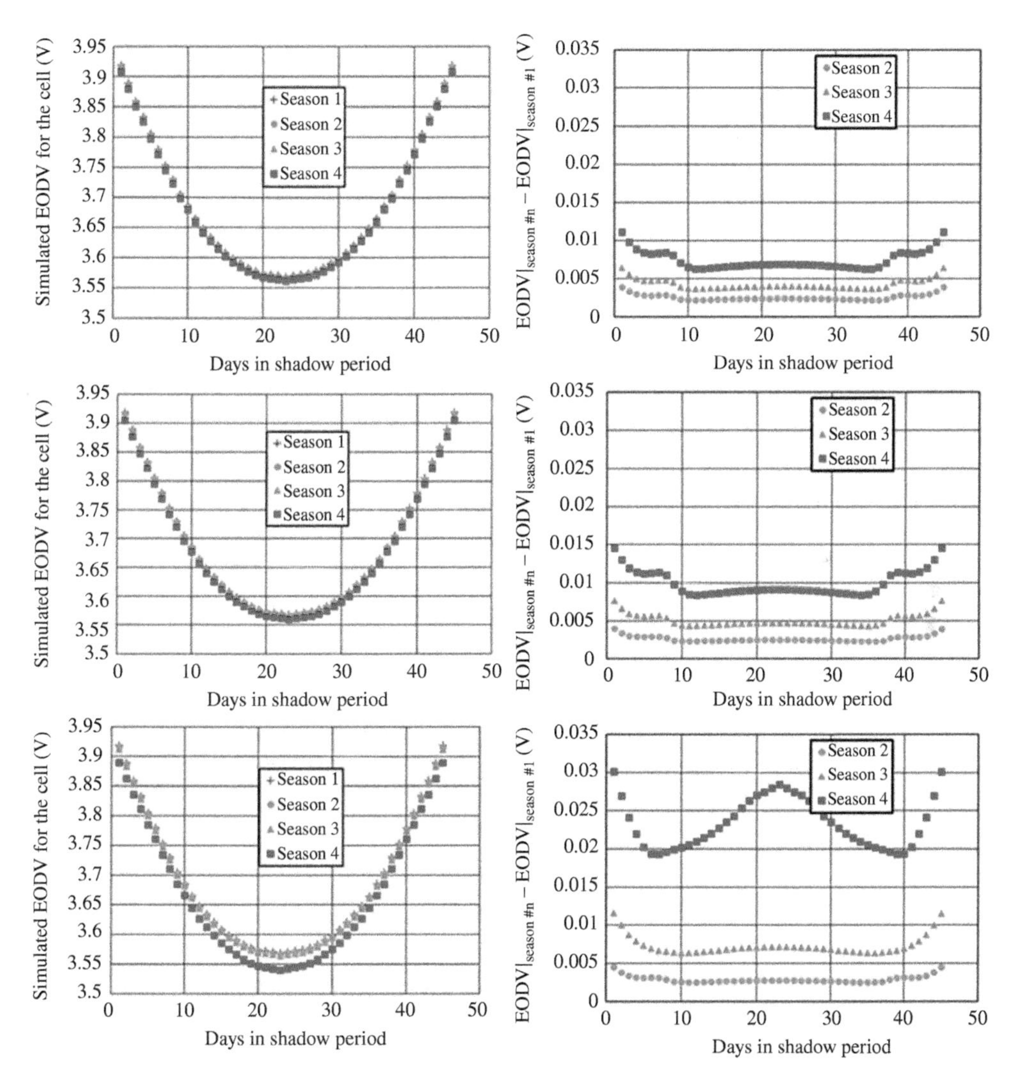

Figure 8.9 Simulated GEO cycle response of single cells cycled under the same operating conditions shown in Figure 8.6: plots (a) to (c) show the end of discharge voltage (EODV) corresponding to four seasons under three hypothetical aging scenarios: (a) minimal aging, (b) degradation controlled by ohmic resistance build-up, (c) accelerated aging under mixed-mode degradation. Plots (d) to (f) show the difference in the EODV for each day from what it was in Season 1, as the cell ages across Seasons 2 to 4. Increase in ΔEODV across days 1–10 is predominantly an ohmic drop whereas that across days 15–30 indicates transport limitations.

results for GEO-cycles for cells cycled under the same conditions shown in Figure 8.6. There are several competing degradation mechanisms that occur within cells as they age. This example shows the simulated EODV for a cell with minimal degradation (Figure 8.9a) and two other hypothetical scenarios where: (i) the degradation is dominated by resistance buildup and resulting ohmic losses (Figure 8.9b) and (ii) a mixed-mode degradation where multiple fade mechanisms including ohmic drop and transport limitations result in accelerated aging (Figure 8.9c). Upon close inspection, there is little difference between Figure 8.9a to c except for an additional drop of about 20 mV in the EODVs observed on a few days across the four seasons simulated.

A closer examination of how the EODV for a given day changes across different seasons (Figure 8.9e and f) allows quantification of the various limiting factors. For scenario (i) (Figure 8.9b), ΔEODV due to ohmic resistances is observable during the first few days of each season and remains fairly constant across the season (Figure 8.9e), whereas for scenario (ii) (Figure 8.9c), where there are multiple fade mechanisms at play, the ohmic drop during days 1–10 is visible but ΔEODV during days 15–23 are more pronounced (Figure 8.9f). These correspond to transport limitations that control cell performance at lower voltages.

Estimates on battery life corresponding to the number of seasons the cells can survive before the requisite ampere-hour throughput forces the EODV below acceptable limits can be calculated. Figure 8.10 shows such model predictions made using the empirical model, as well as the physics-based model for the three scenarios described in Figure 8.9. For cases where the degradation is controlled by resistance build-up only (Figure 8.9d and e), the extrapolation made using the empirical model matches predictions made using a physics-based model.

However, for the case involving competing degradation modes (Figure 8.9f), the physics-based model shows a slightly higher rate of degradation past Season 10, which becomes progressively worse. As noted in this example, the models that can track multiple, competing degradation

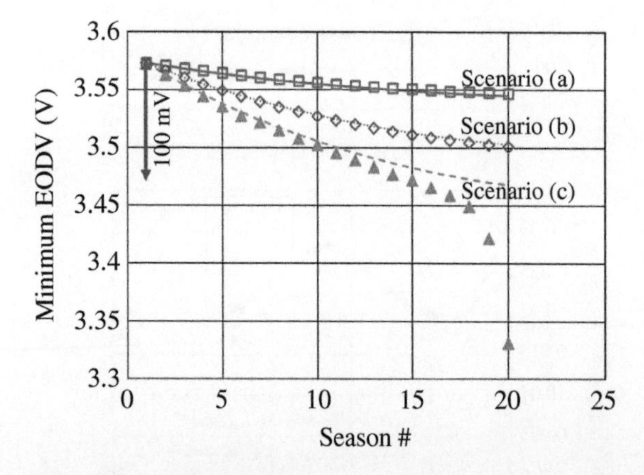

Figure 8.10 LCT cell life estimates. Based on hypothetical end-of-life criterion of a 100 mV drop in the minimum EODV compared to Season 1, when the cell is subject to GEO cycles under the three scenarios outlined on Figure 8.9a to c. The symbols are predictions made using the physics-based model and the lines are estimates based on the empirical model. For the milder degradation under scenarios (a) and (b), results from both models agree, and the predictions show that the cell will last well beyond 20 seasons. For the rapid degradation under Scenario (c), the empirical model projects end-of-life at 20 seasons whereas the physics-based model that accounts for competing degradation modes predicts end-of-life at 15 seasons and a rapid decline in cell performance shortly thereafter. Note that in this example, experimental data from only the first four seasons was used to calibrate degradation parameters in both models.

modes and determine the controlling mechanism for capacity fade under specified operating conditions but are essential when making predictions about long-term performance based on limited test data. Obtaining confidence intervals for the parameters using standard statistical tools allow estimation of the likelihood of a given cell meeting the target performance after the stipulated cycling protocol.

8.11 Summary

This chapter describes satellite LIB LCT best practices commonly used to qualify space LIB-based EPS. Although resource demands associated with long-term LCT can be significant, program investments in conducting cell and/or battery LCT have been shown to be a critical part of meeting spacecraft program mission assurance requirements. Best practices include implementing LCT activities with approved test plans, procedures, and continuous stakeholder engagement throughout the duration of the test program. The space industry commonly conducts real-time LCT activities to verify on-orbit service life of a specific Li-ion cell design. These cell data are then used to assess the life expectancy of space LIBs manufactured from those cells. Ground-based predicts of space LIB service life can be achieved by using a combination of real-time and accelerated LCT data. Quantitative models relating real time to accelerated LCT data can be used to further increase confidence in LIB service life estimates. Finally, real-time cell and battery LCT data are also used as inputs to EPS power budget, energy balance, and LIB reliability analyses.

References

1. Bucher, A.W. (2001). Test Like You Fly. In: *2001 IEEE Aerospace Conference Proceedings* (Cat. No. 01TH8542), March 2001.
2. White, J.D. (2010). Test Like You Fly: Assessment and implementation process, Aerospace Report No. TOR-2010 (8591)-6 (18 January 2010).
3. Reynolds, M.T. (1996). *Test and Evaluation of Complex Systems*. New York: Wiley.
4. Zimmerman, A.H. (2009). *Nickel-Hydrogen Batteries: Principles and Practice*. EL Segundo, CA: The Aerospace Press.
5. Dunlop, J.D., Gopalakrishna, R.M., and Yi, T.Y. (1993). *NASA Handbook for Nickel-Hydrogen Batteries*, Chapter 7. NASA Reference Publication 1314.
6. Beard, K.W. (ed.) (2019). *Linden's Handbook of Batteries*, 5e, Chapter 32. New York: McGraw-Hill.
7. AIAA S-122-2007 (2007). *Electrical Power Systems for Unmanned Spacecraft*. Reston, VA: American Institute of Aeronautics and Astronautics.
8. United States Space Force Space Systems Command Standard, Lithium-ion Battery Standard for Spacecraft Applications, SSC Standard SSC-S-017, 20 January 2022.
9. ANSI/ASQ Z1.4 *Sampling Procedures and Tables for Inspection by Attributes*. Milwaukee, Wisconsin: American Society for Quality.
10. ASTM E122-17 (2017). *Standard Practice for Calculating Sample Size to Estimate, With Specified Precision, the Average for a Characteristic of a Lot or Process*. West Conshohocken, PA: ASTM International www.astm.org.
11. ISO 3951-2 (2013). *Sampling Procedures for Inspection by Variables*. Geneva, Switzerland: International Organization for Standardization.
12. Amstadter, B.L. (1971). *Reliability Mathematics*. New York: McGraw-Hill.

13 Taheri, P., Hsieh, S., and Bahrami, M. (2011). Investigating electrical contact resistance losses in lithium-ion battery assemblies for hybrid and electric vehicles. *J. Power Sources* 196: 6525–6533.

14 Brown S. (2008). Diagnosis of the lifetime performance degradation of lithium-ion batteries, PhD Dissertation, Kungliga Tekniska Högskolan (KTH) Royal Institute of Technology, Stockholm.

15 Preger, Y., Barkholtz, H.M., Fresquez, A. et al. (2020). Degradation of commercial lithium-ion cells as a function of chemistry and cycling conditions. *J. Electrochem. Soc.* 167: 120532.

16 Juarez-Robles, D., Jeevarajan, J., and Mukherjee, P.P. (2020). Degradation-safety analytics in lithium-ion cells: Part I. Aging under charge/discharge cycling. *J. Electrochem. Soc.* 167: 160510.

17 Lam, L.L. and Darling, R.B. (2015). Determining the optimal discharge strategy for a lithium-ion battery using a physics-based model. *J. Power Sources* 276: 195–202.

18 Bandhauer, T.M., Garimella, S., and Fuller, T.F. (2011). A critical review of thermal issues in lithium-ion batteries. *J. Electrochem. Soc.* 158 (3): R1–R25.

19 Zimmerman, A.H. (2020). Accelerated life factors. In: *NASA Aerospace Battery Workshop*. Huntsville, AL.

20 Jiang, J., Gao, Y., Zhang, C. et al. (2019). Lifetime rapid evaluation method for lithium-ion battery with Li(NiMnCo)O_2 cathode. *J. Electrochem. Soc.* 166 (6): A1070–A1081.

21 McDermott, P. (1981). An improved equation for discharge voltage. In: *1981 Goddard Space Flight Center Battery Workshop*, NASA Conference Publication 2217.

22 Hafen, D. (1981). NiCd battery cycle life prediction for low Earth orbit. In: *1981 Goddard Space Flight Center Battery Workshop*, NASA Conference Publication 2217.

23 Corlone, R. (2010). A comparison of in-orbit telemetry data vs. software predicted behaviour of ABSL lithium-ion battery for NASA Lunar Reconnaissance Orbiter. In: *NASA Aerospace Battery Workshop*. Huntsville, AL.

24 Bailey, P. (2010). Consistent battery terminology and detailed battery models from BOL through EOL. In: *NASA Aerospace Battery Workshop*. Huntsville, AL.

25 Camere, A., Powers, A., Marcoux, L., and Jennings, K. (2013). An empirical model of lithium ion cell degradation. In: *NASA Aerospace Battery Workshop*. Huntsville, AL.

26 Guo, J. and Li, Z. (2015). A Bayesian approach for Li-ion battery capacity fade modeling and cycle to failure prognostics. In: *NASA Aerospace Battery Workshop*. Huntsville, AL.

27 Tiedemann, W. and Newman, J.S. (1979). *Battery Design and Optimization* (ed. S. Gross), 39. Princeton, NJ: The Proceedings of the Electrochemical Society.

28 Halpert, G., Bugga, K., Shevade, A. et al. (2009). A first principles-based Li-ion battery performance and life prediction model based on single particle model equations. In: *NASA Aerospace Battery Workshop*. Huntsville, AL.

29 Prevot, D., Borthomieu, Y., and Cenac-Morthe, C. (2014). Experience and modelling, VES16 lithium-ion cell lifetime tests results & SLIM correlation. In: *10th European Space Power Conference*. Noordwijkerhout, Netherlands.

30 Loche, D. (2002). Towards the use of lithium-ion battery on LEO applications. In: *Space Power, Proceedings of the Sixth European Conference*. Porto, Portugal.

31 Jun, M., Smith, K., Wood, E., and Smart, M. (2012). Battery capacity estimation of low-Earth orbit satellite application. *Int. J. Prognost. Health Manag.* 3. ISSN 2153-2648, 2012 009.

32 Santhanagopalan, S., Zhang, Q., Kumaresan, K., and White, R.E. (2008). Parameter estimation and life modeling of lithium-ion cells. *J. Electrochem. Soc.* 155: A345–A353. https://doi.org/10.1149/1.2839630.

33 Arora, P., White, R.E., and Doyle, M. (1998). Capacity fade mechanisms and side reactions in lithium-ion batteries. *J. Electrochem. Soc.* 145 (10): 3647–3667. https://doi.org/10.1149/1.1838857.

9

Ground Processing and Mission Operations

Steven E. Core, Scott Hull, and Thomas P. Barrera

9.1 Introduction

Satellite and launch vehicle (LV) battery ground operations at the manufacturing, system test, and launch site facilities are critical steps in the battery life cycle. Lithium-ion battery (LIB) power system compliance to Range Safety and Department of Transportation (DOT) requirements during launch site testing, transportation, handling, and storage ground operations ensures safe and reliable launch site processing operations. More specifically, adopting a test-like-you-fly (TLYF) approach to electrical power subsystem (EPS) component ground assembly, integration, and test (AI&T) operations, which reflects actual flight environments and operating conditions, has been shown to be an effective satellite program risk mitigation strategy. This includes examination of all applicable battery on-orbit characteristics to determine the fullest practical extent to which those characteristics can be applied during ground spacecraft AI&T activities. As a result, pre-launch electrical verification checks, hardware inspections, and environmental controls of the integrated satellite EPS become necessary to validate and maintain LIBs state-of-health (SOH) prior to and during launch site ground operations. After a successful launch and spacecraft separation, operating critical EPS flight components in a manner consistent with previous ground testing or simulations increases the likelihood of meeting on-orbit satellite mission objectives. At the end of mission (EOM) operations, passivation of the EPS and safe disposal of the satellite are executed as part of the satellite decommissioning process. This chapter describes the ground, launch site, and mission operations of satellite LIB power systems. The on-orbit operations, monitoring, controls, and management of EPS performance over life are emphasized. A special emphasis on EOM passivation guidelines and requirements of satellite LIB-based EPS is provided.

9.1.1 Satellite Systems Engineering

The goal of the satellite systems engineering function is to ensure that the satellite system is designed, manufactured, and operated in a manner that meets mission-specific objectives within program resource constraints. Satellite systems engineering activities encompass all system life-cycle phases

Spacecraft Lithium-Ion Battery Power Systems, First Edition. Edited by Thomas P. Barrera.

starting from initial mission concept development to EOM satellite decommissioning [1]. Among the most important aspects of satellite systems engineering are defining system and subsystem-level requirements for the purposes of product verification and validation. This includes supporting all aspects of spacecraft interactions with internal subsystems and external ground support systems.

After completion of final satellite product design and manufacturing, the system life cycle proceeds through ground processing, launch preparations, launch, initial activation, on-orbit mission operations, and final satellite decommissioning for the purposes of EOM disposal. Ground processing includes all activities related to integration, testing, verification, and validation of the satellite system to expected environments. This phase of the life cycle also includes mission rehearsals, training (for both nominal and contingency operations), and launch readiness of ground operating personnel and spacecraft crew members supporting unmanned or human-rated (crewed) space missions. During on-orbit mission operations the intended primary mission is supported by mission control personnel to include analyzing mission data, executing necessary system adjustments, and resolving on-orbit anomalies. Near the end of the primary mission, mission extension opportunities may be evaluated to extend baseline mission activities or perform new mission objectives. Satellite systems engineering support concludes with implementing the EOM systems decommissioning and disposal plans [2].

9.1.2 Ground and Space Satellite EPS Requirements

Satellite space system design integration is also a major focus of the satellite systems engineering task. Controlling the satellite's physical, functional, and operational design integration requirements between the ground, launch, and space segment elements ensure hardware and software compatibility across the entire system. Ground and launch interfaces to the satellite system include test facilities, modes of transportation, storage facilities, ground command and control stations, launch processing, launch vehicles, and mission control operations. After satellite launch, interactions between the satellite and space environments become the one of the most critical interfaces impacting the operational satellite system.

As discussed in Chapter 4, the satellite EPS is required to reliably generate, store, manage, and distribute electrical energy to the satellite bus and payload user loads. To meet these requirements, the EPS is designed to operate in ground, launch, and space modes that support pre-launch testing, ground storage, launch and ascent, transfer orbit (GEO and MEO missions), satellite initialization, on-station operations, and decommissioning mission phases.

9.2 Ground Processing

Ground processing of space Li-ion cells and batteries begins at the manufacturing facilities of origin. As such, integrated Li-ion cells and batteries may be stored, handled, and transported in various physical configurations and environmental conditions. Typically, ground processing procedures are unique to the cell and battery chemical, electrical, mechanical, and thermal design features. In space applications, proper LIB ground processing is focused on maintaining LIB SOH by executing validated storage, handling, transportation, and operation procedures. More importantly, utilizing approved LIB ground processing procedures and processes ensures that ground personnel safely store, handle, and transport flight LIBs during all phases of ground operations.

9.2.1 Storage

There are numerous non-operational periods during the ground service life of flight LIBs where proper storage is required to maintain LIB SOH. Subsequent to flight LIB manufacturing and acceptance testing, extended non-operational storage periods ranging from months to years may be necessary due to satellite ground operation workflow constraints. Conservative assumptions are commonly included as part of life-cycle analyses to account for worst-case AI&T durations, ground storage, and launch schedule delays. For example, unexpected events, such as delays with LV readiness, range availability, or weather-induced delays, may result in extended non-operational LIB storage periods. After flight LIBs are delivered from the battery manufacturer to the satellite AI&T facilities, battery storage may be required in preparation for integration into the satellite bus structure. To accommodate extended on-satellite non-operational storage periods, flight LIBs are maintained according to battery manufacturer user manual storage specifications. LIB maintenance during on-satellite storage includes periodic recordable monitoring of available battery-level voltages and temperatures. Voltage and temperature data are monitored to maintain flight LIBs within manufacturer specified state-of-charge (SOC) limits. Adherence to proper storage maintenance procedures mitigates the risk of LIB performance degradation prior to launch.

Although short-term (typically less than approximately 30 days) off-satellite storage of flight LIBs may be accomplished in controlled ambient temperature environments, long-term flight LIB storage in certified cold-storage facilities is a common space industry practice. Certified cold-storage equipment is commonly equipped with continuous voltage, temperature, and humidity monitoring capability. LIB SOC, temperature, and humidity storage requirements are dependent on the LIB cell chemistry type, non-operational aging characteristics, and expected storage period. To maintain safe storage conditions, facility alarm systems may be employed to alert ground personnel of potential out-of-specification battery voltage or temperature conditions. Other LIB storage configuration requirements, such as external battery connector electrical protection (including the use of connector savers), mitigates the risk of external electrical shorting hazards.

9.2.2 Transportation and Handling

US and international transportation authorities have classified non-rechargeable and rechargeable Li-ion cells and LIBs as materials that present chemical and electrical hazards. In the US, Li-ion cells are regulated as hazardous materials under the US DOT Hazardous Materials Regulations Title 49 Code of Federal Regulations (CFR) Part 173 [3]. Section 173.185 addresses all aspects of LIB cell handling, packaging, size limits, labeling, and transport requirements for compliance with DOT regulations. DOT regulations also require that all LIB cells and batteries meet the criteria for shipping requirements identified in the United Nations (UN) Recommendations on the Transport of Dangerous Goods, Manual of Tests and Criteria, Section 38.3, Lithium Metal and Lithium-Ion Batteries [4].

Satellite LIBs are not exempt from complying with applicable local, federal, and/or international transportation requirements. LIB handling and transportation requirements include thermal conditions, handling orientation, transportation shock, and vibration environments that adhere to industry standard protocols for demonstrating compliance to verification requirements. In general, the first time a space-qualified LIB is shipped is after successful completion of manufacturer acceptance testing and product sell-off to the satellite customer. LIBs are typically shipped by ground or air transport from the battery manufacturer to the satellite manufacturer's AI&T facility for further ground processing. In some cases, satellite LIBs may be shipped directly from the battery manufacturer to the launch site satellite processing facility. Similar to LIB shipping container transportation requirements, satellite shipping containers are outfitted with sensors to monitor thermal and mechanical transportation environments. The International Air Traffic Association recommends that LIBs be shipped at a SOC not exceeding 30% of their rated capacity [5]. This guidance is provided to the aerospace industry as a recommendation to mitigate risk associated with safety hazards during air transport of LIBs.

LIB handling procedures are intended to prevent exposing the Li-ion cells to excessive mechanical shock, vibration, and/or uncontrolled temperature environments that could cause damage or result in performance degradation. Flight LIBs are typically packaged in rigid storage shipping containers to prevent damage during handling, transportation, and storage. Temperature, shock, and humidity sensors are commonly incorporated into flight LIB shipping containers to record storage, transportation, and handling environmental conditions. Special LIB handling fixtures specifically designed to facilitate shipping, ground tests, and handling during AI&T operations are commonly employed. Finally, satellite LIB ground processing also requires personnel to be trained and certified in the safe storage, handling, and transportation of flight LIBs.

9.3 Launch Site Operations

Upon completion of manufacturer AI&T activities, the fully-integrated satellite system is containerized and shipped to the launch site by ground, sea, or air transport. During satellite shipping, controlled purge environments are maintained and sensors monitor the environments of the satellite to confirm that transportation loads and environments do not exceed specified design or qualification limits. Delivery of the satellite to the launch site signifies the start of the satellite launch campaign activities. Preparing the satellite for launch includes a variety of satellite payload processing activities, such as satellite receiving and inspection, encapsulation into the LV fairing, and payload mating and integration into the LV upper stage. During this time, the mission operations launch team conducts readiness exercises and launch countdown reviews intended to rehearse launch day activities.

9.3.1 Launch Site Processing

Various types of certified satellite ground facilities are available for commercial and government launch site satellite processing prior to launch. Launch site processing services include satellite integration into the LV, propellant management, storage, handling, and transportation capabilities. Some of the major satellite processing facilities are located at the NASA-Kennedy Space Center, Russian Federal Space Agency – Baikonur Cosmodrome, European Space Agency (ESA) – Guiana Space Center, and Japan Aerospace Exploration Agency – Tanegashima Space Center. The most commonly used US-based launch site facility is the NASA-Kennedy Space Center, which serves as a multi-user spaceport enabling ground processing and launch site operations for various government and commercial customers (Figure 9.1). Other US-based launch range locations include the Eastern Range at Cape Canaveral Space Force Station and the Western Range at Vandenberg Space Force Base. As described in Chapters 3 and 7, Range Safety requirements are baselined in the LIB requirements specification. Demonstration of compliance to LIB-related Range Safety requirements occurs during the LIB qualification, proto-qualification, acceptance testing, or other specified satellite AI&T events. More specifically, LV user guides form the basis for launch provider-unique performance and processing capabilities to support satellite mission-specific requirements. User guide requirements also include satellite integration management, LV-to-satellite (payload) interface specifications, and launch facility-based ground procedural requirements [6–10].

Upon receipt at the launch site processing facility, the satellite is removed from the shipping container, receiving and inspections are performed, and SOH ground testing is conducted. A generalized ground processing flow from Representative satellite and battery launch site arrival through launch day is shown in Figure 9.2. Most satellites are delivered to the launch site with the flight solar arrays and LIBs fully integrated into the satellite bus structure. However, selected

Figure 9.1 NASA Vehicle Assembly Building (VAB) and Launch Control Center at NASA's Kennedy Space Center in Florida, USA. Credit: Ben Smegelsky/NASA.

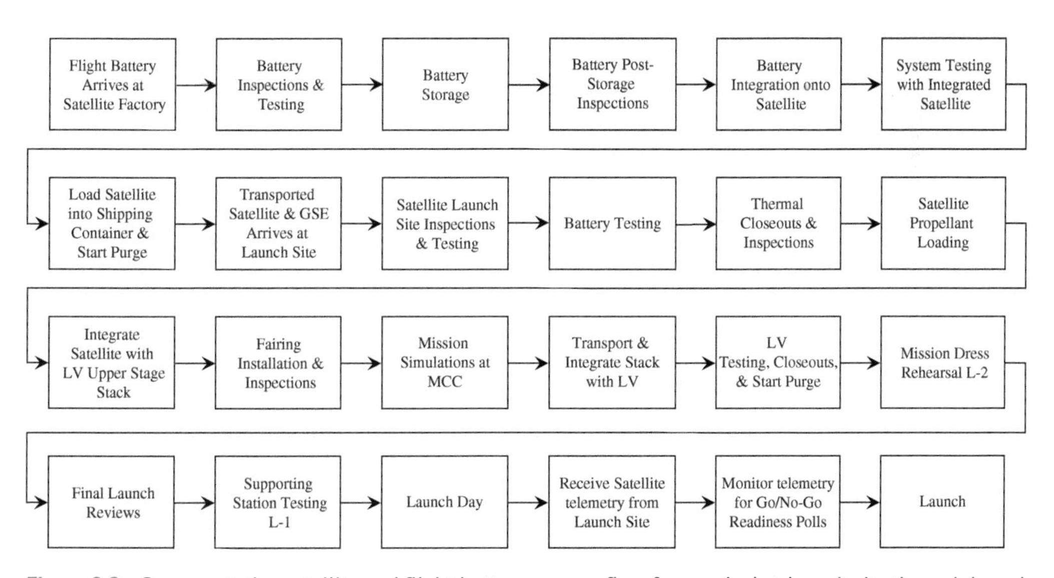

Figure 9.2 Representative satellite and flight battery process flow from arrival at launch site through launch.

programs may choose to integrate flight LIBs at the launch site's ground processing facility. Ground processing also includes provisions for safe handling, transportation, and storage of flight batteries consistent with the Range Safety-approved ground operations and final validation procedures performed at the launch site. Whether integrated into the satellite or maintained in controlled storage, flight LIBs must be handled safely to protect personnel, hardware, and facilities from safety hazards. Proper and safe handling and storage of flight LIBs also mitigates the risk of pre-launch battery performance degradation or damage to battery-level electrical, mechanical, or thermal interfaces.

Prior to encapsulation into the LV fairing, ground support equipment (GSE) is also transported to the launch site processing facility to support final SOH testing on the integrated satellite system. Launch site AI&T activities validate the integrity of system-level interfaces after exposure to transportation loads, including final inspections prior to payload encapsulation into the LV. Launch site satellite EPS SOH testing is supported with electrical GSE (EGSE), as needed, to provide conditioned ground power to the satellite electrical interfaces. These test data are used to simulate satellite loads under normal operating conditions and are compared to satellite manufacturing facility EPS functional test results to validate launch site satellite EPS SOH. EGSE is also used to perform final EPS checkouts of flight battery charge–discharge characteristics, as well as to establish the required minimum pre-launch LIB SOC. Prior to satellite encapsulation, any remaining final inspections are performed, all non-flight covers are removed, and any satellite close-out photographs are acquired for record-keeping purposes. Satellite system close-out photographic and video data facilitate pre- and post-launch anomaly investigations, if needed. Finally, LV purge conditions are established and then monitored after satellite encapsulation prior to launch.

After encapsulation, special purge GSE are used to maintain environmental purge/vent requirements prior to launch, such as maintaining thermal, humidity, and low particle count conditions. Encapsulation includes all ground operations required to integrate the satellite into the launch vehicle's payload fairing (Figure 9.3). After encapsulation into the LVs upper stage, along with the

Figure 9.3 The US–European Sentinel-6 Michael Freilich ocean-monitoring satellite being encapsulated into a SpaceX Falcon 9 rocket payload fairing (Vandenberg Air Force Base, CA, USA). Credit: Randy Beaudoin/NASA.

payload fairing, the integrated spacecraft is stacked on to the LVs booster stage(s) for subsequent roll-out to the launch pad to continue with launch operations.

Prior to the day of launch, the launch procedure and associated sequence-of-events (SOE) are rehearsed with all launch-day personnel to meet TLYF guidelines, using previously established go/no-go launch commit criteria (LCC) and scripted polls between the test site, launch site, and mission control personnel. Program-approved go/no-go LCC are used to ensure that the LV and satellite payload are ready for launch. LCC violations require real-time engineering evaluations and disposition during the pre-defined launch window. Both nominal and contingency operations are rehearsed using established voice communications network protocols and validated telemetry and command (T&C) data links. All T&C data are transmitted to/from the satellite via LV umbilical access ports.

9.3.2 Pre-Launch Operations

Prior to launch operations, an independent review and audit may be performed by key mission stakeholders to ensure that all tailored, program-specific, launch-site, mission, and on-station recommended operations procedures (ROPs) and console manuals (CM) are fully validated and verified. ROPs and CMs are verified against a flight-like satellite simulator that models as best as possible the dynamic response characteristics of satellite performance, preferably with high-fidelity models that attempt to simulate varying on-orbit environmental conditions or operational states. The previous flight history of EPSs that operate with similar designs are also used in the development and verification of ROPs. Any test anomalies or failures are investigated and resolved, satellite simulator models are updated, and mission-relevant procedures are updated and validated as necessary to meet launch readiness requirements. As part of the final validation process, the mission team may attend program-specific mission design reviews to demonstrate that their suite of mission operation products, analysis tools, and associated GSE are validated to meet program-specific mission requirements. The mission team may prepare and present evidence of their design readiness to an independent review team composed of mission-experienced engineers, along with program and mission leadership, and customers. As part of the final certification process, the mission team conducts a mission-specific Flight Readiness Review (FRR) for final launch readiness evaluation.

Just prior to launch day operations, the fully integrated satellite is validated and deemed to be ready-for-launch in a flight-like and powered configuration. This includes the use of all required launch-site GSE, launch procedures, launch operations personnel, and data/voice communications networks. Typically, two days before launch (L-2), a final validated and released version of the launch procedure is executed and rehearsed with live satellite T&C signals transmitted across the wire harness umbilical(s) between the LV and ground systems, and is then distributed to mission operations personnel for monitoring and trending purposes. The launch procedure includes all stages of ground operations leading up to launch, including the launch countdown SOE, as the LV and satellite are chronologically prepared and monitored for launch operations by executing critical commands, and then monitoring the satellite's response to those commands with associated telemetry parameters. Live telemetry data are typically compared to LCC or alarm limits, which monitor the satellite SOH against a minimal list of key performance parameters, typically referred to as the go/no-go criteria. The ultimate goal is to identify and monitor a minimal list of critical SOH parameters that sufficiently confirm the satellite SOH for launch readiness, while avoiding any unnecessary launch delays. For this reason, the launch procedure is scripted and closely monitored for any deviations. After the satellite is encapsulated within the payload fairing, the LV is delivered to the launch pad.

Vertical integration checkouts continue during pre-launch testing, along with electrical, mechanical, and thermal interface SOH testing. SOH diagnostics include LV and satellite payload power-up subsystem functional checks. Satellite EPS SOH diagnostics include monitoring and trending of the EPS bus, battery, module, and/or cell-level voltages. Other LIB SOH telemetry includes temperature, current, heater, and relay status. After completion of closeout operations, including the removal of any non-flight covers, the encapsulated satellite payload thermal environment is typically controlled with a GSE-supplied conditioned dry air or gaseous nitrogen purge system. The roll-out and on-launch pad purge is used to actively and thermally manage the EPS battery and other component temperatures to meet specified pre-launch performance requirements.

9.3.3 Launch Operations

After successful completion of the launch procedure rehearsal and launch site testing, including resolution of any identified problem reports, launch operations can proceed to the day of launch (L-0). The flight dynamics and LV teams identify candidate launch days, as well as designate specific times to open and close any narrow launch window constraints within each launch day, to ensure the LV can ultimately place the satellite within its desired orbital parameters while under nominal conditions. The start time and expected duration of each step of the launch procedure is planned into all launch operations and must factor in mandatory hold periods. During the launch procedure execution, particularly during the final countdown period just moments prior to the launch, the LV and satellite payload LCC are closely monitored by the mission team. Examples of EPS-specific LCC parameters include subsystem launch configuration, EPS relay status, power consumption, bus voltage, battery voltage, cell or module voltages, battery temperatures, and SOC. The satellite is commonly powered by EGSE via the LV umbilical connection until typically within 15 minutes prior to launch. After this countdown event, the umbilical is disconnected and the EPS batteries begin to discharge as they provide full power to the satellite subsystems. After launch and during ascent, satellite SOH cannot be monitored until after the initial acquisition-of-signal (AOS) is received following the LV separation on-orbit, where in-view ground station(s) are able to acquire and distribute satellite telemetry to the ground mission operations team. With a successful satellite AOS event, the phase of the program progresses from launch operations to on-orbit mission operations.

9.4 Mission Operations

Mission operations begin after launch upon initial AOS of on-orbit satellite telemetry. Satellite antennae are designed to communicate with the ground station network, such that telemetry data are received at the ground-based Mission Control Center (MCC). Located in the MCC, the mission operations team is trained and certified in all of the nominal and contingency operating procedures available for execution on-orbit (Figure 9.4). Ground controller commands issued from the MCC are received at the satellite for on-board processing by the flight computers. After AOS, the mission team initially executes procedures in a scripted SOE to check the post-launch satellite SOH. Next, the ground mission operations team performs in-orbit testing (IOT) of the bus and payload subsystems [11]. The SOE includes operating procedures to perform subsystem functional testing, swap antenna ground stations, deploy solar array mechanisms, perform transfer orbit operations, validate on-orbit performance, and establish the final satellite configuration prior to customer delivery on-station.

Throughout the service life of the satellite, key EPS performance data are monitored and trended to maintain satellite performance requirements. Battery performance parameters are especially monitored during eclipse and sunlight on-orbit periods. Typical EPS SOH telemetry includes bus

Figure 9.4 The NASA Johnson Space Centers MCC ISS flight control room flight controllers support the Expedition 37 mission. Credit: NASA.

voltage, bus unit current, solar array current, solar array voltage, battery voltage, battery current, cell voltage, battery temperature, and fault mode indicators. As discussed in Chapter 4, satellite EPS battery telemetry points are generally limited to battery terminal voltage, cell or module voltage, charge–discharge current, and temperature measurements. Due to these EPS telemetry limitations, key battery SOH quantities such as battery resistance, capacity, depth of discharge (DOD), and SOC must be calculated by ground operations personnel. Other EPS-level performance data are monitored, trended, and summarized during periodic SOH assessments in order to facilitate real-time operational decision making. Periodic on-orbit EPS SOH analyses include, but are not limited to, comparisons of on-orbit EPS performance versus ground power margin analysis predictions, out-of-family data trending, and other operations performance analyses. In the event of excessive battery degradation or observed cell failures over life, the satellite operators may choose to reduce satellite loads to meet LIB performance requirements during eclipse seasons and/or develop battery charge management controls to mitigate battery degradation rates.

After completion of IOT, a mission summary and parameters handbook is typically prepared and delivered to the customer as part of the satellite commissioning process. The handbook includes subsystem performance data, flight hardware descriptions, and as-executed IOT operations summaries to be used for reference and trending purposes over the life of the satellite.

9.4.1 GEO Transfer Orbit

Post-launch, geosynchronous orbit (GEO) satellites enter a transfer orbit period where the satellite apogee is raised from a low Earth orbit (LEO) to a GEO orbit. The time required to raise the satellite orbit depends on the transfer orbit trajectory and method of transfer via the satellite propulsion subsystem design. During the transfer orbit period, MCC satellite telemetry data are received and analyzed by the ground mission team to monitor and trend satellite SOH. For EPS, this includes the subsystem configuration status, bus voltage, bus unit current, solar array current, solar array voltage, battery voltages, battery current, cell voltages, battery temperatures, and power-related fault protection status. After confirming a nominal SOH, transfer orbit operations include: (i) initialize and perform bus subsystem functional tests, (ii) configure the satellite bus for on-orbit operations, and (iii) perform orbit-raising operations to complete bus IOT operations. In addition to satellite payload deployments, solar array deployments represent the most significant transfer orbit event for the EPS and may impart high mission current loads on the flight batteries. Solar array deployment enables full satellite power capability for all bus and payload subsystems. Battery charging rates are also fully enabled due to the increase in available solar array current.

Payload IOT operations are initialized and performed at the designated test orbital slot to perform payload subsystem functional tests and to configure the satellite payload for on-orbit operations, upon which any final station changes can then be performed prior to customer commissioning to signify the start of on-station operations.

To automate the telemetry data monitoring process, ground systems may implement a system of alarm limits that are established in advance for specific transfer orbit operations. The alarm limits bound minimum and maximum telemetry values before a specific alarm limit is triggered and the satellite operator is notified. Alarm limits may include a particular type of performance or status data, such as battery temperatures, battery voltages, EPS unit on/off states or power latch configuration states, solar array temperatures, bus unit currents or voltages, payload unit currents or voltages, and battery charge/discharge states or currents. Alarm thresholds may be established at specific minimum and maximum values to permit time for ground operator intervention, which ensures sufficient energy is available for satellite user load consumption. For example, satellite operators monitor battery temperature limits during charge and discharge operations in order to avoid permanent damage to the battery cells. Adjustments may be required during battery charge management and heater control operations to maintain nominal battery conditions over the planned service life of the satellite.

9.4.2 GEO On-Station Operations

Upon completion of final station change relocations to the on-station orbital slot, a formal satellite commissioning occurs between the transfer orbit mission team and the on-station operations team. This satellite commissioning process typically includes a review and performance assessment that demonstrates the successful completion of: (i) transfer orbit operations, including bus and payload IOT operations, (ii) on-orbit satellite performance requirements validation, (iii) ground system performance requirements validation, (iv) resolution and closure of any on-orbit issues, problem reports, and/or anomalies, (v) delivery of satellite database updates, and (vi) delivery of IOT operations report(s). The commissioning process is intended to demonstrate compliance to key performance requirements and budgets, such as EPS power budgets, while meeting on-station mission life predictions.

Subsequent to IOT checkout activity completion, the satellite mission transitions to a nominal on-station operations phase. For Earth-orbiting satellites and planetary spacecraft, real-time mission operations include execution of the mission plan, payload operations, and continued maintenance of satellite SOH [12]. Once the satellite begins a full operational status, the EPS batteries continue to support full satellite power demands during eclipse periods. A special focus of ground control monitoring of satellite EPS includes analyzing and trending battery and solar array SOH versus pre-launch power budget and operational life predictions. As described in Chapter 3, LEO and GEO satellite EPS battery DOD, charge–discharge cycles per year, and solstice durations are significantly different. Battery SOH monitoring is therefore analyzed in terms of expected electrical performance under on-orbit mission-specific operational conditions.

During on-station operations, the satellite operator monitors and trends the satellite subsystem performance and SOH. All subsystem SOH telemetry data are monitored, trended, and analyzed from BOL through EOM operations. These generalized on-station mission operations may need to be tailored as the satellite performance changes with real-time on-orbit operating and environmental conditions. Unplanned changes to on-orbit satellite performance may occur due to seasonal variations in solar radiation, micrometeoroid and orbital debris (MMOD) events, and/or unpredicted anomalies or failures that could degrade on-station flight hardware performance.

Depending upon the available satellite-specific EPS telemetry, various engineering data parameters are monitored and trended on a daily basis. For GEO during satellite eclipse seasons, SOH monitoring includes EPS bus voltage, battery/module/cell voltages, battery temperature, SOC, and capacity (if available). Satellite load currents are also monitored to ensure that the battery can support those loads during eclipse periods, particularly during GEO peak eclipse day(s) that can last up to 72 minutes in daily duration. For example, at the end of each Spring and Fall GEO eclipse seasons, the battery and EPS performance data are trended and compared to prior Spring and Fall eclipse seasons in order to identify any excessive performance differences that could indicate premature wear-out or off-nominal aging characteristics. As discussed in Chapters 2 and 3, on-orbit LIB resistance, capacity and SOC can be estimated from battery voltage telemetry data.

9.4.3 On-Orbit Maintenance Operations

Throughout the duration of a mission, on-orbit satellite SOH maintenance requires ground operators to perform routine corrections and adjustments to EPS operations. Assessing EPS performance deviations between pre-launch ground analysis predictions versus real-time on-orbit performance are commonplace for on-orbit satellite systems. EPS SOH deviations may result from routine changes to on-orbit operational modes, impacts from relevant on-orbit environments, and unexpected hardware anomalies. For example, modified EPS load management operations may be required due to an on-orbit EPS component anomaly that results in a loss of hardware redundancy. A failed battery cell (or battery module) or solar array string will require ground operators to assess impacts to EPS SOH to maintain satellite operational objectives. Alternatively, variations in thermal environments may necessitate periodic changes to battery or system-level heater set-points during GEO solstice or eclipse periods.

Evaluating on-orbit LIB performance during the satellite mission requires real-time measurements of battery voltage, current, and temperature telemetry data. As discussed in Chapter 5, the NASA ISS 30s1p LIB power system real-time telemetry data includes battery voltage, individual cell voltages and temperatures, battery baseplate temperature, and battery charge–discharge current (Figure 9.5). The ISS LIB charge current rate is based on individual cell voltage telemetry. Every LEO orbital period, each battery cell is charged to a specified end-of-charge voltage (EOCV) limit (nominally 3.95 V per cell). Charge current is stepped down as battery cells reach the specified EOCV limit. Once each battery cell reaches a full SOC, individual cells are bypassed until all cells reach a full SOC that maintains battery SOH [13]. On-orbit LIB data for a 155-minute highly elliptical orbit (HEO) eclipse period for the Van Allen Probes satellite 8s1p LIB is shown in Figure 9.6. Subsequent to deep LIB DOD eclipse periods, a two-step, constant-current taper charge protocol was used to optimize battery SOC prior to the next eclipse period.

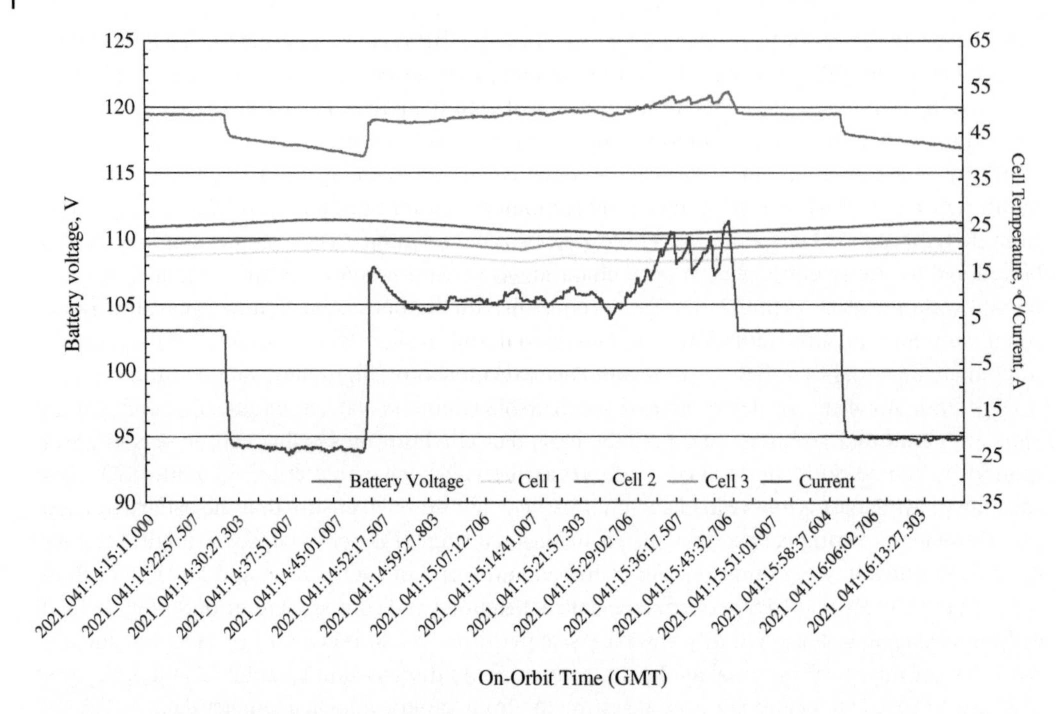

Figure 9.5 Representative NASA ISS 30s1p LIB 3A1 power channel on-orbit voltage, charge–discharge current, and average cell temperature telemetry data for a single LEO orbital period. Credit: Courtesy of NASA.

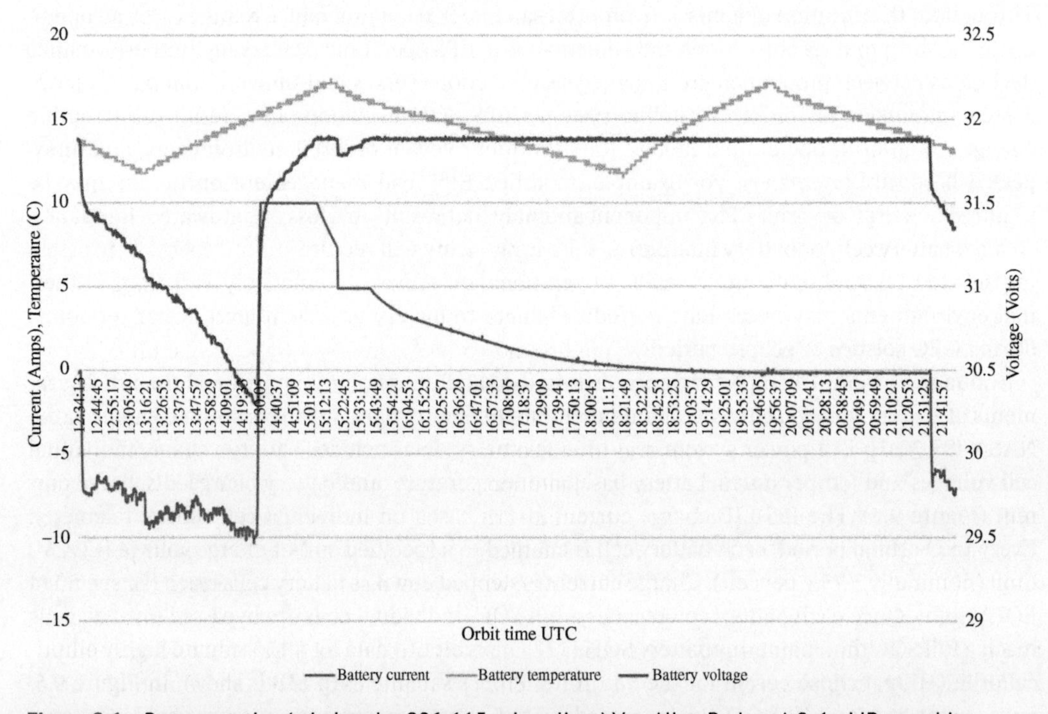

Figure 9.6 Representative (mission day 206, 115-min eclipse) Van Allen Probes-A 8s1p LIB on-orbit voltage, current, and temperature telemetry data. Credit: Courtesy of The Johns Hopkins University APL.

Ideally, on-orbit battery maintenance includes comparing flight battery performance to ground predictions, either by analysis or by test, of a ground battery cycled in flight-representative conditions. These SOH assessments enable adjustments of battery utilization, as necessary, to meet mission requirements. For example, industry flight experience has shown that on-orbit satellite power duty cycle characteristics may differ from ground test power duty cycle values used to qualify the flight LIB [14]. As a result, ground operators commonly quantify and trend on-orbit battery deviations from the ground battery qualification life cycle test conditions. This analysis includes charge–discharge rates, DOD, EOCV, end-of-discharge voltage, and temperature during all satellite sunlight and eclipse periods. On-orbit flight experience has also shown that satellite ground operators prioritize battery capacity and resistance data to assess SOH and predict remaining service lifetime.

Although most satellite EPS telemetry budgets do not include LIB capacity and resistance measurement data, ground operators routinely estimate on-orbit battery capacity and resistance trends using available on-orbit battery voltage, current, and temperature telemetry data. Other benefits of trending on-orbit battery SOH data include the early detection of battery underperformance characteristics that may lead to on-orbit anomalies. For example, if the on-orbit battery SOH data trends predict anomalous performance in advance of the next eclipse period, proactive measures to develop tailored battery management operational procedures can be implemented to meet satellite-critical EPS user load requirements. EPS load management operations may include user load shedding options to manage LIB SOH.

9.4.4 Contingency Operations

Throughout their entire mission lifetime, all on-orbit satellites are vulnerable to space environmental hazards and anomalies resulting from hardware or software faults. Errors in ground operator commanding may also contribute to the likelihood of an on-orbit satellite anomaly or failure. While certain anomalies may have benign impacts to satellite operation, other anomalies may result in significant hardware or software failures that can cause loss of a critical satellite function or entire mission. As discussed in Chapter 4, autonomous fault protection is incorporated into satellite EPS architectures to reduce the risk of losing satellite function and to maximize the likelihood of survival.

9.4.4.1 Safe Mode

Satellite safe (or survival) mode is an on-orbit contingency condition resulting from an on-board processor fault detection response. Safe mode events may be caused by an anomaly in a critical spacecraft subsystem, operator error during ground control commanding, or some other safety threat. These system anomalies may cause loss of satellite control, attitude, or orientation resulting in a reduction of solar illumination time on the solar array. Usually during safe mode contingencies, the satellite autonomously re-orients to a fixed sun-pointing attitude to maximize on-sun time to the solar arrays. Under satellite safe mode conditions, battery energy must be conserved to maintain a positive energy balance in order to maintain critical satellite unit functionality.

To preserve battery SOC and EPS SOH, the satellite may autonomously begin sequentially powering off non-essential user loads (load shedding), which decreases battery energy demand, thus providing additional time for ground control personnel to detect and correct the anomalous condition. Load-shedding operations may be controlled autonomously by the satellite flight computers or by manual commands from ground controllers. The sequence of powering off non-essential user loads is based on real-time mission analysis of satellite orientation, battery SOC, and sun-search capability necessary for survival. Non-essential satellite user loads may include selected payload units, heaters, and bus units deemed non-critical for safe on-station operations. Although safe mode may cause the satellite to be operated as a single-string system, sufficient power must be available to mission-critical systems that enable ground controllers to communicate with onboard processors, analyze real-time telemetry data, maintain thermal stability, assess satellite SOH, and perform recovery operations.

Battery capability to supply power to the spacecraft in safe mode is dependent on remaining battery capacity and the duration of the safe mode contingency period. As discussed in Chapter 4, a worst-case energy balance analysis estimates the rate of battery SOC decrease during a spacecraft safe mode contingency period. Spacecraft batteries are typically sized and analyzed to survive a worst-case safe mode condition, which is considered to occur at the end of eclipse when the battery is at its lowest operational SOC. To survive the worst-case safe mode scenario, battery reserve energy margins may decrease sufficiently, resulting in a negative energy balance spacecraft condition. An unsuccessful safe mode response results if mission ground control operators cannot restore nominal spacecraft operations and reacquire sufficient solar array illumination in time to return the EPS to a power-positive energy state. In the event of a failed safe mode response, the battery may completely discharge in a manner that depletes the battery cells of all usable energy.

9.4.4.2 Dead Bus Survival

As described in Chapter 4, a dead bus is an anomalous condition where the primary bus voltage of a spacecraft collapses to any voltage below the minimum operational bus voltage level (including zero volts) for any time interval exceeding normal or survival bus undervoltage limits. The most common causes for a dead bus condition include safe mode contingencies, component hardware single-point failures, flight software anomalies, or ground operator errors [15]. If a safe mode contingency occurs at the beginning of or during an eclipse period, the time to discharge the battery to a low-voltage condition will be shortened, leaving less time for ground controller intervention. In all cases, the discharged battery will cause a low-voltage or zero-volt bus condition. For many satellite systems, safe mode contingencies will initiate autonomous user load shedding, which may result in a reduction of satellite component temperatures. Degraded satellite thermal control and loss of heater power reduces the ability to thermally manage payload units, critical bus unit electronics, propulsion subsystem propellant lines, and batteries. Loss of thermal control may cause the battery temperatures to decrease below the lower operating or qualification temperature control limits. Under worst-case dead bus conditions, the satellite LIB may be completely discharged to a zero-volt condition and exposed to extremely low environmental temperatures.

A dead bus event is an extremely serious condition, requiring that every possible effort be available to recover the satellite EPS as soon as possible. In addition to any long-term loss of commercial revenue or scientific data, a drifting satellite left on-orbit is a potential collision risk to other spacecraft because it lacks the means to be disposed of in a controlled manner according to the mission plan. In addition, while an unresolved dead bus event, by definition, may no longer permit safe passivation of the EPS, it also prevents passivation of propulsion subsystem pressure vessels, presenting the long-term risk of the satellite breaking up and creating many more pieces of orbital debris. More importantly, a dead bus condition will prevent the satellite from being decommissioned.

9.4.4.3 Dead Bus Recovery

Operationally, recovering from a dead bus event requires that the satellite EPS has the hardware and software capability to autonomously return to normal mission operations when sufficient sunlight is reacquired by the solar arrays. As discussed in Chapter 4, adherence to dead bus recovery requirements results in an EPS architecture designed for autonomous bus voltage and safe battery recovery from dead bus conditions. This design capability includes recovery from EPS operating conditions where the battery voltage and temperature are below specified minimum operating limits. If the cause of a dead bus event is a hardware anomaly, satellite recovery to a nominal level of on-orbit operation may require hardware reconfiguration to mitigate the risk to repeating the anomaly. Restoring bus power is dependent on EPS architecture characteristics, fault protection hardware topology, and fault management software design.

Flight LIB recovery is also dependent on EPS design features such as the capability to autonomously recharge in an offline mode. As discussed in Chapter 7, LIBs are generally intolerant to overdischarge conditions, which are known to cause a permanent loss in capacity and degraded cycle life performance. In addition, recharging LIBs under cold temperature conditions require additional protocols to avoid irreversible lithium plating of battery cell electrodes [16]. To avoid the negative effects of high-rate charging of low-voltage LIBs at low temperatures, battery recharge rates are maintained below nominal recharge rates until the LIB voltage exceeds the minimum bus voltage limits. Systematic restoration of thermal control is critical to avoid thermal overstress of the battery and other satellite hardware. Restoring satellite thermal control includes reactivating primary heater power to all bus subsystem critical components, including the battery's active thermal control system and battery charging capability. Under all circumstances, timely action by ground operators to recover and return to full mission operations is critical because extended time in a dead bus condition increases the likelihood for permanent battery and other satellite hardware damage due to extended low-temperature exposure below specified voltage limits.

9.4.5 End-of-Life Operations

Since the retirement of Ni-H_2 space battery technology in the mid-2000s, an increasing number of government and commercial satellite systems have been launched with LIB-based EPS. To date, the on-orbit satellite industry experience has demonstrated that space-qualified LIBs are meeting on-orbit service life requirements. As a result, many satellite manufacturers and operators have implemented mission life extension strategies aimed at extending on-orbit satellite service life. Satellite mission extension strategies include analyzing opportunities to adjust how the on-orbit LIB is operated. Extending satellite end-of-life (EOL) operations may require ground operators to perform mission corrections, station changes, or modified operational adjustments via ground commands, flight software updates, or other operational modifications. For example, as the satellite approaches the end of its service life, solar array and LIB performance will have gracefully degraded during the mission as predicted by EOL analyses. For example, on-orbit satellite experience has demonstrated that solar array performance gracefully degrades due to accumulated in-service exposure to mission-specific orbital environments.

However, LIB in-service performance degradation is largely due to the combined effects of calendar life and charge–discharge cycling on battery capacity loss. Long-term LIB operation at elevated environmental temperatures will also decrease usable LIB capacity. Near EOL, a GEO satellite operator may trend battery performance data over prior eclipse seasons to compare season-to-season differences and monitor for performance degradation trends or any anomalous signatures. Ground operator intervention may be required due to battery aging effects near EOL to ensure that battery capacity is maintained to meet mission requirements. Battery DOD is managed via user load management, EOCV, or other EPS-level adjustments. Operational adjustments may be required during GEO eclipse seasons near EOL to ensure that the battery can support satellite user loads with adequate margin. Load management strategies may include satellite bus and payload user load-shedding to manage battery voltage and DOD. Other operational EPS-level strategies may include disabling certain fault protection responses, widening fault protection thresholds, and actively managing satellite user loads to maintain greater battery and bus voltage margins.

Due to degraded battery performance at EOL, GEO commercial satellites may no longer be able to support nominal payload operations during eclipse operations. Under these circumstances, the satellite operator may reduce payload capability by diverting payload traffic to other satellite(s) in their on-orbit fleet and/or turn-off some or all of their revenue-generating payload units. Station-keeping operations are still required to maintain the satellite within its intended orbital slot, including inclined-orbit operations to extend the mission life of on-station operations. Even without carrying any payload traffic, operators may choose to maintain that satellites orbital slot and associated frequency spectrum for future satellites, or potentially preserve that satellite as a spare on-orbit asset within their fleet, albeit with limited revenue-generating capability. However, all satellites must eventually be decommissioned when they can no longer be economically or safely operated in their intended orbital slot. Examples include satellites without sufficient unit redundancy or on-orbit performance, such that they are potentially only one failure away from being uncontrollable, which would seriously jeopardize the integrity or safety of the nearby on-orbit satellite community. It is also possible for a satellite to reach the EOM with all other hardware still viable, but forced to end the mission due to insufficient propellant to both extend the mission and perform a responsible disposal while maintaining adequate safety or propellant uncertainty margins.

9.5 End-of-Mission Operations

EOM is generally defined as when a satellite or LV: (i) completes the tasks, functions, and/or useful life for which it has been designed, other than its disposal, (ii) becomes non-functional as a consequence of failure, or (iii) is permanently halted through a voluntary decision. A primary objective of EOM disposal is the process of safely moving a satellite to an orbit or trajectory considered acceptable for limiting the generation of new orbital debris. Advance planning for safe Earth-orbiting satellite and LV orbital stage disposal is a key element for successful EOM operation activities. Passivation is the process of minimizing any stored energy on-board the spacecraft at the EOM. Since the consequences of not passivating the spacecraft can lead to a spacecraft break-up, which is a leading contribution to the orbital debris environment, decommissioning plans, including passivation, are generally linked closely to orbital debris limitation policies. For example, NASA and the Inter-Agency Space Debris Coordination Committee (IADC) guidelines specify that all on-board satellite or orbital stage sources of energy must be depleted or safed when they are no longer required for mission operations or post-mission disposal [17, 18].

As part of the broadcast licensing process of the desired frequency spectrum, the satellite design and on-orbit operations must adhere to international standards, regulations, and telecommunications law intended to maintain the integrity, reduce the risk, and preserve the safety of the on-orbit satellite community. For example, international regulating agencies, such as the Federal Communications Commission (FCC) and the International Telecommunication Union (ITU), mandate to satellite operators that satellites be safely and sufficiently decommissioned in a controlled fashion to avoid the potential of creating on-orbit debris.

The most common on-board satellite components containing stored energy at the EOM are found in the propulsion (residual propellants, pressure vessels, and unfired pyrotechnic devices), electrical power (rechargeable and non-rechargeable battery energy storage devices), and attitude control (reaction wheel) subsystems. Hence, satellite passivation requires incorporating design features and operational capabilities into the satellite design to enable safe and reliable disposal at the EOM.

Planning for the EOM requires a different approach than designing for successful spacecraft operations. Once the mission is over, the task shifts to making the spacecraft safe from break-up in order to preserve the space environment for current users and future missions. This alternative perspective is necessary during the EPS design phase, as well as during ground-based mission operations as part of the decommissioning process. Since spacecraft telemetry and SOH status will no longer be available after decommissioning, intervention by ground operators is no longer an option to prevent unexpected problems. The final state of the EPS, especially batteries, needs to be as stable as possible and for as long as possible.

9.5.1 Satellite Disposal Operations

There are several acceptable options for post-mission disposal, the most effective of which is generally driven by the operational orbit region. Satellites operating in the GEO region are usually raised into a higher circular retirement orbit (re-orbited), typically at least 300 km above the operational GEO arc (nominally 35 786 km), so that they will not interfere with operational missions for at least 100 years. There is a corresponding storage orbit for lower altitude missions beginning above 2000 km, though it is generally more efficient and widely preferred to lower the orbit (de-orbited) of most LEO missions at the EOM, eventually resulting in re-entry. Disposal by re-entry can involve either controlled maneuvers to target an unpopulated location or simply lowering the orbit to a degree that uncontrolled re-entry occurs within 25 years after the EOM. The expected risk to the population from surviving debris is usually the deciding factor when designing for controlled versus uncontrolled re-entry. Satellites disposed of using targeted re-entry do not require passivation, since most satellite subsystems are needed to execute the disposal process.

Satellite EOM disposal options are commonly established early in the design process to ensure that the necessary hardware and software features are incorporated into the EPS architecture. For example, solar array or battery disconnect relay architectures will depend on which passivation techniques are chosen to meet EOM disposal requirements. It can be advantageous, however, to design the EPS with multiple passivation options, in order to enable alternative or redundant disposal approaches that can be considered later in the mission. For example, the Van Allen Probes mission was designed for a passivation method that was only effective if re-entry took place within six months after the EOM. When the remaining propellant was used to extend the mission, passivation could no longer be accomplished because the orbit could no longer be reduced for prompt re-entry. A relatively easy design change early in the mission design could have preserved the option for both the extended mission and full passivation. Early planning for disposal and passivation must be consistent with internationally accepted requirements, guidelines, and best practices in order to preserve the orbital environment [19].

9.5.1.1 LEO Disposal Operations

The high on-orbit spatial density of satellite and other objects (orbital debris) in near-Earth altitudes underscores the importance of formulating optimal disposal strategies and options. Satellite mission designers have determined that the most effective means to dispose of a LEO satellite is via a controlled (or targeted) re-entry method. A targeted re-entry involves maneuvering the satellite for re-entry over a specific area, which also reduces the risk of human casualties. The most significant benefit of a targeted re-entry is the complete removal of the satellite from orbit soon after the

EOM. This minimizes the need for EOM passivation design features and best preserves the orbital environment. If targeted re-entry is employed, EPS functionality is needed during controlled re-entry for commanding, attitude control, and telemetry transmission, so that passivation operations are not implemented. If the disposal approach includes leaving the satellite on-orbit, whether as part of storage operations in a disposal orbit or as a consequence of an anomaly preventing de-orbit operations, passivation is performed whenever possible, and may be required by applicable national or international standards.

9.5.1.2 GEO Disposal Operations

The high altitudes of GEO satellites make targeted re-entry disposal operationally impractical and cost ineffective. Instead, designers and operators implement GEO EOM planning, which includes maneuvering the satellite into a higher-altitude orbit (super-GEO) where the satellite does not interfere with other satellites or orbital stages. ISO 26872 specifies requirements for the EOM disposal of a satellite operating at GEO altitudes to ensure final disposal configuration characteristics and verify that sufficient propellant is reserved for any planned retirement orbit maneuvers [20]. IADC guidelines further identify a GEO-protected region (at least GEO $\pm$200 km) which an EOM satellite should also avoid to further protect the operational GEO environment from orbital debris for extended periods (typically up to 100 years) of time. EOM disposal of satellite and orbital stages between LEO and GEO should be carefully designed to minimize interference with medium Earth orbit (MEO)-based global navigation satellite systems, such as the US Global Positioning System (GPS), European Union Galileo, and Russian Global Navigation Satellite System (GLONASS) networks.

9.5.2 Passivation Requirements

At the EOM, all Earth-orbiting satellites left in storage orbits must be passivated. Satellite passivation is the act of permanently depleting, irreversibly deactivating, or making safe all on-board sources of stored energy capable of causing an accidental break-up or explosion that could generate orbital debris. Other benefits of passivating the EPS include elimination of any potential radio frequency (RF) interference, stored mechanical energy in reaction wheels, the potential for heaters to over-heat and rupture pressure-sensitive components like pressure vessels, and ensuring that the satellite cannot become re-activated later by any unexpected mechanism. NASA maintains a detailed history of Earth-orbiting satellite fragmentations documenting satellite break-ups since the Transit 4A Able-Star vehicle break-up in June 1961. NASA has documented 241 satellite break-ups, which includes an assessment of the primary causes (where able to be determined) between 1961 and 2018 [21]. These data indicate that over 66% of break-ups are due to a combination of satellite propulsion subsystem failures and deliberate actions related to national security activities (Figure 9.7).

While only 10 on-orbit satellite break-ups are known during that period to have been caused or initiated by batteries, these incidents generated over 1000 pieces of debris large enough to be able to be tracked by radar (Table 9.1). Based on the satellite launch dates, it is likely that all 10 satellite batteries were either Ni-Cd or Ni-H_2 type battery power systems. Within that time frame, the last reported satellite break-up caused by a battery failure was the Defense Meteorological Satellite Program (DMSP) Block 5D-2 incident in February 2015, involving a satellite that had launched in 1995 [22]. Analysis indicated that the DMSP event was likely due to a short circuit in the battery charger, which caused an explosion of one of the Ni-Cd power system batteries. This damage propagated to a larger break-up, resulting in over 200 tracked pieces of debris being generated. The ESA analysis of orbital fragmentation history also indicated that most satellite EPS break-up events

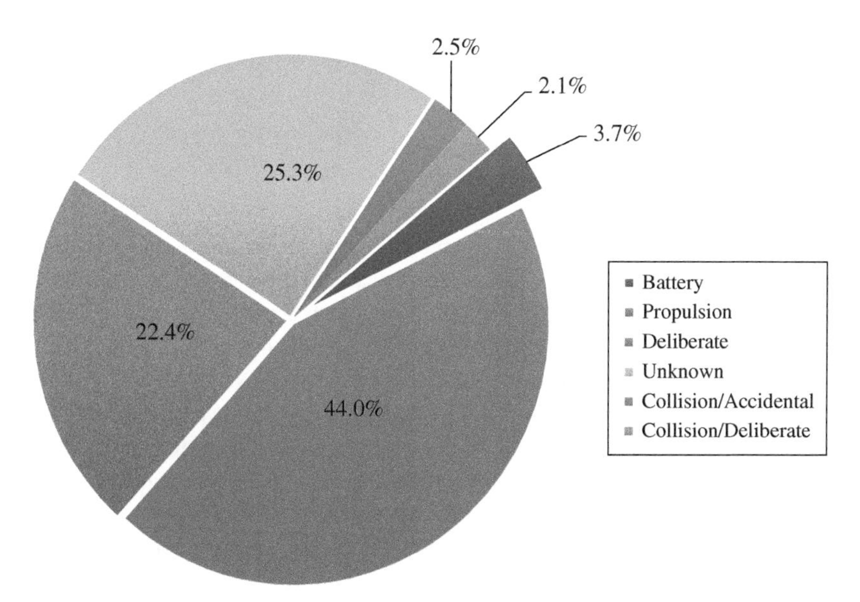

Figure 9.7 Causes of known satellite break-ups in Earth orbit between 1961 and 2018.

Table 9.1 History of on-orbit satellite breakups caused by batteries.

Satellite name	Launch date	Breakup date	Debris generated
COSMOS 839	8 July 1976	29 September 1977	70
EKRAN 2	20 September 1977	23 June 1978	5
COSMOS 880	9 December 1976	27 November 1978	49
COSMOS 1275	4 June 1981	24 July 1981	479
COSMOS 1375	6 June 1982	21 October 1985	61
COSMOS 1691	9 October 1985	22 November 1985	21
NOAA 8	28 March 1983	30 December 1985	5
COSMOS 1823	20 February 1987	17 December 1987	150
DMSP 5D-2F13	24 March 1995	3 February 2015	236
NOAA-16	21 September 2000	25 November 2015	458

occurred due to overcharging and subsequent explosion of batteries [23]. Although only a small fraction of recorded satellite break-ups are battery-related, an increased understanding of safely passivating high specific energy LIB power systems is of great importance to the continued preservation of the near-Earth space environment [24]. While LIB behavior has been studied over the course of up to a decade, very long-term storage (such as a century or longer in super-GEO orbit) might present failure mechanisms that have not yet been considered.

9.5.2.1 United States Passivation Guidance

General guidance on passivation of satellites owned by or regulated by the United States is found in the US Government Orbital Debris Mitigation Standard Practices (ODMSP) [25]. This document includes qualitative and quantitative goals for limiting orbital debris generation from intentional release, accidental break-ups, collisions, and long-term accumulation. It is employed both by agencies that directly launch satellites and by those that license and regulate LV and satellite operations. The description most applicable to EPS passivation is found in Section 2.2: "All on-board sources of stored energy of a satellite or upper stage should be depleted or safed when they are no longer required for mission operations or post-mission disposal." The phrase "or safed" refers to a process sometimes called "soft passivation."

While NASA requirements are not directly applicable to most US space missions, they are a good example of how the ODMSP is applied. NASA requirements to control the generation of orbital debris and mitigate its growth are intended to preserve the near-Earth space environment from risks to human life and satellite missions [26]. The procedural requirements in NPR 8715.6 are applied to the mission management, and technical requirements are found in NASA-STD-8719.14. With the exception of planetary missions other than to Mars, NASA programs are required to conduct technical assessments of, and to limit, the potential to generate orbital debris during deployment, mission operations, and at the EOM. NASA passivation requirements are consistent with guidelines published by the IADC, UN Committee on the Peaceful Uses of Outer Space, and in compliance with the ODMSP [27]. IADC guidelines commonly serve as a basis for the development of technical standards on space debris mitigation measures.

9.5.2.2 International Passivation Guidance

International standards used by various non-US space agencies provide additional references for how other nations approach the act of satellite passivation. To meet the ESA standard for the technical requirements on space debris mitigation for agency projects, ESA adopted the ISO 24113 standard for space debris mitigation. Regarding the EOM disposal of satellite and LV orbital stages, ISO 24113 states that all remaining on-board sources of stored energy shall be permanently depleted or safed in a controlled sequence. This requirement is well aligned with the NASA, IADC, and UN guidelines for satellite passivation.

9.5.3 Satellite EPS Passivation Operations

The high specific energy of satellite EPS LIBs creates a potentially hazardous condition for a passivated satellite at the EOM. To eliminate sources of stored energy, NASA passivation requirements specify that the preferred form of EPS passivation is disconnection of the solar array output from the satellite bus at the EOM. Electrical isolation of the solar array from the EPS bus prevents current from inadvertently recharging or overcharging the satellite LIBs after the EOM, as well as ensuring complete passivation of all electrically-powered components on the satellite. For example, a decommissioned satellite with the solar arrays left connected, even while performing no active attitude control, is intermittently exposed to sunlight as the satellite tumbles, allowing intermittent battery recharging. Degraded or failed satellite power circuits may also cause inadvertent LIB recharging if EPS charging circuits are not disconnected. If disconnecting the solar array is not possible, NASA passivation requirements allow for battery disconnection from the EPS bus charging circuits and other approaches intended to prevent catastrophic LIB cell safety hazards. IADC Space Debris Mitigation guidelines also recommend that batteries should be designed and manufactured to prevent break-ups and that battery charging lines should be deactivated during EOM operations.

There are two basic approaches employed by satellite designers to demonstrate compliance to passivation requirements. The first method is to implement an EOM "hard passivation" plan by depleting all onboard sources of stored energy and disconnecting all energy generation sources. This is the preferred approach, but if satellite hard passivation cannot be implemented, then satellite passivation may be achieved by mitigating onboard sources of energy to a sufficiently low level of risk that cannot result in an explosion or deflagration large enough to release orbital debris or break-up the satellite. This second method of satellite passivation is often referred to as soft passivation.

Satellite passivation, especially the EPS, typically requires permanently disabling the automated onboard fault detection and correction mechanisms designed into the satellite to provide reliable on-orbit operations. By definition, passivation is the very opposite of reliable on-orbit operations, such that the software and hardware safety net designed to prevent a premature EOM must be defeated, while retaining those design features that prevent overcharging the battery. Generally, specific software commands are executed for disabling these capabilities, but those commands must be designed so that, once sent, the fault protection hardware and software will not reverse them. Particularly in the case of soft passivation, the EPS may be designed to autonomously restart when the voltage drops below a certain minimum threshold. As such, the battery charge may increase as the satellite tumbles, resulting from the inadvertent re-illumination of the solar arrays. If this EPS restart event should inadvertently re-enable the fault correction capabilities, large portions of the satellite may be able to return to safe-mode operations, including full battery charging and the subsequent defeat of the passivation process.

9.5.3.1 Hard Passivation Operations

The preferred option is to deplete all on-board EPS sources of energy via physical disconnection of all energy generation and storage sources. The physical disconnection of the solar array and/or battery components of the satellite EPS is referred to as hard passivation. Hard passivation of solar array current sources can be accomplished by either physically shorting (shunting) the array output or disconnecting the solar array output from the satellite EPS bus. Permanently disabling the satellite solar arrays results in the remaining bus and payload user loads discharging and eventually depleting the connected battery energy storage source. Safe battery charge depletion may also include powering on active heater loads for the remaining operational life of the satellite. Hard passivation of the solar array is a preferred EOM satellite configuration, since reducing battery recharge rates or increasing battery discharge rates may not be sufficient to completely deplete stored battery energy during long periods of on-orbit time [28].

The second option to achieve satellite EPS hard passivation is to physically disconnect the batteries from the EPS bus. Battery discharge as a result of disconnection from the EPS bus charging circuit, such as via relays and/or activating cell bypass switches, will fully passivate the battery [29]. For example, the small satellite European Student Earth Orbiter (ESEO) 6p6s LIB contained a solid-state isolation switch designed to isolate the EPS bus from the satellite user loads for safe ground operations, resulting in EPS passivation [30]. The EOM planning for the ESEO LEO satellite included depletion of the LIB, followed by commanding the isolation switch to open to hard-passivate the ESEO EPS. To comply with NASA passivation requirements, the NASA Solar Dynamics Observatory (SDO) GEO satellite also employed a battery disconnect relay control function to implement hard passivation of the EPS LIB [31].

One major challenge to implementing hard passivation is that it can represent a single point of failure for the mission, thus reducing satellite reliability, which is usually avoided as part of best engineering design practices. It would generally be unwise to incorporate any possibility for a single ground command that could result in accidentally and prematurely ending the mission. With adequate design considerations, however, hard passivation can be employed, while observing adequate mission assurance practices. One approach, employed by the SDO mission, was to provide the relays for disconnecting the LIB from the satellite power bus. However, during on-orbit operations, the commands to actuate those relays were part of the forbidden command database. At the eventual EOM, a new permitted command database will be uploaded to the satellite that will allow these relays to be armed and actuated, along with other disposal and passivation operations.

In general, satellite EPS designers are averse to including relay switches that may inadvertently disconnect the solar arrays and/or batteries, creating a single-point failure that could lead to premature loss of the mission. To mitigate this risk, certain EPS designs may include robust fault-tolerant architectures that protect the satellite from unintentional activation of EOM passivation functions. This may include implementation of parallel relay drivers with independent commanding features to avoid the risk of single ground commands from inadvertently disconnecting the batteries. Alternatively, solar array or battery disconnect commands may be forbidden in the satellite flight software until needed and uploaded by ground operators at EOM.

Another approach to safely incorporating hard passivation is to require a complex set of maneuvers and commands to be executed in a specific SOE in order to achieve passivation. Most satellites have the ability to disconnect individual solar array strings from the power generation input, in case of damage or degradation during the mission that could adversely affect the EPS. Passivation plans for Aqua and Aura satellites involved a complex SOE, where the satellite must be maneuvered to rotate the solar arrays away from the Sun, thus disconnecting most of the solar array strings. Power from the remaining hard-wired strings must then be shunted by an unusual, non-standard configuration of the EPS, which would never be used during the mission. The Ni-H_2 batteries would then be depleted by maintaining the satellite in this specific configuration with the aid of the reaction wheels. By first disabling the fault protection mechanisms, the shunting configuration and disconnection from the solar array would then become permanently invoked after the batteries are depleted. If only a portion of this SOE were to be initiated prematurely, there would be many opportunities for ground intervention in order to recover the satellite before premature passivation could occur.

9.5.3.2 Soft Passivation Operations

To maintain reliability and mission assurance requirements, satellite EPS designers may choose to not permit the capability to disconnect solar arrays from the batteries or batteries from the satellite bus. The IADC guidelines and ODMSP recognize this issue, and allow for the possibility to safe the EPS, sometimes referred to as soft passivation. Techniques are commonly employed to limit onboard energy sources to acceptable levels so as to meet the intent of industry passivation requirements. Best engineering design practices may be used to control onboard sources of stored energy to a level sufficiently low to reduce the risk of an explosion large enough to release orbital debris or break-up the satellite. For example, batteries may be discharged partially, in conjunction with implementing power safing controls intended to avoid conditions that could lead to a catastrophic failure.

Soft passivation techniques for LIBs typically include discharging the batteries to a safe SOC and voltage level. As discussed in Chapter 7, reducing LIB SOC decreases the rate and magnitude of energy release that could result from catastrophic cell failure events [32]. As discussed in Chapter 3, on-orbit satellite LIBs experience capacity loss due to time on-orbit and charge–discharge cycling operating conditions. Hence, the amount of usable stored energy in the satellite LIBs is expected to be lower at EOM compared to BOL launch conditions. Over extended periods of time, battery self-discharge will also contribute to lowering the SOC. Decreasing the available EOM energy of the satellite LIB also has the advantage of reducing the risk of a catastrophic battery failure hazards (such as thermal runaway) resulting from MMOD impacts. A comparison between hard and soft passivation approaches for a satellite EPS is given in Table 9.2.

Since soft passivation leaves the EPS in a unique non-operational configuration, a careful and thorough engineering assessment must be performed to consider all possible failure scenarios, along with the resultant battery cell responses associated with those failures. For example, when

Table 9.2 EPS hard and soft passivation approaches and states.

EPS component	Hard passivation	Soft passivation	No passivation
Battery	Physically disconnected No further charging	Reduced SOC Maintained	Nominally operational
Battery charge controller	Disabled	Reduced SOC Commanded	Nominally operational
Solar array	Physically disconnected	Partially disconnected Partially shunted Not tracking the sun	Nominally operational

the attitude control subsystem is disabled as part of passivation, the satellite will likely tumble for at least some of the time, during which the battery radiator will occasionally face the Sun, potentially resulting in abnormally high temperatures. Partial discharges may also eventually lead to the development of short circuits within the battery cells. Due to anomalous resistive cell internal short circuits, continued uncontrolled LIB recharging may induce a catastrophic thermal runaway condition which damages the satellite structure and releases debris objects.

Soft passivation conditions need to be self-sustaining and safe over a long period of time, with no ground monitoring or intervention available if any unexpected conditions should occur. Passivated satellites can be left in orbit for up to 25 years in LEO, or for centuries after GEO mission disposal. Off-nominal conditions are very likely to occur during such long time periods, making fail-safe designs challenging. Unplanned availability of power to the EPS after passivation may result in the possibility of EPS reconfiguration and partial EPS reactivation, such as due to single event effects (upsets), small particle orbital debris penetrations, or simple wear-out of parts forced to function for significantly longer than their mission design life. In order to perform passivation at EOM, it is typically necessary to defeat the automated fault detection and correction mechanisms that are normally relied upon during normal on-station operations. Therefore, designing a truly safe soft passivation approach includes more than simply commanding to a lower battery charge current.

9.5.3.3 LV Orbital Stage EPS Passivation Operations

LV batteries are inherently passivated by self-discharge and attached loads, and thus traditionally do not need intentional passivation. Whether non-rechargeable primary cells or rechargeable secondary cells, LV upper stages do not have the means to recharge the battery on-orbit, so they remain in a discharge state throughout the mission. Passivation of other sources of stored energy on LVs (propellants or pressurants) might still be performed if the stage is to be left in orbit after its mission. More recently, certain LV upper stages perform controlled re-entry maneuvers, thus precluding the need for passivation.

9.6 Summary

Ground, launch site, and mission operations of integrated LIB-based EPS systems have leveraged past on-orbit operational experience and lessons learned from heritage rechargeable battery EPS satellite systems. Systems engineering best practices include verification of flight LIB interfaces during bus AI&T, adherence to manufacturer-recommended LIB transportation, handling, and storage requirements, performing pre-launch electrical verification checks and hardware inspections, and following TLYF principles. Ground processing of flight LIBs in a TLYF manner is necessary to demonstrate flight-like ground testing of expected EPS orbital operational conditions and to verify flight commands, sequencing, and telemetry databases. During on-orbit satellite operations, regular trending of LIB voltage, temperature, and user load performance data permits ground operators to assess EPS SOH status and identify performance changes over mission life. After completion of on-station operations, passivation of the EPS and safe disposal of the satellite are essential to controlling orbital debris and meeting international guidelines during spacecraft decommissioning. Although LIB-based EPS have not yet experienced on-orbit break-ups that directly create orbital debris, their higher specific energy may represent a considerable risk of that mechanism potentially occurring in the future. For these reasons, it is essential that the satellite LIBs (and solar arrays) be made as inert as possible at the EOM in order to prevent other components from generating debris or RF interference. Passivation can be interpreted as the opposite of high reliability mission management, which requires a different engineering approach in order to achieve effective EOM designs and operational execution.

References

1. *NASA Systems Engineering Handbook*, NASA SP-2016-6105 (2017).
2. *European Cooperation for Space Standardization, Space Engineering: System Engineering General Requirements*, ESA-ESTEC, ECSS-E-ST-10C, 15 February 2017.
3. US Department of Transportation Hazardous Materials Regulations, Title 49, *Code of Federal Regulations*, Part 173, Section 173.185.
4. *Manual of Tests and Criteria, United Nations Recommendations on the Transport of Dangerous Goods*, Part III, Section 38.3, 7e (2019).
5. International Air Transport Association (2020). *Lithium Battery Guidance Document, Transport of Lithium Metal and Lithium-Ion Batteries.*
6. *Delta-IV Launch Services Users Guide* (2013). United Launch Alliance.
7. Atlas, V. (2010). *Launch Services Users Guide*. March: United Launch Alliance.
8. *Space Launch System Mission Planners Guide, NASA ESD 30000* (2018).
9. *Falcon Users Guide*. SpaceX (2020).
10. *Ariane 5 User's Manual* (2016). Arianespace.
11. Chatel, F. (2011). Ground segment. In: *Spacecraft Systems Engineering*, Chapter 14, 4e (ed. P. Fortescue, G. Swinerd and J. Stark), 467–494. Wiley.
12. Sorenson, T.C. (2011). Mission operations. In: *Space Mission Engineering: The New SMAD*, Chapter 29 (ed. J. Wertz, D. Everett and J. Puschell), 903–936. Microcosm Press.
13. Dalton, P., Bowens, E., North, T., and Balcer, S. (2019). International space station lithium-ion battery status. In: *NASA Aerospace Battery Workshop*. Huntsville, AL.
14. International Organization for Standardization (2020). *Space Systems, Space Batteries, Guidelines for In-Flight Health Assessment of Lithium-ion Batteries*, ISO/TR 20891, 1e.
15. Landis, D.H. (2018). *Dead Bus Recovery Handbook for Earth Orbiting Spacecraft*, The Aerospace Corporation, TOR-2018-00319.
16. Crompton, K.R. and Landi, B.J. (2016). Opportunities for near zero volt storage lithium ion batteries. *Energy Environ. Sci.* 9: 2219.
17. NASA-STD-8719.14B (2019). *Process for Limiting Orbital Debris.*
18. Inter-Agency Space Debris Coordination Committee (IADC) (2007). *IADC Space Debris Mitigation Guidelines*, IADC-02-01, Rev. 1.

19 International Organization for Standardization, ISO 24113 (2019). *Space Systems – Space Debris Mitigation Requirements*, 3e.

20 ISO 26872:2019 (2019). *Space Systems – Disposal of Satellites Operating at Geosynchronous Altitude.*

21 NASA/TM-2018-220037 (2018). *History of On-Orbit Satellite Fragmentations*, 15e. NASA Orbital Debris Program Office.

22 NASA Orbital Debris Program Office (2015). *Orbital Debris Quarterly News: Recent Breakup of a DMSP Satellite*, vol. 19, Issue 2. https://orbitaldebris.jsc.nasa.gov/quarterly-news/pdfs/odqnv19i2.pdf (accessed 11 September 2020).

23 ESA Space Debris Office (2020). *ESA's Annual Space Environment Report*, GEN-DB-LOG-00288-OPS-SD, Issue 4.0.

24 Aouizerate, M., Samaniego, B., Bausier, F., and Nestoridi, M. (2019). Spacecraft Battery Passivation, ESPC 2019 Juan-les-Pins, France.

25 NASA Orbital Debris Program Office. *US Government Orbital Debris Mitigation Standard Practices.* https://orbitaldebris.jsc.nasa.gov/library/usg_od_standard_practices.pdf (accessed 10 September 2020).

26 NPR 8715.6B (2017). *NASA Procedural Requirements for Limiting Orbital Debris and Evaluating the Meteoroid and Orbital Debris Environments.*

27 United Nations Office for Outer Space Affairs. *Space Debris Mitigation Guidelines of the Committee on the Peaceful Uses of Outer Space*, adopted by the United Nations General Assembly in its Resolution A/RES/62/217 of 22 December 2007. https://www.unoosa.org/pdf/publications/st_space_49E.pdf (accessed 10 September 2020).

28 Hull, S. (2011). End of mission considerations. In: *Space Mission Engineering: The New SMAD, Chapter 30* (ed. J. Wertz, D. Everett and J. Puschell), 937–946. Microcosm Press.

29 Bausier, F., Nestoridi, M., Samaniego, B. et al. (2017). Spacecraft electrical passivation: From study to reality. In: *ESPC 2016, E3S Web of Conferences*, vol. 16, 13002. http://dx.doi.org/10.1051/e3sconf/20171613002.

30 Tambini, A., Antonini, F., DeLuca, A. et al. ESEO Power System. In: *2019 European Space Power Conference (ESPC)*, vol. 2019, 1–7. Juan-les-Pins, France: http://dx.doi.org/10.1109/ESPC.2019.8932086.

31 Johnson, N.L. (2007). The disposal of satellite and launch vehicle stages in low Earth orbit. In: *2nd International Association for the Advancement of Space Safety Conference*, 14–16.

32 Jeevarajan, J.A., Duffield, B., and Orieukwu, J.C. (2015). Safety of lithium-ion cells at different states of charge. In: *Space Safety Is No Accident* (ed. T. Sgobba and I. Rongier), 131–134.

Appendix A

Terms and Definitions

Acceptance Test

A non-destructive test or series of tests and inspections conducted on typically 100% of flight hardware to demonstrate that it is free of workmanship defects and meets design and performance requirements. The test demonstrates that design and performance requirements are met during or after environmental test exposures equal to flight predicted plus a required margin or more severe conditions to ensure adequate screening of defects.

Battery

An assembly of battery cells, cell banks, or modules electrically connected to provide the desired voltage, current, and capacity capability. The battery may also include one or more assemblies, such as electrical bypass devices, charge control electronics, heaters, temperature sensors, thermal switches, and thermal control elements, which are attached to the battery or cells. In a battery, the cells and assemblies are often enclosed within a case or other structural element to enable attachment to the spacecraft structure. Connectors and harnesses are included for power and telemetry command and sense.

Battery Capacity

The battery (or cell) capacity is the total number of ampere-hours (Ah) that can be withdrawn from a fully charged battery under specified conditions of discharge. It is the integral of current over discharge time. Integration limits are from the start of discharge to either the minimum cell voltage, minimum power subsystem battery-voltage limit, the point at which the first cell reaches the lower cell voltage limit, or the defined time duration is achieved.

Battery Energy

The battery (or cell) energy is the total number of watt-hours (Wh) that can be withdrawn from a fully charged battery under specified conditions of discharge. It is the integral of the product of discharge current and voltage over discharge time. Integration limits are from the start of discharge to

Spacecraft Lithium-Ion Battery Power Systems, First Edition. Edited by Thomas P. Barrera.

either the minimum cell voltage, minimum power subsystem battery-voltage limit, the point at which the first cell reaches the lower cell voltage limit, or when the defined time duration is achieved.

Bus Voltage

The nominal DC voltage at the spacecraft electrical node closest to the power sources where power is controlled and made available to the user load equipment.

Cell Bank

A grouping of two or more interconnected cells in a parallel arrangement, into a single mechanical and electrical unit.

Cell Bypass Switch

A device used to provide an alternate parallel low-impedance path around a cell or bank.

Cell Formation

One or more charge–discharge cycles that create the initial solid-electrolyte interphase on the electrode surface just after electrolyte is added to the cell.

Cell Lot

All cells manufactured in a single or multiple production run manufactured from the same anode, cathode, electrolyte, and separator material sublots with no change in processes, drawings, or tooling, and within a defined production run or time period.

Cell Lot Acceptance Test

A series of performance, manufacturing quality, and safety tests performed on a sample of cells from a lot that has passed flight screening tests. Its purpose is to determine performance margins, manufacturing quality, and safety margins after rigorous environmental exposure, cycle life, and end-of-life conditions.

Cell Screening

A series of inspections and non-destructive tests performed on a lot of cells to identify statistical outliers and assess performance uniformity of the lot.

C-Rate

The discharge or charge current for a battery (or cell), in amperes, expressed as a multiple of the rated capacity in ampere-hours.

Commercial Off-the-Shelf (COTS) Cells

COTS Li-ion cells are mass-produced, intended for terrestrial use, and procured directly from cell manufacturers or from third party distributors. Examples include 18650, 21700, and 26650-sized cylindrical Li-ion cells.

COTS Cell Lot

A group of cells manufactured within one day as verified by the individual cell lot date code stamp. The cell lot is to be procured and delivered as one group of cells and subsequently stored and tested together to maintain a single lot definition. The cell lot may be procured directly from the cell manufacturer or through a retail establishment with traceability to a cell manufacturer.

Current Interrupt Device (CID)

A built-in non-resettable pressure-activated switch that disables the electrical path through a cell when the cell internal pressure exceeds a specified threshold value.

Dead Bus

An anomalous condition where the primary bus voltage of a spacecraft decreases to any voltage below the minimum operational bus voltage level (including zero volts) for any time interval exceeding normal or survival bus undervoltage limits.

Dead Bus Recovery

A sequence of events that would normally begin prior to the complete loss of a spacecraft's main bus power and ends when the spacecraft has been restored to a thermally and power-safe state of operation or has transitioned to safe disposal operations.

Dead Bus Survival

Actions that serve to maximize the likelihood of recovery from a dead bus event. These actions may include: autonomous primary power restoration, load shedding, heater shedding, battery depletion prevention, enabling battery recovery heaters, incremental battery recharge, and small thermostatic heaters for thermally sensitive mission critical loads.

Depth-of-Discharge (DOD)

The ratio of the quantity of electricity (in amp-hours or watt-hours) removed from a battery (or cell) on discharge to its rated capacity or energy.

Design Mission Life

The contractually required period of on-orbit time over which the spacecraft must meet all of its performance and safety requirements.

Design Reference Case (DRC)

Representative mission or set of operational conditions that are used in an analytical or simulation setting to show that a design meets or exceeds its performance requirements.

Destructive Physical Analysis (DPA)

The process of disassembling, testing, and inspecting battery (or cell) components to verify internal design, materials, construction, and workmanship. DPA is often used effectively in post-mortem analysis to examine design features, workmanship characteristics, or discover process defects for identification of production lot problems not detected during normal screening tests.

Electrical Power Subsystem (EPS)

Set of all equipment, wiring, and EPS-controlling software whose task is the generation, storage, management, and distribution of electrical energy to the input power terminals of the spacecraft user loads.

Electromagnetic Compatibility (EMC)

EMC is the condition that exists when various electronic devices are performing their functions according to design in a common electromagnetic environment.

Electromagnetic Interference (EMI)

EMI is electromagnetic energy that interrupts, obstructs, or otherwise degrades or limits the effective performance of electrical equipment.

End-of-Mission (EOM)

When a spacecraft either: (i) completes the tasks, functions, and/or useful life for which it has been designed, other than its disposal, (ii) becomes non-functional as a consequence of failure, or (iii) is permanently halted through a voluntary decision.

Energy Balance

Balance between energy generated (or stored) and energy consumed (or wasted) over a defined orbital period.

Energy Storage

Devices that store energy for use in powering the spacecraft loads during orbital periods (such as eclipse) when the output of the power generation element is insufficient to meet the overall load demand.

External Short Circuit (ESC)

A direct electrical connection between the external terminals of a battery (or cell) that provides an unintended path for current flow between the terminals or to ground.

Fault Management

Process of detecting and reacting to the occurrence of a fault or anomaly, whether in hardware or software.

Foreign Object Debris (FOD)

A substance, debris, or article that is foreign to the cell or battery.

Ground Support Equipment (GSE)

Non-flight items used in the ground testing of the EPS as integrated into the spacecraft either at the spacecraft manufacturer facility, launch site, or other location.

Hard Passivation

Type of spacecraft passivation referring to complete and permanent electrical disconnection of the solar arrays from the power bus coupled with complete discharge of the batteries at the end of the mission. This absolute approach is often not practical, resulting in methods that could be considered soft passivation.

Internal Short Circuit (ISC)

A direct electrical connection between the cathode material, anode material, and/or metal current collectors internally in a cell that provides an unintended path for current flow. Manufacturing defects, foreign objects, debris, and/or lithium deposition are potential causes of internal short circuits.

Life Cycle Test (LCT)

Real-time life testing of battery service life expectancy conducted on a representative sample of cells, cell banks, or batteries configured in a flight-like manner under simulated on-orbit mission-specific operating temperatures, charge-control profiles, and electrical loads.

Lithium-Ion Cell

A rechargeable electrochemical cell in which the positive and negative electrodes are both intercalation compounds constructed with no metallic lithium in either electrode. An Li-ion cell contains an assembly of electrodes, separators, electrolyte, safety devices, case, and terminals. Various form factors exist for the case such as prismatic, cylindrical, and elliptical-cylindrical. A Li-ion cell is a subassembly of a Li-ion battery.

Load

Device or unit that consumes electrical power provided by the spacecraft EPS.

Load Margin

Ratio of the power margin to the contingent load power expressed as a percentage.

Lock-Up (or Latch-Up)

A condition wherein the solar array is operated at a point well below its maximum power point such that the solar array output power is insufficient to power the load and recharge the battery.

Module or Battery Module

A grouping of interconnected cells in a series and/or parallel arrangement into a single mechanical and electrical unit. The battery module may also include one or more assemblies, such as electrical bypass devices, charge control electronics, heaters, temperature sensors, thermal switches, and thermal control elements that are attached to the battery or cells. In a battery the cells and assemblies are often enclosed within a case or other structural element to enable attachment to a spacecraft structure. Connectors and harnesses are included for power and telemetry command and sense. Several battery modules can then be connected in series or parallel to form a battery unit.

Native Object Debris (NOD)

A substance, debris, or article native to the battery (or cell) design.

Normal Operation

A range of operational states of the spacecraft that exist or occur by design and in which the spacecraft spends the majority of its time, according to the expectations of the mission designers and planners.

Operating Voltage

A range of voltage bound by an upper limit and a lower limit within which charging and discharging of a battery (or cell) is specified.

Operational States

All foreseeable and intentional combinations of states, modes, or conditions within the EPS hardware and software.

Overcharge

Charging of a battery (or cell) after all of the active material has been converted to the charged state. Charging continued after 100% SOC, full-charged state, or upper battery (or cell) voltage limit is exceeded.

Overdischarge

Discharge of a battery (or cell) below the manufacturer's recommended lower discharge voltage limit..

Passivation

The process of removing stored energy from a spacecraft, which could credibly result in eventual generation of new orbital debris after disposal. This includes removing energy in the form of electrical, pressure, mechanical, or chemical energy.

Positive Temperature Coefficient (PTC)

An internal protective safety device used in some COTS 18650 Li-ion cells to provide safety hazard protection against high currents or short circuits external to the cell.

Power Budget

Method of accounting for the spacecraft's electrical loads and losses associated with these loads.

Power Generation

All equipment involved in the generation of DC power for use by the loads and for charging the energy storage devices.

Power Margin

Difference between contingent source power and contingent load power.

Proto-qualification Test

Proto-qualification is a modified qualification approach where design verification testing is performed at amplitudes and/or durations reduced from qualification levels and with analyses to the qualification requirements. Proto-qualification test hardware is considered flight-worthy.

Qualification Test

Qualification is the process that demonstrates that the design (including parts and materials), manufacturing processes, and acceptance programs produce products that meet specification requirements throughout their design life. Qualification is established by demonstrating margins relative to the product's design conditions to account for expected variations in parts, material properties, dimensions, manufacturing processes, and operating conditions. Qualification tests include destructive and abusive tests to determine performance, manufacturing quality, and safety margins.

Rated (or Nameplate) Battery Capacity/Energy

The minimum ampere-hours (or watt-hours) of a battery (or cell) specified at the beginning of life (BOL) for a defined range of charge control conditions, discharge load conditions, temperature profile, and minimum discharge voltage recommended by the cell manufacturer.

Regulated Bus

One whose voltage is controlled by means of a closed-loop negative feedback control scheme.

Safe Mode

A temporary state of minimized satellite operations that is transitioned into as a result of an autonomously irreconcilable spacecraft fault or safety threat. This state discontinues mission payload operations, configures satellite assets for sufficient power collection and minimal power usage, maintains equipment health and command and telemetry communications with the satellite ground segment, and yields fault/safety threat corrective action and operating state control authority to the satellite ground segment.

Service Life

The service life of a battery, battery module, or battery cell starts at cell formation and continues through all subsequent ground fabrication, acceptance testing, handling, storage, transportation, testing preceding launch and all phases of on-orbit mission operations.

Single Point Failure

Single component, wiring, or connector failure, software, or computer failure that results in the permanent loss of the spacecraft's ability to perform its primary mission for the intended design mission life span.

Soft Passivation

Type of spacecraft passivation referring to the reduction and control of generated and stored energy at the end of the mission to an extent, confirmed by analysis or test, that minimizes the risk of self-induced breakup of the spacecraft during post-mission storage or orbit decay. By contrast, complete disconnection and discharge could be considered as hard passivation.

Spacecraft

An integrated set of subsystems and units, including their software, capable of supporting a specified vehicle designed for travel or operation outside the Earth's atmosphere. Types of spacecraft include launch vehicles, astronaut crew transfer vehicles, cargo supply transport vehicles, satellites, and planetary rovers, landers, and orbiters.

Specific Energy

The ratio of the energy output of a battery (or cell) to its mass expressed in units of Wh/kg.

Specific Power

The ratio of the power delivered by a battery (or cell) to its mass expressed in units of W/kg.

State-of-Charge (SOC)

The available capacity in a battery (or cell) expressed as a percentage of rated capacity.

Thermal Runaway (TR)

Rapid self-sustained heating of a battery cell driven by exothermic chemical reactions of the materials within the cell. Thermal runaway is generally evidenced by a rapid increase in temperature and pressure and a decrease in cell voltage. Leakage, venting, smoke, fire and/or explosion of a battery cell can occur during thermal runaway.

Vent

A safety design feature of a battery (or cell) that activates to relieve excessive internal pressure.

Index

Note: Page number followed by '*f*' refer to figures and '*t*' refer to tables.

Spacecraft Lithium-Ion Battery Power Systems, First Edition. Edited by Thomas P. Barrera.

m

n

o

p